DATE DUE

# Biotechnology
# for
# Biological
# Control
# of
# Pests and Vectors

Editor

## Karl Maramorosch, Ph.D.

Robert L. Starkey Professor of Entomology
Department of Entomology
Rutgers — The State University
New Brunswick, New Jersey

CRC Press
Boca Raton Ann Arbor Boston London

**Library of Congress Cataloging-in-Publication Data**

Biotechnology for biological control of pests and vectors/editor,
  Karl Maramorosch.
      p.   cm.
   Includes bibliographical references and index.
   ISBN 0-8493-4836-6
   1. Vector control—Biological control.   2. Pests—Biological
control.   3. Biological pest control agents—Biotechnology.
I. Maramorosch, Karl.
RA639.3.B57   1991
632′.95—dc20                                                      90-25420
                                                                      CIP

Direct all inquiries to CRC Press, Inc., 2000 Corporate Blvd., N.W., Boca Raton, Florida 33431.

© 1991 by CRC Press, Inc.

International Standard Book Number 0-8493-4836-6

Library of Congress Card Number 90-25420
Printed in the United States

# PREFACE

Chemical pesticides have been used successfully for many decades, but their effect on groundwater pollution, residues on food crops, effects on nontarget organisms, and the development of resistance to chemicals have focused attention at alternative control measures. It has been estimated that despite the large amount of chemical pesticides — 500 million kilograms annually — used worldwide to control pests, nearly 50% of the world's food production has been lost to insects, diseases, weeds, microorganisms, birds, and rodents. Recent developments in biotechnology influenced the development of novel biopesticides that will play an important role in future pest management systems. These new biopesticides include bacteria, viruses, nematodes, fungi, and protozoa. Entomopathogenic bacteria, especially *Bacillus thuringiensis* (*Bt*) and *B. sphaericus* (*Bs*) are of special interest because they can be mass produced by industrial companies and, in fact, *Bt* sales have already amounted to $50 million per year worldwide. Entomopathogenic viruses also can be mass produced and genetically manipulated. Epizootics caused by viruses in forest insects have frequently resulted in nearly complete destruction of insect pest populations. Baculoviruses, including nuclear polyhedrosis and granulosis viruses, are the primary targets for biotechnological improvement aimed at increased pathogenicity and effectiveness. Nematodes that are used against soil insects are exempt from Environmental Protection Agency registration in the U.S. and are therefore favored by industrial companies as alternative biopesticides. Biotechnological modifications of fungi and protozoa, specifically microsporidia, have lagged behind the other biopesticides and have not been dealt with in the present volume.

Effective methods of pest and vector control are urgently needed for the expanding human populations. New methods must be compatible with environmental quality. Novel strategies in combating invertebrate pests and carriers of pathogens have focused on microbial and viral biological control agents that can now be manipulated by genetic engineering so as to become more target oriented and more virulent. Successful use of several microbial preparations applied against target pests in agriculture and against vectors in medicine has prompted active research into enhancement of epizootics for long-term pest management strategies. The manipulation of invertebrate pathogens and cells, using recombinant DNA and cell fusion technology to improve crop production and human health, hold considerable promise for the future. Biotechnology permits the modification of microbial and viral insecticides and of plants in ways that have never before been possible. Traditional approaches to agriculture and the management of the major crop plants are being dramatically changed by biotechnology. The use of genetically altered biopesticides and genetically modified plants that are self-protected against insect pests, as well as the improved control of vectors of pathogens, have potential for broad-reaching effects that will promote human health in the next century.

The release of genetically engineered microbial and viral pesticides has raised environmental concerns among the public, and to a lesser extent, among scientists. The safety of bioengineered pesticides has been of concern because genetically engineered bacteria, nematodes, and viruses released in the environment can reproduce in nature and thus differ from chemical pest control agents. Based on present experience, the genetically engineered biopesticides have proven safe, but constant vigilance is required to protect the environment and prevent hazardous events from happening. The public became acutely aware of the environmental damage that resulted from the excessive use of chemical pesticides that entered the food chain and affected not only nontarget and beneficial organisms, but humans as well. Therefore special precautions have to be taken to avoid any unnecessary hazards when bioengineered microbial and viral pesticides are being released in the environment. Deliberate release into the environment of microorganisms and baculoviruses containing recombinant DNA raised controversies because of the difficulties encountered in the assessment of environmental risks. Risk assessment must consider the survival, multiplication, dispersal, and effect on the environment. In the absence of any of these four individual probabilities, there would be no risk involved but accurate prediction of the four events is very difficult. Although the fears of the public might seem exaggerated, environmental consequences could be serious if genetically modified biocontrol agents would affect the environment. Adequate education of the public in industrialized and in less developed countries is needed so that actual facts can be understood, exaggerated fears removed, and genetic engineering of viral and microbial pesticides exploited without endangering the environment. New genetically engineered pesticides will benefit society and provide assurance that environmental concerns have been taken into account. The timing of the expected benefits will depend on the training, abilities, and facilities of scientists and on available funding. The final results will benefit humanity in developed and less developed countries in the 21st century.

Interdisciplinary research and significant genetic engineering advances have been achieved in America, Europe, and Asia. Active researchers who are in the forefront of this field in Belgium, China, Guatemala, Japan, the Philippines, Singapore, Thailand, and the U.S. have been invited to describe the techniques, the state-of-the-art areas that require further attention, current regulation, safety precautions, and the range of impacts of genetically engineered organisms in the environment. Biochemical and genetic mechanisms of gene variation and transfer, transposable elements, virus-mediated transfer, applications of genetically engineered bacteria, nematodes, and viruses to biological pest control, bioengineering of plants, new intriguing biomedical applications, mass production procedures, cell fusion, regulatory aspects assuring the safety of microbial and viral insecticides, and potential hazards of biocontrol agents are covered in the seven sections of this treatese.

This book brings together for the first time the large body of information and the significant achievements accomplished in laboratories in America, Europe, and Asia. The information is presented by the foremost authorities, who contributed on basic and applied concepts. Their international experience has been combined so that the latest developments in this fascinating and rapidly expanding field are presented in a comprehensive manner. The large body of information brings into sharp focus anticipated new directions in pest and vector control and basic microbial and viral biology. It is hoped that this book will set the benchmark for future research and serve for years to come as a standard source of information for research workers and students of biological control, microbiology, virology, and molecular biology. The worldwide growth of biotechnology is unique in the history of biology and the combined efforts of eminent contributors to discuss and evaluate new information will hopefully benefit all who are interested in biotechnology.

The editor wishes to express his sincere gratitude to the contributors for their effort and care with which they have prepared their chapters and to CRC Press for their part in indexing, proofreading, and other aspects of producing the volume.

**Karl Maramorosch**

# THE EDITOR

**Karl Maramorosch, Ph.D.,** is Robert L. Starkey Professor of Microbiology and Professor of Entomology at Rutgers — The State University of New Jersey, New Brunswick, New Jersey.

Dr. Maramorosch graduated from the Agricultural University of Warsaw, Poland, summa cum laude and obtained his Ph.D. degree in 1949 from Columbia University in New York. His scientific career began at Rockefeller University where he worked on plant viruses and insect vectors for 12 years. Afterwards he became Program Director of Virology and Insect Physiology at the Boyce Thompson Institute, and since 1974 he has been Distinguished Professor at Rutgers University, where he became the Robert L. Starkey Professor in 1983. He is a Fellow and former Recording Secretary and Vice-President of the New York Academy of Sciences, Fellow of the American Association for the Advancement of Science, the Entomological Society of America, the American Phytopathological Society, President of the International Association for Medicinal Forest Plants, Honorary Fellow of the Indian Virological Society, and a member of the Tissue Culture Association, the Harvey Society, the Society for Invertebrate Pathology, and several other professional organizations.

He was the winner of the 1980 Wolf Prize in Agriculture and many other awards, including the Ciba Geigy Award in Entomology, the Jurzykowski Award in Biology, the AIBS Award of Distinction, the Cressy Morisson Prize of the N.Y. Academy of Sciences, the Waksman Award, and the AAAS-Campbell Award. He was nominated by the Entomological Society of America for the National Medal of Science and was the 1990 Founders Lecturer in Australia of the Society for Invertebrate Pathology. He is an elected member of the Leopoldina, the oldest European Academy of Sciences, and Fellow of the Indian National Academy of Sciences.

Dr. Maramorosch served twice as visiting Fulbright professor in Yugoslavia and as visiting professor in P. R. China, the U.S.S.R., The Netherlands, Germany, Japan, India, and Poland. He has consulted for the Food and Agriculture Organization of the United Nations in the Philippines, India, Nigeria, Thailand, Sri Lanka, and Mauritius, and he accepted short-term assignments from the Rockefeller Foundation, the Agency for International Development of the U.S. State Department, the U.S. Department of Agriculture (FERRO), and the Ford Foundation in Mexico, India, Kenya, and the Philippines. He speaks fluent German, Romanian, and several Slavic languages, including Polish and Russian. He has edited more than 50 volumes on viruses, vectors, plant diseases, and tissue culture and he is the author or co-author of more than 500 research papers. His major research interests include comparative virology, invertebrate cell culture, parasitology, diseases caused by spirochetes, viroids, and spiroplasmas, biotechnology, and international scientific cooperation.

# CONTRIBUTORS

**Michael J. Adang, Ph.D.**
Associate Professor
Department of Entomology
University of Georgia
Athens, Georgia

**Spiros N. Agathos, Ph.D.**
Assistant Professor
Department of Chemical and Biochemical
  Engineering, and
Waksman Institute
Rutgers — The State University
Piscataway, New Jersey

**Clayton C. Beegle, Ph.D.**
Office of Pesticide Programs
Environmental Fate and Effects Division
Environmental Fate and Ground Water
  Branch
U.S. Environmental Protection Agency
Washington, District of Columbia

**Colin Berry, Ph.D.**
Postdoctoral Research Fellow
Institute of Molecular and Cell Biology
National University of Singapore
Singapore
Republic of Singapore

**Lee A. Bulla, Jr., Ph.D.**
Professor
Department of Molecular Biology
University of Wyoming
Laramie, Wyoming

**Pepito Q. Cabauatan**
Senior Research Assistant
Department of Plant Pathology
International Rice Research Institute
Manila
Philippines

**Eddie W. Cupp, Ph.D.**
Professor of Medical-Veterinary
  Entomology
Department of Entomology
The University of Arizona
Tucson, Arizona

**Robert M. Faust, Ph.D.**
National Program Leader
Plant Protection Institute
Agricultural Research Service
U.S. Department of Agriculture
Beltsville, Maryland

**Arnold S. Foudin, Ph.D.**
Deputy Director
Biotechnology Permits
Biotechnology, Biologics and
  Environmental Protection
Animal and Plant Health Inspection
  Service
U.S. Department of Agriculture
Hyattsville, Maryland

**Roy L. Fuchs, Ph.D.**
Monsanto Agricultural Company
St. Louis, Missouri

**Hiroyuki Hibino, Dr. Agric. Sci.**
Plant Virologist
Department of Plant Pathology
National Agriculture Research Center
Kannondai
Japan

**John Hindley, Ph.D.**
Institute of Molecular and Cell Biology
National University of Singapore
Singapore
Republic of Singapore

**Anne K. Hollander**
Associate
The Conservation Foundation, and
Biotechnology Project Manager
U.S. Environmental Protection Agency
Washington, District of Columbia

**Theodore M. Klein, Ph.D.**
Biological Products Division
E. I. du Pont de Nemours & Company
Experimental Station
Wilmington, Delaware

**Ricardo Luján, Ph.D.**
Professor and Director
Centro de Investigaciones en
 Enfermedades Tropicales
Instituto de Investigaciones
Universidad del Valle de Guatemala
Guatemala City
Guatemala

**Susan C. MacIntosh**
Senior Biochemist
Ento Tech, Inc.
Davis, California

**Pal Maliga, Ph.D.**
Professor of Genetics
Waksman Institute
Rutgers — The State University
Piscataway, New Jersey

**Karl Maramorosch, Ph.D.**
Department of Entomology
Rutgers — The State University
New Brunswick, New Jersey

**Pamela C. Marrone, Ph.D.**
President
Ento Tech, Inc.
Davis, California

**Jun Mitsuhashi, Dr. Agric.**
Tokyo University of Agriculture and
 Technology
Tokyo
Japan

**Coreen Oei**
Institute of Molecular and Cell Biology
National University of Singapore
Singapore
Republic of Singapore

**Somsak Pantuwatana, Ph.D.**
Professor
Department of Microbiology
Faculty of Science
Mahidol University
Bangkok
Thailand

**Marnix Peferoen, Ph.D.**
Research Manager
Plant Genetic Systems N.V.
Ghent
Belgium

**George O. Poinar, Jr., Ph.D.**
Department of Entomological Sciences
University of California
Berkeley, California

**Kathleen C. Raymond, Ph.D.**
Temporary Assistant Professor
Department of Molecular Biology
University of Wyoming
Laramie, Wyoming

**Robert I. Rose, Ph.D.**
Larus, Inc.
Baltimore, Maryland

**Sivramiah Shantharam, Ph.D.**
Senior Biotechnologist
Biotechnology, Biologics, and
 APHIS Environmental Protection
U.S. Department of Agriculture
Hyattsville, Maryland

**Steven R. Sims, Ph.D.**
Research Specialist
Monsanto Agricultural Company
St. Louis, Missouri

**Terry B. Stone**
Senior Research Biologist
Monsanto Agricultural Company
St. Louis, Missouri

**Nitaya Thammapalerd**
Associate Professor
Department of Microbiology and
 Immunology
Faculty of Tropical Medicine
Mahidol University
Bangkok
Thailand

**Savanat Tharavanij, Ph.D.**
Professor and Head
Department of Microbiology and
  Immunology
Faculty of Tropical Medicine
Mahidol University
Bangkok
Thailand

**Herman Van Mellaert, Ph.D.**
Business Development Manager
Plant Genetic Systems N.V.
Ghent
Belgium

**H. Alan Wood, Ph.D.**
Virologist
Boyce Thompson Institute for Plant
  Research
Ithaca, New York

**Yu Ziniu**
Associate Professor
Department of Soils and Agrochemistry
Huazhong Agricultural University
Wuhan
China

# TABLE OF CONTENTS

*Genetic Engineering of*
*Bacterial Biocontrol Agents*

Chapter 1

# *BACILLUS THURINGIENSIS* INSECTICIDAL CRYSTAL PROTEINS: GENE STRUCTURE, ACTION, AND UTILIZATION

Michael J. Adang

## TABLE OF CONTENTS

# I. INTRODUCTION AND BACKGROUND

*Bacillus thuringiensis* is a bacterium that produces toxic protein crystals during sporulation. The specific toxicity of these crystals against pest insects provides the basis for the use of this organism as a biological insecticide. Thousands of *B. thuringiensis* strains have been isolated from soil and other natural environments colonized by insects. Most are active against larval stages of Lepidoptera (butterflies and moths), while some kill species of Diptera (specifically mosquitoes and flies), Coleoptera (e.g., Colorado Potato Beetle), or both Lepidoptera and Diptera. Products based on *B. thuringiensis* have been registered as pesticides in the U.S. since 1961. As with other biological pesticides, *B. thuringiensis* offers a number of advantages over chemical pesticides, including lack of polluting residues, high specificity against target insects and safety to nontarget organisms, and low registration costs. Its broadest uses are on food crops and in forestry where its safety and specific action are desirable. Disadvantages of *B. thuringiensis* include a history of control failures and lack of potency against certain key insect pests. Where susceptible insects are target pests and safety is paramount, *B. thuringiensis* species are valuable pest control tools.

The pathogenicity of the numerous *B. thuringiensis* varieties for insects is caused by insecticidal crystals produced during spore formation. The proteinaceous crystals are aggregates of subunits called insecticidal crystal proteins (ICPs), δ-endotoxins, or Cry proteins. Insects feeding on plants ingest crystals and a combination of high pH and proteases in the insect's midgut dissolve the crystals and activate their subunits to insect-toxic proteins. Activated toxins bind to sites in the midgut, causing cell lysis, paralysis of the insect's gut, and mortality in several days.

*Cry* gene products determine the potency and toxic spectra of each *B. thuringiensis* strain against certain insect species. Progress made in identifying and characterizing *cry* genes has resulted in new insecticidal genes for pest insect control and encouraged research investigating the relationship between gene structure and action on insects. The ease of genetically manipulating *B. thuringiensis* insecticidal genes in bacteria and plants makes them ideal for future development.

Recent reviews have focused on *B. thuringiensis* biology, insecticidal genes, and commercialization as a biopesticide. Aronson et al.,[1] Whiteley and Schnepf,[2] and Andrews et al.[3] give useful overviews of this important organism and its insecticidal proteins. Tojo[4] reviewed the mode of action of Cry proteins. Höfte and Whiteley[5] focused on insecticidal gene structure, diversity, and regulation. Wilcox et al.[6] and Carlton[7] discuss the genetic manipulation of insecticidal genes to improve their qualities as microbial pesticides. This author will use *cry* gene structure and Cry protein mode of action as a framework for presenting examples of research targeted at making *B. thuringiensis* a better biological insecticide. Also covered are cases where *cry* genes are manipulated for expression in microorganisms and plants to produce new products for insect control.

# II. STRUCTURE OF *B. THURINGIENSIS CRY* GENES

Since the insecticidal activity of *B. thuringiensis* is determined by the *cry* genes within the host bacterium, cloning and sequencing *cry* genes is an obvious key to understanding this relationship. A second key is to study the insecticidal activity of individual *cry* gene products expressed in *B. thuringiensis* or *Escherichia coli*. Based on comparisons of nucleotide sequences and insecticidal spectra, Höfte and Whiteley[5] organized a classification system for the 42 sequenced *cry* genes. Fourteen *cry* genes were designated holotypes and placed into four classes; I, Lepidoptera-specific; II, Lepidoptera- and Diptera-specific; III, Coleoptera-specific; and IV, Diptera-specific. Table 1 summarizes their *cry* gene classification system. Table 1 also includes *cry* genes reported in 1990 (*cryIE, cryIF, cryIIIB,* and

**TABLE 1**
**Crystal Protein Protoxins and Toxins**

| Cry[a] protein | Target insects[b] | Protoxin size (kDa) | Toxin size (ca. kDa) | Refs. |
|---|---|---|---|---|
| CryIA(a) | L | 133 | 60—70 | 9—12, 49, 52 |
| CryIA(b) | L | 131 | 60—70 | 137—140 |
| CryIA(c) | L | 133 | 60—70 | |
| CryIB | L | 138 | 60—70 | |
| CryIC | L | 135 | 60—70 | |
| CryID | L | 132 | 60—70 | |
| CryIE | L | 130—135 | 60—70 | |
| CryIF | L | 130—135 | 60—70 | |
| CryIIA | L/D | 71 | 51 | 17—19 |
| CryIIB | L | 71 | 49 | |
| CryIIIA | C | 73 | 55—65 | 25 |
| CryIIIB | C | 73 | ? | |
| CryIIIC | C | 74 | ? | |
| CryIVA | D | 134 | 65—75 | 26, 31, 36, 37, 53 |
| CryIVB | D | 128 | 65—75 | |
| CryIVC | D | 78 | 58? | 38 |
| CryIVD | D | 72 | 38 | 31, 53 |
| CytA[c] | D/cytol | 27 | 22—25 | 31, 53, 141, 142 |
| CytB | D/cytol | 25, 28 | 23 | |

[a]  Refer to Höfte and Whiteley[5] for references to published nucleotide sequences except for *cryIIIB*.[23] Sequences for *cryIE*,[63] *cryIF*,[143] and *cryIIIC*[144] are not reported.

[b]  L = Lepidoptera, D = Diptera, C = Coleoptera, cytol = cytolytic.

[c]  Knowles et al.[142] propose the cytolytic genes from *B. thuringiensis* subsp. *israelensis* and *morrisoni* PG14 remain as designated, *cytA*, while genes from subsp. *darmstadiensis* and *kyushiensis* be designated *cytB*. This is based on the lack of antigenic homology between the groups.

*cryIIIC*). This organization of *cry* genes facilitates structural and functional comparisons of known genes and proteins while allowing flexibility for novel gene types as they are discovered.

## A. *cryI* GENES

The *cryI* genes were the earliest to be characterized and the first to be genetically modified. *CryIA(a)*, *cryIA(b)*, and *cryIA(c)* genes were formerly called 4.5-, 5.3-, and 6.6-kb genes, based on the location of a HindIII site within the coding region and a second 5' flanking HindIII site.[8] They are over 80% homologous and were placed together in a subgroup. The other members of the *cryI* group; *cryIB*, *cryIC*, and *cryID* are slightly less related or conserved. The nucleotide sequences of *cryIE* and *cryIF* are not yet published. *CryI* gene products kill numerous lepidopteran larvae. The *cryI* genes encode 130- to 140-kDa protoxins which are processed by trypsin-like enzymes in the midgut to ~60-70 kDa toxins (Table 1). Deletion analysis of *cryIA* genes[9-11] and amino acid sequencing of activated toxins[12] demonstrated that the toxin region is located in the N-terminal half of the protoxin. The C-terminal regions of CryI proteins have the greatest number of conserved amino acids. In contrast, amino acid differences in the N-terminal half are not randomly distributed, but are concentrated in subregions or hypervariable regions that determine insecticidal specificity.[13]

## B. *cryII* GENES

Isolates of several *B. thuringiensis* subspecies, recognized by toxicity to Lepidoptera and the formation of bipyramidal crystals, contain flattened cuboidal-shaped crystals and kill aquatic Diptera. Generally, the dipteran activity of these strains is low compared to subsp. *israelensis* and easily overlooked. Subspecies *kurstaki* strain HD-1 is a typical dual-action insecticidal strain that produces both crystal types. Strain HD-1 bipyramidal crystals have 131- to 133-kDa CryIA protein subunits (formerly called P1 proteins) while cuboidal crystals have 65 kDa CryII protein subunits (P2 mosquito factor).[14] Other strains of this type are found in subspecies *thuringiensis, tolworthi, kenyae,* and *galleriae*.[5,15,16] Note that subspecies *israelensis* and *morrosoni* PG14, which are highly toxic to Diptera, do not contain *cryII* genes.

*CryII* genes have been cloned from subsp. *kurstaki* strains HD-1[17] and HD-263.[18] These genes, designated *cryIIA* and *cryIIB,* have open reading frames which encode 633 amino residues and peptides of 71 kDa molecular mass. *CryIIA* encodes the "P2" insecticidal protein or "mosquito factor". CryIIA and B proteins are identical for 87% of their amino acid residues, yet their amino acid differences affect insect toxicity. For example, CryIIA insecticidal protein expressed in *Bacillus megaterium* was toxic to *Heliothis virescens, Lymantria dispar,* and *Aedes aegypti*.[18] However, CryIIB protein expressed in *E. coli* was toxic to *M. sexta* but not *A. aegypti*.[17] Widner and Whiteley[19] defined the mosquito specificity domain of CryIIA protein. They exchanged DNA fragments between *cryIIA* and the highly homologous *cryIIB* (87% amino acid homology) to identify the insect specificity domain. The region defining mosquito specificity in CryIIA is in a 76 amino acid segment between residues 307-382. These studies indicate CryIIA has a bifunctional host range that includes Lepidoptera and Diptera, while CryIIB is active only against Lepidoptera, and that residues in a 76-amino acid domain determine dipteran toxicity.

The unique physical and biochemical characteristics of bipyramidal and cuboidal crystals have facilitated the purification and characterization of their subunit peptides. Nicholls et al.[16] dissolved strain HD-1 crystals using a combination of alkaline pH (9.5) and a reducing agent which leaves cuboidal crystals intact. Subsequently, insoluble cuboidal crystals were dissolved by high alkaline pH[12] yielding CryII protein. The authors used this method to analyze the cuboidal crystal composition of five strains. Subspecies *kurstaki* HD-1, *thuringiensis* HD-117, and HD-770 had crystals with 63-kDa CryII proteins, while subsp. *galleriae* strains HD-29 and HD-916 contained 61-kDa CryII proteins. Analysis of peptide fragments supported their observation that the 63-kDa and 61-kDa peptides were distinct proteins. Crystals with 63-kDa subunits were 10-fold more toxic against *Pieris brassicae* and also more toxic to *A. aegypti* and *Anopheles gambiae*. Nicholls et al. treated solubilized CryII proteins with *P. brassicae* gut extract and chymotrypsin to produce protease resistant cores of 51 kDa and 49 kDa, respectively. N-terminal amino acid sequencing of the first 10 residues of the 51-kDa peptide shows that it aligns with amino acids 146-155 of CryIIA. This indicates activated CryIIA toxin has 145 residues removed from the N-terminus of the protoxin. This may correspond to the 50-kDa peptide observed by Widner and Whiteley[17] on immunoblots of solubilized HD-1 cuboidal proteins.

## C. *cryIII* GENES

The discovery of *B. thuringiensis* strains active against Coleoptera enhanced commercial development of this organism as a bioinsecticide. Four beetle-active strains are reported; subsp. *tenebrionis,*[20] subsp. *san diego,*[21] strain EG2158,[22] and a new isolate of subsp. *tolworthi*.[23] The beetle-active insecticidal protein (CryIIIA) coding sequence is identical for the first three, while the subsp. *tolworthi* isolate harbors a closely related gene, *cryIIIB*.[23] The *cryIIIA* gene encodes a 644 amino acid, 73-kDa protein. Like other crystal protein genes, *cryIIIA* is expressed in sporulating cells. Unlike CryI proteins, the translated peptide is not what is isolated from native crystals. During the process of crystal production and subsequent release, proteolysis converts the 73-kDa peptide to the major crystal subunit of 67 kDa and a minor 56-kDa component.[24,25] Carroll et al.[25] also found that trypsin or gut enzyme treatment produced a 55 to 56-kDa peptide equal in toxicity to the native 67-kDa crystal protein. N-terminal amino acid sequencing showed that 145 amino acid residues had been cleaved from the N-terminus of the native 644-residue peptide. This resulted in a 55-kDa peptide beginning with asparagine at position 159. Adang and Sekar (unpublished observation) have found that removing four amino acids from the C-terminus destroys toxicity. Alignment of the *cryIIIA* coding region with the *cryI* genes shows both of these results are not unexpected. The natural 3'-terminus of *cryIIIA* corresponds closely to the known end of the toxin regions of several dipteran and lepidopteran-active crystal protein genes.[5,26]

## D. *cryIV* AND *cyt* GENES

*B. thuringiensis* subsp. *israelensis* produces multiple crystals of different sizes. This crystal complex is insecticidal to larvae of mosquitoes and blackflies. Major peptide subunits are 28 kDa, 65 kDa, and 130 kDa.[27] There has been controversy over which subunit has mosquitocidal activity. First, bioassays are complicated by the filter feeding behavior of mosquitoes which makes particulate insecticidal proteins more toxic than solubilized proteins. Second, each subunit has been shown to have some amount of larvicidal activity, but individually they are less potent than native crystals.[28,29] Most likely, the composite insecticidal activity is the result of the synergistic action of several proteins.[30]

The similarities, i.e., structural relationships, between subsp. *israelensis* crystal proteins were initially measured by immunological assays and they by cloning and sequencing individual genes. Antigenic analysis of the crystal peptides demonstrated that 28-, 65-, and 130-kDa peptides were immunologically distinct.[31] Cloning and nucleotide sequence analysis has identified five distinct gene types from *B. thuringiensis* subsp. *israelensis*. Four of these are designated *cryIV* genes due to their general conversation of structure with other *cry* genes.[5] The fifth gene, designated *cyt*, encodes the cytolytic toxin.

*CryIVA*[32,33] and *cryIVB*[26,33-35] encode 130-kDa insecticidal proteins which are toxic to mosquito larvae. Similar to *cryI* genes, which also encode 130-kDa proteins, they have an insecticidal protease resistant core of 65-72 kDa.[31,36] Sequential deletions from the N- and C-termini of *cryIVB* indicate 459 amino acids of the 1136 amino acid peptide can be removed from the C-terminus, and 38 residues could be removed from the N-terminus without a significant loss of larvicidal activity.[26,36,37] Chungjatupornchai et al.[26] point out that residues 619 to 633 match comparable amino acids in CryIA(b) known to constitute the end of the toxic fragment. CryIVA and cryIVB have nearly identical 3' halves. Overall there is little amino acid homology with the *cryI* lepidopteran active genes, but there are potentially significant stretches of homology and a conservation of hydrophobic and hydrophilic amino acids.

*CryIVC*[38] encodes a dipteran-toxic peptide of 72 kDa. The gene product accumulates in *E. coli* and *B. subtilus* as a protein with a molecular size of approximately 58 kDa. *CryIVC* (called ORF1 in Thorne et al., 1989) is followed by a second open reading frame (ORF2). ORF1 and ORF2 correspond to a rearranged 130-kDa type gene, with ORF1 being the N-terminal toxic region and ORF2 the C-terminus.[5] Delecluse et al.[37] determined that a *cryIVB* gene is located adjacent to ORF1 in subsp. *israelensis*. The gene products encoded by ORF1 and ORF2 act in a synergistic manner to broaden their toxic range. For instance, CryIVB protein kills *A. aegyptii* and *An. stephensi* larvae, while CryIVB plus the ORF1 gene product (CryIVC) are required to kill *Culex pipiens* larvae.

*CryIVD*, cloned and sequenced by Donovan et al.,[39] also encodes a 72-kDa protein. Interestingly, it is not related to *cryIVC*, but has homology in limited regions to *cryIIB* and *cryIIIA*. The 72-kDa protein is degraded by proteases to peptides of 38 kDa.[31] Yamamoto et al.[28] found peptides of this molecular size to be toxic to mosquitoes.

The $M_r$ 28-kDa crystal protein gene, or *cyt* gene, is unique among the *B. thuringiensis* insecticidal genes. It is not genetically related to *cry* genes at the nucleotide or amino acid levels. It is activated to a 22- to 25-kDa toxin by proteases. Alone, it has low toxicity to mosquitoes but synergizes the 68- and 130-kDa toxins,[41] indicating a role in the insect toxicity of subsp. *israelensis* crystals. Once activated, it is hemolytic and cytolytic.[42]

# III. ACTION OF CRY PROTEINS

When a susceptible lepidopteran larvae ingests CryI insecticidal crystals, a series of molecular events begins in the insect's midgut that leads to physiological changes and mortality. The onset of symptoms is surprisingly rapid. Generally, the midgut is paralyzed, feeding ceases within several minutes, and the insect dies in several days. Cytotoxic changes include swelling of the goblet cells and columnar cell microvilli followed by the formation of blebs or protrusions. These protrusions are sloughed into the lumen accompanied by cell lysis. Lüthy and Ebersold[43] describe the pathological effects of toxemia in more detail. The following discussion summarizes the known process of Cry protein toxicology, including crystal solubilization, toxin recognition of midgut proteins, and primary intracellular effects.

## A. PROTEOLYSIS OF CRY PROTEINS

All insecticidal crystals require solubilization in the insect midgut. This process is most thoroughly characterized by CryI proteins. Lepidoptera, the target insects of CryI proteins, have highly alkaline midguts of pH 10 or greater and a complement of digestive proteinases.[44-46] In the midgut environment, crystals dissolve releasing 130- to 140-kDa protoxin peptides. Trypsin-type proteases cleave protoxins to 60- to 65-kDa toxins.[47,48] Choma et al.[49] reported that protoxin is successively cleaved in 10-kDa increments from the C-terminus to the N-terminus stopping at a 67-kDa toxin. In most cases, CryI protein amino acid sequence determines the protease stable toxin size and not the actual protease acting on the protoxin.

Some interesting variations are known among the Cry proteins in this process of toxin activation. For example, not all proteolysis of Cry proteins occurs in the insect midgut. Andrews et al.[50] demonstrated that *B. thuringiensis* strains high in intracellular proteases convert protoxin to toxin which is packaged in the crystal concurrently with protoxin. Sekar et al.[24] observed a similar result for the coleopteran-active CryIII crystals, where in addition to the expected 73-kDa translation product, a 65-kDa peptide was noted. Further experiments by Sekar[51] and Carroll et al.[25] showed this to be a result of posttranslational protease modification. Also, for certain Cry proteins, the nature of protease activation determines target insect specificity. Haider et al.[52] found that a subsp. *aizawai* CryI protein can be differentially processed by *P. brassicae* enzymes to a 55-kDa lepidopteran-specific toxin or by dipteran *A. aegypti* enzymes to a 52-kDa mosquito-active toxin. *B. thuringiensis* subsp. *israelensis* potency against mosquitoes is affected by protease cleavage. Pfannenstiel et al.[53] digested subsp. *israelensis* mosquitocidal Cry protein with chymotrypsin to yield protease-resistant domains. The domains were equally as toxic as native crystals to *C. quinquefasciatus* larvae, but the toxicity of the chymotrypsin-digested Cry protein was reduced five-fold or more for *A. aegypti* larvae. They present a model where covalently attached amino sugars are required for toxicity for *A. aegypti* larvae but not *C. quinquefasciatus* larvae. These variations from the accepted CryI protein model of toxin processing indicate that, for some Cry proteins, midgut digestive enzymes are important in determining toxin specificity.

Haider and Ellar[54] measured the effects of protease processing on CryI protein specificity using cell cultures. To accomplish this, they cloned a *cry* gene from *B. thuringiensis* var. *aizawai* and expressed the 130-kDa protoxin in *E. coli*. Trypsin digestion of purified protoxin produced a 55-kDa peptide toxic to *Choristoneura fumiferana* CF-1 cells and not *A. albopictus* cells. If the protoxin was first treated with trypsin and then *Aedes* gut proteases, the resulting 53-kDa protein lysed dipteran cells, but had no effect on lepidopteran CF-1 cells.

**TABLE 2**
*In Vitro* **Assays for CryI Toxin Action**

| Toxin effects measured | Refs. |
|---|---|
| **Artificial Phospholipid Membranes** | |
| $^{14}$C-Sucrose release | 56 |
| Binding, increased light scattering, $^3$H-alanine and $^3$H-uracil release | 57 |
| 86Rb$^+$-K$^+$ release | 58 |
| Cation-selective channels | 78 |
| **Cultured Cells** | |
| Cytotoxicity, ionic effects | 145 |
| Release of intracellular markers | 77 |
| Toxin-receptor binding, cytoxicity, osmotic protectants | 68 |
| Toxin-receptor binding, cytoxicity, *N*-acetylgalactosamine protection | 146 |
| K$^+$-ATPase inhibition | 147 |
| **Brush Border Membrane Vesicles** | |
| High-affinity binding | 60—66 |
| Ion-amino acid cotransport | 74—76 |
| Putative receptor identification | 66 |
| Alkaline phosphatase inhibition | 81 |
| **Isolated Midguts** | |
| Toxin disrupts short-circuit current | 71, 72 |
| Ba$^{2+}$ and Ca$^{2+}$ neutralize toxin effect on short circuit current | 73 |
| Enzyme release | 148 |
| Toxin binding to histological sections | 149 |

Their study indicated a role of insect gut proteases in determining cell toxicity and demonstrated the utility of cell culture in investigating toxin action. A limitation of cultured cell assays is that most insect cell lines are not derived from midgut epithelial tissue and most likely do not have the same membrane peptides. Particularly, currently available cell culture assays do not quantiatively reflect *in vivo* insecticidal activity.[101]

## B. TOXIN BINDING TO TARGET CELLS

A characteristic feature of Cry proteins is their specificity for certain insects. Lüthy[55] and others postulated the role of receptors in *B. thuringiensis* toxin specificity, but only recently have experiments demonstrated their presence. Artificial phospholipid membranes, cultured cells, and membrane vesicles isolated from midguts have been employed as models to study toxin binding (Table 2).

Studies with artificial membranes show an affinity of CryI toxins for cell membranes. At high concentrations, CryI toxins are able to directly bind to phospholipid membranes without the presence of receptors.[56-58] Like cell culture assays, the concentrations required are approximately 1000-fold greater than what is needed for *in vivo* toxicity. However, binding can be accelerated by incorporating brush border membranes from susceptible insects into the artificial membrane bilayers.[58] More information is needed to compare toxin-membrane interactions in reconstituted membranes vs. brush border membrane vesicle (BBMV) and midgut assays (discussed below).

Brush border membrane vesicles (BBMV) have been used as tools to analyze insect midgut transport for 10 years.[59] Purified BBMV provide a method for presenting membrane surface molecules to CryI toxins *in vitro* for the purpose of characterizing toxin binding sites. Hoffman et al.[60,61] and Van Rie et al.[62,63] published a series of papers in which they demonstrate high affinity binding sites for various CryI toxins in target insect midguts utilizing BBMV. In their experiments $^{125}$I-labeled toxin is incubated with BBMV and free toxin separated from bound toxin by centrifugation or filtration. By the addition of unlabeled homologous toxin or heterologous toxins (i.e., other CryI toxins) the affinity and specificity of binding sites is determined. First, they[60,61] demonstrated that *B. thuringiensis* CryI toxins recognize high-affinity receptors in the insect midgut and that binding specificity is central to the host range of *B. thuringiensis* toxins. Specifically, *B. thuringiensis* var. *thuringiensis* 4412 toxin (CryIB) bound with high affinity ($K_d$ = 46 n$M$) to BBMV isolated from highly susceptible *P. brassicae* larvae.[60] Hoffman et al.[61] then extended the correlation by selecting two toxins, CryIA(b) and CryIB that have different insect toxicities. BBMV were prepared from the tobacco hornworm, *M. sexta,* or larvae of the cabbage butterfly, *P. brassicae.* Two toxins, CryIA(b) and CryIB were studied for binding to BBMV. Toxins lethal to the respective insects bound to brush border sites with a high affinity. For example, CryIA(b) kills *M. sexta* and *P. brassicae* and binds with high affinity and saturably to sites on both insects. In contrast, CryIB toxin from subsp. *thuringiensis* 4412 kills *P. brassicae,* but not *M. sexta,* and binds with high affinity and saturably only to *P. brassicae* vesicles. Van Rie et al.[62] used this assay to test three toxins; Bt2[CryIA(b)], Bt3[CryIA(a)] and Bt73[CryIA(c)], against vesicle preparations of *M. sexta* and *H. virescens.* The three toxins have similar toxicity to *M. sexta* but differ in their potency against *H. virescens.* Both insect vesicles had saturable, high-affinity sites for each toxin; however competition studies revealed heterogeneity in the *H. virescens* binding sites. Van Rie et al. propose that *M. sexta* has a binding site population that binds all CryIA toxins and a second population that binds Bt3 toxin. *H. virescens* has three populations of binding sites which bind one or multiple toxin types. As Wolfersberger[64] notes, heterologous competition experiments become complex and difficult to interpret in insects such as *S. litoralis* and *H. virescens* that have multiple populations of binding sites. In spite of the difficulty in assigning toxin types to binding sites there is a clear relationship between insect toxicity and binding site specificity.

The recognition of binding sites may not be sufficient to ensure toxicity. For example, CryIC toxin binds with high affinity ($K_d$ = 22.4 n$M$) to midgut BBMV from *H. virescens,* but the toxin is not active against larvae.[63] Two anomalies are reported with CryIA(c) toxin and BBMV binding. The affinity of CryIA(c) toxin for *P. brassicae* midgut vesicle sites is greater than that of CryIA(b) toxin, yet its potency is 400-fold less.[65] Also, Garczynski et al.[66] found that CryIA(c) toxin binds with high affinity to *S. frugiperda* BBMV, but the toxin at a dose of 2000 ng/cm$^2$ does not have larvicidal activity. Overall, lepidopteran larvae contain multiple binding sites for toxins in their epithelial membranes. For some toxins there is a quantitative correlation between site number and toxicity. However, the above examples also illustrate cases where the correlation between binding and toxicity is lacking. The investigation of Cry toxin binding sites is at an early stage. Binding sites are important, but are only part of the insecticidal process.

The nature of CryI toxin receptors has not been determined. One approach to identifying receptors, called ligand blots, involves the separation of proteins from target cell or membrane by SDS-PAGE and transfer to membrane filters. Candidate binding proteins are identified by hybridization with labeled ligand. This method has identified receptors for growth factors, and other biologically active compounds whose effects are receptor mediated. Refer to Soutar and Wade[67] for a review of this method and its applications. Haider and Ellar[68] probed protein blots of CF-1 cells with iodinated subspecies *aizawai* CryI toxin and identified 68-kDa and 120-kDa peptides as candidate toxin receptors. Given the success of toxin/brush border vesicle binding experiments, efforts to identify and clone toxin binding proteins will be renewed. Putative toxin receptor proteins will need to be incorporated into functional assays to demonstrate their roles in toxemia.

## C. EFFECTS OF CRY TOXINS ON CELL MEMBRANES

The lepidopteran midgut has a mechanism for secreting $K^+$ from the hemolymph to the gut lumen.[69,70] *B. thuringiensis* toxins disrupt $K^+$ transport (measured by short-circuit current) across isolated midgut.[71,72] Divalent cations, $Ba^{2+}$ and $Ca^{2+}$, prevent and cause a reversal of short circuit current in toxin-treated *M. sexta* midguts.[73] A related effect is seen on the uptake of amino acids by the brush border. Amino acids move from the midgut lumen into intestinal cells via a cotransport mechanism with $K^+$. Using brush border membrane vesicles as an *in vitro* assay for amino acid/$K^+$ cotransport, Sacchi et al.[74] showed that low concentrations of *B. thuringiensis* HD-1 toxins disrupt amino acid uptake by BBMV. Their data suggest this is due to increased membrane permeability to $K^+$ as opposed to a direct effect on the amino acid symport protein. Incubation of larval *M. sexta* vesicles with Bt73 toxin [CryIA(c)] resulted in a dose-dependent increase in both $K^+$ and $Na^+$ ions.[75] The effect of toxins on ion-amino acid cotransport correlates with toxicity against insects.[76] A disruption of the normal cation flux by Cry toxins is sufficient to perturb the midguts' ion potential and lead to cell lysis.[44,64]

A two-step model (toxin binding to a cell receptor and subsequent pore formation) was proposed for the mechanism of action of *B. thuringiensis* toxins by Knowles and Ellar.[77] Tissue culture cells (CF-1) were treated with activated toxin and the release of $^{86}Rb$, [$^3H$]uridine, or $^{51}Cr$ measured. Pore sizes of 0.5 to 1 n$M$ were estimated by adding molecules of known radii to the outside of cells. This suggests that after binding to a receptor in the cell membrane *B. thuringiensis* toxins generate "holes" or "pores" in the cell membrane. The minimal cross-sectional area for a membrane pore permeable to both $K^+$ and $Na^+$ must be approximately 17 Å, and a membrane pore permeable to $K^+$ ions can have a cross-sectional area of 9 Å.[75] The formation of pores allows a net inflow of ions and water which leads to colloid-osmotic lysis. Haider and Ellar[57] studied the interaction of liposomes with CryI toxins. They found that CryI toxins can integrate into phospholipid vesicles and act directly as a pore. After toxin binds/integrates into these membranes, pores are created allowing ions to leak across. Slatin et al.[78] observed, by a voltage clamp method, that CryIA(c) and CryIIIa toxins form cation selective pores in artificial membranes. For each *in vitro* system utilized (artificial phospholipid membranes, BBMV, cultured cells and isolated midgut) there are unresolved interactions regarding the relationships between toxin type, nature of the receptors, pore size, and the involvement of $Na^+$ and $K^+$, but overall, these *in vitro* studies point to the creation of pores in brush border membranes as central to the action of *B. thuringiensis* toxins. Parker et al.[79] note the common occurrence of pore formation among toxins of very different activities. They suggest this is a consequence of the requirement of water-soluble proteins to turn inside-out on insertion into membranes. An efficient method of forming the pore is by oligomerization of toxins, because it reduces the conformational requirements used by a monomer for pore formation. The net result is a hydrophobic outer surface and a charged water-filled pore.

The observation that *B. thuringiensis* toxins create pores does not eliminate the possibility that they may have other direct actions. For example, CryI toxin inhibits a $K^+$-ATPase in insect cells and an alkaline phosphatase located in midgut apical membranes.[80,81] Finally, insects have been shown to recover from subacute doses of toxin.[82] This response may differ between species and be a factor in insecticidal activity. It will be of great interest to combine our understanding of Cry toxin action on cell membranes with further molecular studies on intracellular effects.

## IV. MODIFICATION AND USE FOR INSECT CONTROL

The bulk of the commercial *B. thuringiensis* market is controlled by three producers: Abbott Laboratories (North Chicago, IL), Solvay (Brussels, Belgium), and Sandoz (Basel, Switzerland). Opportunities for discovering novel strains and genetically improving existing strains have encouraged more companies to develop *B. thuringiensis* biopesticides. Entotech Division of Novo Laboratories, Mycogen, and Ecogen are new entrants to the agricultural *B. thuringiensis* market. The dominant commercial product since the 1970s is based on subsp. *kurstaki* strain HD-1 (serotype 3a3b). It is effective on over 200 crops and against more than 55 lepidopteran species. It comprises one of the products of choice against forest insects, spruce budworm, and gypsy moth. *B. thuringiensis* is increasingly used for early season control of *Heliothis* species on cotton, where worms are resistant to pyrethroid insecticides, and on vegetables against diamondback moths, which are resistant to all known chemical controls.

*B. thuringiensis* subsp. *israelensis* was isolated from diseased mosquito larvae.[83] With support of the World Health Organization, this bacterium was registered and a biopesticide produced to control mosquitoes and blackflies, important dipteran vectors of human diseases. More recently, the discovery of subsp. *tenebrionis* by Krieg et al.[20] and subsp. *san diego* by Herrnstadt et al.[84] broadened the host range for biopesticides to include Coleoptera. The first use of this strain is for the control of Colorado potato beetles. In some areas of the U.S., Colorado potato beetles are resistant to all chemical pesticides and alternative methods of control are needed. Mycogen Corp. is marketing a product M1™, and other companies (Abbott Laboratories, Sandoz, and Ecogen) are near commercial production with similar formulations.

Considering the range of insects killed, why do *B. thuringiensis* and other biopesticides constitute only 0.5% of the pesticide market?[7] Part of the answer lies in the commercial perception of disadvantages. *B. thuringiensis* biopesticides are more expensive than conventional insecticides and are short-lived. This short life is environmentally attractive, but results in increased applications and higher cost to the farmer. They are specific for some insects making broad insect control more difficult than with chemical pesticides. If a farmer has several classes of insects attacking his field at one time, several microbial pesticides are needed. They generally are toxic to young larvae, so careful monitoring is necessary, and *B. thuringiensis* treatments must adequately cover plant parts eaten by the pest. Some of these limitations can be overcome through improvements in formulation and application of genetic technologies.

## A. IMPROVEMENT OF *B. THURINGIENSIS* BIOPESTICIDES

It is generally accepted that the use of *B. thuringiensis* and other biopesticides will increase greatly in the next 5 to 10 years. However, these projections depend on predicted improvements in insect control with *B. thuringiensis* biopesticides. *B. thuringiensis* HD-1 products of today are more potent than those of 5 years ago. This is mostly due to improved manufacturing and formulation. The challenge for a new series of *B. thuringiensis* biopesticides will be to accomplish improved pest control while maintaining the desirable qualities of *B. thuringiensis,* particularly its safety to nontarget organisms. Insect pests commonly cited as targets for new biopesticides are *Heliothis* species on cotton, European corn borer, and corn rootworms. The development of new ways of delivering Cry proteins to pest insects using transgenic plants and microorganisms spurs the search for new strains and insecticidal genes.

The process of *B. thuringiensis* discovery and improvement depends heavily on the tools of molecular biology and the knowledge of *cry* genes and how they work. This applies to strain isolation as well as to the genetic modification and transfer of insecticidal genes. An advantage of *B. thuringiensis* is the decades of experience in fermentation, processing and formulation that allow the delivery of a stable biopesticide to the plant surface. Also important is the experience entomologists have in the use of this bacterium in pest management practices. Wilcox et al.[6] and Shieh[85] emphasize the need for improved spore/crystal formulations that provide better coverage and are more stable on plant surfaces. These practical considerations must be integrated into strategies for *B. thuringiensis* improvement, because the commercial success of this research will be determined by pest control in the field.

Common approaches are emerging for the discovery and development of improved *B. thuringiensis* strains. They can be considered a continuum which should provide novel biopesticides for many years. Each of these steps will be discussed below.

## 1. Nonrecombinant Methods for Improving Biopesticides

Biopesticide products developed using methods other than recombinant DNA encounter fewer regulatory hurdles than products employing those technologies. Faced with the common commercial pressures of today this tends to bias near-term research and development towards natural isolates and nonrecombinant technologies.

### a. Strain Isolation and Screening

The discovery of novel *B. thuringiensis* strains and insecticidal genes provides a basis for new biopesticides. Systematic searches for novel *B. thuringiensis* strains are routine in many industrial laboratories. These screening programs are proprietary, but the *B. thuringiensis* literature suggests several approaches to isolate and identify strains with unique biological specificities. Success is likely to come from a combination of approaches and technologies.

The source of samples is of obvious importance in the isolation of *B. thuringiensis* strains. Historically, new isolates are most often found in diseased and dead insects. *B. thuringiensis* was first isolated by Ishiwata[86] in Japan, where it was a pathogen in a silk worm colony. Next it was reported by Berliner[87] in Germany in a flour moth infestation. This trend continued with the isolation of the commercial strain *B. thuringiensis* subsp. *kurstaki* HD-1 from a mass-reared colony of pink bollworms.[88] More recently, subsp. *tenebrionis* was found in a dead common meal beetle (*Tenebrio molitor*). Studies investigating the ecological distribution of *B. thuringiensis* now direct our attention to the isolation of *B. thuringiensis* from natural environments, particularly soil and grain dusts. Researchers utilize this bacterium's ability to form spores and their incumbent heat resistance.[39,89] Ohba and Aizawa[90] collected 136 soil samples from throughout Japan, heat-treated suspensions at 65°C for 30 min, plated the mixtures on nutrient agar, and chose colonies for examination based on morphology. They obtained an average of 50 *B. thuringiensis* – *Bacillus cereus* colony types per gram of soil, and of these 2.7% produced crystal inclusions. Of the 189 isolates from this study, 48 were *Lepidoptera* toxic, 20 were mosquito toxic, and 121 were toxic to neither.[90] A study of the distribution of *B. thuringiensis* in soils in the U.S. utilized an enrichment medium[91] containing polymyxin B and penicillin to reduce the number of non-*Bacillus* bacteria in samples.[92] In a 2-year evaluation of 115 fields, they found *B. thuringiensis* in 7% of the soils tested where it comprised 0.5% of *Bacillus* species. Travers et al.[93] describe a novel method of enriching for *B. thuringiensis*. They selectively inhibit the germination of *B. thuringiensis* spores by sodium acetate while allowing the germination of other *Bacillus* species. This is followed by pasteurization at 80°C to kill non-*B. thuringiensis* organisms and plating on rich medium. They report that 20 to 96% of the colonies picked from agar plates were crystal-forming *Bacillus* species. Relative to its occurrence in soil, *B. thuringiensis* abounds in grain elevator dusts.[94] They analyzed 20 samples of settled dust and 53 of respirable dust from 4 grain elevators and found *B. thuringiensis* in 55% of the settled samples and 17% of the respirable samples. Using this information on *B. thuringiensis* occurrence, Donovan et al.[39] isolated a beetle-toxic strain from soybean grain dust in Kansas. The above studies demonstrate that *B. thuringiensis* is found in many ecological niches, most often, in relatively low numbers and can be isolated using standard microbiological enrichment techniques.

After a panel of strains is isolated the task is to identify those strains harboring insecticidal proteins that may have potential for commercialization. Some biotypes of *B. thuringiensis* have unique crystal shapes which are discernable under phase contrast microscope observation. For instance, coleopteran-active cuboidal crystals are easily distinguished from the more common lepidopteran-active bipyramidal crystals. Insect bioassays are the foundation for strain evaluation. Unfortunately, the maintenance of insect colonies is expensive and bioassays are tedious. The simplest method is to grow *B. thuringiensis* to sporulation and use a crude spore/crystal mixture in feeding tests.[95] Others have dissolved insecticidal crystals in alkali containing reducing agents and tested insects against solubilized protoxins.[96,97] Alternatively, insect cell cultures have been used to evaluate Cry protein toxicity (reviewed by Johnson).[98] Cell culture assays have been adapted for rapid screening using 96-well micro titer plates and a micro plate reader.[99] Recently, an assay was developed that uses 1-μl samples and can detect 24 pg of toxin.[100] The same cautions hold for using cell cultures to evaluate toxin activity as for toxin binding, i.e., *in vitro* potency may not relate to insecticidal strength.[101]

Most insecticidal crystals consist of several toxic protein subunits, and crystal potency results from their combined action. This complicates basing strain selection on bioassay potency. For example, *kurstaki* HD-1 has three *cryIA* genes and two *cryII* genes. These multiple proteins have widely different insecticidal specificities that contribute to the insecticidal profile of strain HD-1.[102,103] Strains *B. thuringiensis* subsp. *israelensis* and *morrosoni* PG-14 need the combined action of several insecticidal proteins to kill mosquitoes optimally.[41] These examples illustrate how a strain may have marginal activity against an important pest yet merit further study using individual insecticidal proteins.

Novel strains and insecticidal genes may be revealed through a combination of molecular techniques. One method is to use DNA hybridization of gene-specific probes to search for homologous crystal protein genes in unknown strains. Kronstad et al.[8] first used a *cry* gene probe to demonstrate the diversity and distribution of insecticidal genes among *B. thuringiensis* strains. As more *cry* genes were cloned and sequenced the probes became more specific and useful as screening tools. Prefontaine et al.[104] used oligonucleotide probes specific for *cryIA(a)*, *cryIA(b)*, and *cryIA(c)* lepidopteran-active genes to determine the distribution of these gene types among 13 *B. thuringiensis* serotypes. In a similar study, Visser[105] cloned restriction fragments homologous to 4 *cryI*-type genes and screened DNA from 25 strains (18 serotypes) by Southern hybridization analysis. By comparing homologous fragments in test strain DNA with the expected size fragments in the cloned genes he was able to, first, identify strains harboring the four gene types and, second, where the size of the homologous fragment differed, identify potential *cry* gene variants. Isolating *B. thuringiensis* DNA from a few strains is a manageable task, but handling hundreds or thousands of isolates will require simplified DNA extraction methods. However, after DNAs are isolated, hybridization assays are relatively rapid. Also, strategies using polymerase chain reaction (PCR) can be adapted to identify and clone novel *cry* genes. These methods, particularly when enhanced with PCR, will be useful in selecting variant *cry* genes with increased activity against pest insects.

A second approach to identifying novel insecticidal proteins uses antibodies against individual crystal components. Krywienczyk et al.,[106,107] Lynch and Baumann,[108] and Smith,[109] used polyclonal antisera to compare crystal serotypes among the lepidopteran-active *B. thuringiensis* strains. Each study contributed to the understanding that *B. thuringiensis* crystals contain serologically distinct components and that often a strain's crystal serotype does not correspond to its flagellar antigen serotype. Polyclonal antisera have recognized limitations for protein screening because they discriminate between Cry proteins with major antigenic variation, but not between more closely related proteins.

In contrast to polyclonal antisera, monoclonal antibodies have the specificity to distinguish closely related Cry proteins. Huber-Lukac et al.[110] first used monoclonal antibodies to study k-1 (*kurstaki*-1; see Krywienczyk et al.[106,107]) type crystal proteins and found similar crystal antigens in other subspecies. Monoclonal antibodies are best used when the gene product they are binding to is characterized. Adang et al.[111] used a monoclonal antibody which binds to CryIA(c) toxin and not CryIA(a) or CryIA(b) to isolate the *cryIA(c)* gene from a subsp. *kurstaki* HD-1 plasmid library. Höfte et al.[102] used monoclonal antibodies that react with CryIA, CryIB, and CryIC lepidopteran-active proteins to measure their distribution among *B. thuringiensis* strains. They found that CryIA proteins are the most widespread of the Cry proteins among the serovars. They also observed that CryIB, the "type" protein of subsp. *thuringiensis* strain 4412, and CryIC, a subsp. *aizawai* and *entomocidus Spodoptera*-toxic protein are relatively restricted in distribution. Monoclonal antibodies have the specificity and, when used in ELISA tests, the speed necessary for use in large screening programs.

### b. B. thuringiensis *Plasmid Curing and Exchange*

There are other methods for improving *B. thuringiensis* biopesticides that employ biotechnology and recombinant DNA, yet the final product is not considered "genetically engineered" by regulatory agencies. The process of plasmid exchange allows the transfer of plasmids between related *Bacillus* species.[112] This process is useful in the improvement of *B. thuringiensis* biopesticides. Insecticidal crystals constitute approximately 30% of the total weight of sporulating cells and there is probably little opportunity for changing crystal yield per cell. Due to the differences in Cry protein action, crystals frequently contain ineffective proteins against a particular species. Plasmid exchange allows a greater amount of the desired Cry protein to accumulate in the strain's crystals. This occasionally results in increased effective range or potency.[7] Though not well documented, patent applications claim that the combination of *cry* gene types in a strain can have surprising synergistic activity or toxicity to a pest not killed by parental gene types. More expected is the result where a lepidopteran gene and coleopteran-active gene are harbored in the same bacterium, producing a strain with combined insect toxicity. This permits the fermentation and marketing of a single strain. Biopesticides produced by conjugation of *B. thuringiensis* are now being field tested and will be registered for use in 1990.

### c. Nonliving Delivery Systems for Cry Proteins

In most cases, the *B. thuringiensis* spore itself is not critical to insect toxicity. Exceptions are with marginally susceptible insects, such as meal moth larvae, *Plodia interpunctella* and *Ephestia kuhniella*. This permits the development of Cry protein-based insecticides which are independent of the normal host bacterium. An advantage of this type of biopesticide is that insecticidal gene systems can be used that will not replicate in the environment.

Mutants of *B. thuringiensis* that do not produce spores may be used for the production of wild-type or genetically modified insecticidal proteins. This would permit the manufacture of a product containing insecticidal crystals but not viable bacteria or spores, obviating the concern regarding the deliberate release of genetically engineered organisms.

Scientists at Mycogen developed a system, called MCAP™, of encapsulating *B. thuringiensis* proteins within a protective bacterial cell wall coat.[113] They express insecticidal protein in *Pseudomonas fluorescens* bacteria which are killed after fermentation. Dead bacteria/insecticidal protein sprayed on plants is more stable to environmental factors than wild-type insecticidal crystal, yet the gene encoding the protein does not replicate. MCAP™ biopesticide containing a CryIA toxin has been field tested for the past two summers with encouraging results. Future tests with USDA/APHIS and EPA approval will utilize genetically modified insecticidal proteins. Gaertner[113] reviews the MCAP™ and other cellular delivery systems for insecticidal proteins in more detail.

### 2. Molecular Modification of Insecticidal Genes

The genetic modification of insecticidal proteins is a tool to dissect insect physiological processes as well as a means to increase the potency and host range of biopesticides. To meet the applied goal of improving biopesticides, techniques are needed to transfer engineered *cry* genes from *E. coli* back into *B. thuringiensis,* other microbes, and plants. In most cases, this means reintroduction of engineered genes into *B. thuringiensis* where combinations of genes can be used in their natural host. Most cloning and genetic engineering is done using plasmid vectors so systems for plasmid transfer into *B. thuringiensis* are most desirable. Early methods using phage and protoplasts for DNA transfer into *B. thuringiensis* were complicated and inefficient. Heirson et al.[114] reported plasmid DNA-mediated transformation of vegetative cells at efficiencies up to $10^{-3}$ transformants per viable cell or $10^4$ transformants per μg of DNA. Using electroporation, plasmids, sometimes harboring insecticidal genes, were transferred back into *B. thuringiensis* at frequencies of greater than $10^7$ transformants per microgram DNA.[115-117] Baum et al.[118] report the development of shuttle vectors based on replication origins originating from *B. thuringiensis* plasmids. The development of efficient transformation methods and plasmid shuttle vectors overcomes an impediment to *B. thuringiensis* genetic engineering.

Success in the rational design of improved toxins relates to knowledge of protein structure and mode of action. To date the improvements are modest. First, the *CryIA* genes were each truncated from the 3'-end to produce 1.8-kb coding regions that expressed insecticidal toxin.[9,11,119] This is analogous to the process of protoxin activation by midgut proteases. The comparison of proteins with similar functions but different structures assists in designing experiments to modify activity. Ge et al.[13] combined information on insecticidal specificity and gene homologies to identify the specificity domain in a CryIA toxin. Their hypothesis was that the hypervariable region in CryIA toxins was the region that determined the range of insect toxicity. This allowed them to construct and express hybrid genes in *E. coli.* Bioassays on expressed crystal protein resolved the specificity domain to a region spanning 118 amino acids (residues 330-450 of CryIA(a). Honee et al.[120] linked two *cry* genes that individually encode proteins with distinct insecticidal toxicities. They showed that the insecticidal profile of the composite gene product spanned the toxic range of both parental *B. thuringiensis* proteins. Rusche et al.[101] used random nucleotide mutagenesis to generate a panel of mutant *cry* genes. Among 15,000 candidates, several of the modified genes were "up" mutants and, when expressed in bacteria, killed insects more effectively.

The best example of *cry* gene modification is described in the plant section below where *cry* gene coding regions are engineered to increase protein expression in plants. Each of these experiments represents the biorational design for better pest control. These are early results and later experiments are likely to have more dramatic impact and may yield biotoxins active against pests not now susceptible to *B. thuringiensis* insecticidal proteins.

### 3. Alternate Delivery in Epiphytes/Endophytes

Genetic engineering technology allows the use of microorganisms that multiply on (epiphytes), or in (endophytes) plants to produce insecticidal proteins. This approach can overcome problems of insufficient coverage and persistence while providing season-long insect control. *Pseudomonas fluorescens, Pseudomonas cepacia,* and *Clavibacter xyli* are plant-associated bacteria that have been used for the expression of insecticidal proteins.[121-123] At Monsanto Company, a *B. thuringiensis* insecticidal gene was cloned into a corn root colonizer, *P. fluorescens,* using transposase deletion derivatives of the transposon Tn5 containing a *cry* gene.[121] This biopesticide was insecticidal to lepidopteran insects in growth chamber tests. When met with pending litigation from environmental advocacy groups, field trials with this organism were cancelled. Subsequently, a sensitive marker system was developed for following fluorescent pseudomonads in the environment.[124]

Crop Genetics International (CGI) developed a novel biopesticide by inserting a *B. thuringiensis cry* gene into the chromosome of *Clavibacter xyli* subsp. *cynodontis.*[122] *C. xyli* is a coryneform bacterium that inhabits the xylem of bermuda grass. Scientists at CGI found that *C. xyli* can be transferred to corn via seed treatment. After germination, bacteria colonize xylem and express *B. thuringiensis* protein. If sufficient insecticidal protein is produced, it will protect the stem from damage by European corn borers (*Ostrinia nubilalis*). The second small-scale field trial was performed in 1989 and a registered biopesticide is scheduled for the early 1990s.

There is certainly potential for developing numerous biopesticides based on producing insecticidal proteins in plant-associated microorganisms. The major questions to be answered relate to environmental issues of gene stability, microbial persistence, and impact on non-target invertebrates and animals. These issues must be addressed and a certain probability of public acceptance expressed before there is expanded research using *B. thuringiensis* proteins in plant-associated microorganisms.

## B. INSECTICIDAL GENES IN PLANTS

The list of genes encoding proteins conferring insect resistance upon transfer to plants is short. The qualifications of high activity against target insects upon feeding and safety for nontarget invertebrates and animals are met by few known proteins. The candidate insecticidal protein needs to be stable in the plant cell and the insect midgut. Because the exoskeleton, foregut, and hindgut of insects are sclerotized cuticle, the most likely route of toxicity is through the midgut. *B. thuringiensis* insecticidal proteins fit these criteria and experiments began in the early 1980s to develop insect-resistant transgenic plants based on the introduction of *B. thuringiensis* toxin genes.

Critical to transfer of genes into plant chromosomal DNA were discoveries in the field of plant transformation. *Agrobacterium tumefaciens* is a bacterium that transfers a segment of its DNA into the genome of plants. Scientists have used the nucleotide sequences necessary for DNA transfer to design vectors, called Ti vectors, that allow foreign genes to be engineered into crop plants. With the aid of Ti vectors, *B. thuringiensis* genes were introduced into the genomes of plant cells. Tissues harboring the transferred genes were regenerated into whole plants, unchanged except for the introduced genes. Plants with the foreign DNA in their chromosomes were called transgenic. The transfer of *B. thuringiensis* genes using Ti vectors has now been accomplished for a number of dicot species.

*CryI* genes from *B. thuringiensis* subsp. *kurstaki* were expressed in tobacco[125,126] and tomato.[127] Tobacco hornworms were killed after feeding on transgenic plants expressing low levels of *B. thuringiensis* protein. The significance of these papers is that they demonstrate a technology for protection against feeding insects, albeit in the laboratory, with broad application to crop plants. An underlying theme of these papers is that much less *B. thuringiensis* protein is found in plant tissues than expected. Generally, truncated, i.e., toxin-sized *cry* genes are expressed ten times better than full-length *cry* genes. Barton et al.[128] and Adang et al.[129] improved the production of *B. thuringiensis* protein slightly in tobacco and tomato through the use of DNA sequences proximal to the toxin coding sequence that stabilize mRNA and protein in plants. Murray et al.[130] measured the expression of *cry* genes in transgenic tobacco plants and electroporated carrot protoplasts. Low levels of *cryIA(b)* gene expression in plants and carrot cells was due to mRNA instability. The low levels of Cry protein in transgenic plants led to the construction of synthetic genes designed for optimal plant expression.[131,132] Rebuilt *cry* genes eliminate A-T nucleotide stretches found in native *cry* genes and match the codon usage pattern of plant genes. Scientists at Monsanto report that *B. thuringiensis* protein in transgenic cotton using synthetic *cryIA(b)* and *cryIA(c)* genes is 500 to 1000 times higher than reported for native genes.[132] This amount of insecticidal protein is sufficient to kill less suceptible insects like the *Heliothis* complex (tobacco budworms and cotton bollworms) and *Spodoptera litoralis* (cotton leafworm).

Corn that is tolerant to cornborers and rootworms is a major target for plant genetic engineering. However, attempts to use traditional *Agrobacterium* systems on that monocot species have not been successful. In 1990, corn cells were transformed via the particle gun method and fertile plants were regenerated.[133] The particle gun method bombards plant cells with micron-sized metal particles coated with DNA.[134,135] By combining a strong selectable marker (in these reports an herbicide resistance gene) with a particle gun for DNA delivery, plants were recovered that had foreign DNA integrated into corn genomic DNA. This is a major advance because now *B. thuringiensis* and other insecticidal proteins can be moved into this crop plant.

After nearly a decade of research directed towards genetically engineering plants for insect resistance, plants with the desired phenotype are being delivered to agronomists and entomologists for field evaluation. In 1990 field trials, Cry proteins expressed in cotton conferred worm control comparable to weekly insecticide applications.[136]

## C. POTENTIAL IMPACTS OF *CRY* GENE MODIFICATION AND INCREASED ENVIRONMENTAL IMPACT

Insect-resistant plants, recombinant microorganisms with crystal protein genes, and improved potency of crystal protein will contribute to the increased presence of *B. thuringiensis* Cry proteins in the environment. The threat of insects acquiring resistance to *B. thuringiensis* proteins is a major concern for all involved in these technologies. Stone et al. (Chapter 4) review this subject. Steps to delay the onset of resistance are being discussed in academia and industry. Commercial concerns have established a "*B.t.* Management Working Group" to address these issues.

# REFERENCES

1. **Aronson, A. I., Beckman, W., and Dunn, P.,** *Bacillus thuringiensis* and related insect pathogens, *Microbiol. Rev.,* 50, 1, 1986.
2. **Whiteley, H. R. and Schnepf, H. E.,** The molecular biology of parasporal crystal body formation in *Bacillus thuringiensis, Annu. Rev. Microbiol.,* 40, 549, 1986.
3. **Andrews, R. E., Jr., Faust, R. M., Wabiko, H., Raymond, K. C., and Bulla, L. A., Jr.,** The biotechnology of *Bacillus thuringiensis, CRC Crit. Rev. Biotechnol.,* 6, 163, 1987.
4. **Tojo, A.,** Mode of action of bipyramidal δ-endotoxin of *Bacillus thuringiensis* subsp. *kurstaki* HD-1, *Appl. Environ. Microbiol.,* 51, 630, 1986.
5. **Höfte, H. and Whiteley, H. R.,** Insecticidal crystal proteins of *Bacillus thuringiensis, Microbiol. Rev.,* 53, 242, 1989.
6. **Wilcox, D. R., Shivakumar, A. G., Melin, B. E., Miller, M. F., Benson, T. A., Schopp, C. W., Casuto, D., Gundling, G. J., Bolling, T. J., Spear, B. B., and Fox, J. L.,** Genetic engineering of bioinsecticides, in *Protein Engineering: Applications in Science, Medicine, and Industry,* Inouye, M. and Sarma, R., Eds., Academic Press, New York, 1986, 395.
7. **Carlton, B. C.,** Development of genetically improved strains of *Bacillus thuringiensis:* a biological insecticide, in *Biotechnology for Crop Protection,* Hedin, P. A., Menn, J. J., and Hollingworth, R. M., Eds., American Chemical Society, Washington, D.C., 1988, 260.
8. **Kronstad, J. W., Schnepf, H. E., and Whiteley, H. R.,** Diversity of locations for *Bacillus thuringiensis* crystal protein genes, *J. Bacteriol.,* 154, 419, 1983.
9. **Adang, M. J., Staver, M. J., Rocheleau, T. A., Leighton, J., Barker, R. F., and Thompson, D. V.,** Characterized full-length and truncated plasmid clones of the crystal protein of *Bacillus thuringiensis* ssp *kurstaki* HD-73 and their toxicity to *Manduca sexta, Gene,* 36, 289, 1985.
10. **Wabiko, H., Held, G. A., and Bulla, L. A., Jr.,** Only part of the protoxin gene of *Bacillus thuringiensis* subsp. *berliner* 1715 is necessary for insecticidal activity, *Appl. Environ. Microbiol.,* 49, 706, 1985.
11. **Schnepf, H. E. and Whiteley, H. R.,** Delineation of a toxin-encoding segment of a *Bacillus thuringiensis* crystal protein gene, *J. Biol. Chem.,* 260, 6273, 1985.
12. **Nagamatsu, Y., Itai, Y., Hatanaka, C., Funatsu, G., and Hayashi, K.,** A toxic fragment from the entomocidal crystal protein of *Bacillus thuringiensis, Agric. Biol. Chem.,* 486, 6110, 1984.
13. **Ge, A. Z., Shivarova, N. I., and Dean, D. H.,** Location of the *Bombyx mori* specificity domain on a *Bacillus thuringiensis* δ-endotoxin protein, *Proc. Natl. Acad. Sci. U.S.A.,* 86, 4037, 1989.

14. **Yamamoto, T. and McLaughlin, R. E.,** Isolation of a protein from the parasporal crystal of *Bacillus thuringiensis* var. *kurstaki* toxic to the mosquito larva, *Aedes taeniorhynchus, Biochem. Biophys. Res. Commun.,* 103, 414, 1981.

15. **Samasanti, W., Tojo, A., and Aizawa, K.,** Insecticidal activity of bipyrimidal and cuboidal inclusions of δ-endotoxin and distribution of their antigens among various strains of *Bacillus thuringiensis, Agric. Biol. Chem.,* 50, 1731, 1986.

16. **Nicholls, C. N., Ahmad, W., and Ellar, D. J.,** Evidence for two different types of insecticidal P2 toxins with dual specificity in *Bacillus thuringiensis* subspecies, *J. Bacteriol.,* 171, 5141, 1989.

17. **Widner, W. R. and Whiteley, H. R.,** Two highly related insecticidal crystal proteins of *bacillus thuringiensis* subsp. *kurstaki* possess different host range specificities, *J. Bacteriol.,* 171, 965, 1989.

18. **Donovan, W. P., Dankocsik, C. C., Pearce, G. M., Gawron-Burke, M. C., Groat, R. G., and Carlton, B. C.,** Amino acid sequence and entomocidal activity of the P2 crystal protein, *J. Biol. Chem.,* 263, 561, 1989.

19. **Widner, W. R. and Whiteley, H. R.,** Location of the dipteran specificity region in a lepidopteran-dipteran crystal protein from *Bacillus thuringiensis, J. Bacteriol.,* 172, 2826, 1990.

20. **Krieg, V. A., Huger, A. M., Langenbruch, G. A., and Schnetter, W.,** *Bacillus thuringiensis* var. *tenebrionis:* a new pathotype effective against larvae of Coleoptera, *Z. Angew. Entomol.,* 96, 500, 1983.

21. **Herrnstadt, C., Soares, G. G., Wilcos, E. R., and Edwards, D. L.,** A new strain of *Bacillus thuringiensis* with activity coleopteran insects, *Bio/Technology,* 4, 305, 1986.

22. **Donovan, W. P., Dankocsik, C., and Gilbert, M. P.,** Molecular characterization of a gene encoding a 72-kilodalton mosquito-toxic crystal protein from *Bacillus thuringiensis* subsp. *israelensis, J. Bacteriol.,* 170, 4732, 1988.

23. **Sick, A., Gaertner, F., and Wong, A.,** Nucleotide sequence of a coleopteran active toxin gene from a new isolate of *Bacillus thuringiensis* subsp. *tolworthi, Nucleic Acids Res.,* 18, 1305, 1990.

24. **Sekar, V., Thompson, D. V., Maroney, M. J., Bookland, R. G., and Adang, M. J.,** Molecular cloning and characterization of the insecticidal crystal protein gene of *Bacillus thuringiensis* var. *tenebrionis, Proc. Natl. Acad. Sci. U.S.A.,* 84, 7036, 1987.

25. **Carroll, J., Li, J., and Ellar, D. J.,** Proteolytic processing of a coleopteran-specific δ-endotoxin produced by *Bacillus thuringiensis* var. *tenebrionis, Biochem. J.,* 261, 99, 1989.

26. **Chungjatupornchai, W., Höfte, H., Seurinck, J., Angsuthanasombat, C., and Vaeck, M.,** Common features of *Bacillus thuringiensis* toxins specific for Diptera and Lepidoptera, *Eur. J. Biochem.,* 173, 9, 1988.

27. **Ibarra, J. E. and Federici, B. A.,** Isolation of a relatively non-toxic 65-kilodalton protein inclusion from the parasporal body of *Bacillus thuringiensis* subsp. *israelensis, J. Bacteriol.,* 165, 527, 1986.

28. **Yamamoto, T., Iizuka, T., and Aronson, J. N.,** Mosquitocidal protein of *Bacillus thuringiensis* subsp. *israelensis:* Identification and partial isolation of the protein, *Curr. Microbiol.,* 2, 279, 1983.

29. **Hurley, J. M., Bulla, L. A., and Andrews, R. E.,** Purification of the mosquitocidal and cytolytic proteins of *Bacillus thuringiensis* subsp. *israelensis, Appl. Environ. Microbiol.,* 53, 1316, 1987.

30. **Yu, Y. M., Ohba, M., and Aizawa, K.,** Synergistic effects of the 65- and 25-kilodalton proteins of *Bacillus thuringiensis* strain PG-14 (serotype 8A:8B) in mosquito larvicidal activity, *J. Gen. Appl. Microbiol.,* 33, 459, 1987.

31. **Pfannenstiel, M. A., Couche, G. A., Ross, E. J., and Nickerson, K. W.,** Immunological relationships among proteins making up the *Bacillus thuringiensis* subsp. *israelensis* crystalline toxin, *Appl. Environ. Microbiol.,* 52, 644, 1986.

32. **Ward, E. S. and Ellar, D. J.,** Nucleotide sequence of a *Bacillus thuringiensis* var. *israelensis* gene encoding a 130 kDa δ-endotoxin, *Nucleic Acids Res.,* 15, 7195, 1987.

33. **Sen, K., Honda, G., Koyama, N., Nishida, M., Neki, A., Sakai, H., Himeno, M., and Komano, T.,** Cloning and nucleotide sequences of the two 130 kDa insecticidal protein genes of *Bacillus thuringiensis* var. *israelensis, Agric. Biol. Chem.,* 52, 873, 1988.

34. **Tungpradubkul, S., Settasatien, C., and Panyim, S.,** The complete nucleotide sequence of a 130 kDa mosquito-larvicidal δ-endotoxin of *Bacillus thuringiensis* var. *israelensis, Nucleic Acids Res.,* 16, 1637, 1988.

35. **Yamamoto, T., Watkinson, I. A., Kim, L., Sage, M. V., Stratton, R., Akanda, N., Li, Y., Ma, D. P., and Roe, B. A.,** Nucleotide sequence of the gene coding for a 130-kDa mosquitocidal protein of *Bacillus thuringiensis israelensis, Gene,* 66, 107, 1988.

36. **Pao-intara, M., Angsuthanasombat, C., and Panyim, S.,** The mosquito larvicidal activity of 130 kDa δ-endotoxin of *Bacillus thuringiensis* var. *israelensis* resides in the 72 kDa amino-terminal fragment, *Biochem. Biophys. Res. Commun.,* 153, 294, 1988.

37. **Delecluse, A., Bourgouin, C., Klier, A., and Rapaport, G.,** Specificity of action on mosquito larvae of *Bacillus thuringiensis israelensis* toxins encoded by two different genes, *Mol. Gen. Genet.,* 214, 42, 1988.

38. **Thorne, L., Garduno, F., Thompson, T., Decker, D., Zounes, M., Wild, M., Walfield, A., and Pollock, T.,** Structural similarity between the Lepidoptera- and Diptera-specific insecticidal endotoxin genes of *Bacillus thuringiensis* subsp. *kurstaki* and *israelensis, J. Bacteriol.,* 166, 801, 1986.

39. **Donovan, W. P., Gonzalez, J. M., Gilbert, M. P., and Dankocsik, C.,** Isolation and characterization of EG2158, a new strain of *Bacillus thuringiensis* toxic to Coleopteran larvae, and nucleotide sequence of the toxin gene, *Mol. Gen. Genet.*, 214, 365, 1988.

40. **Chillcott, C. N. and Ellar, D. J.,** Comparative toxicity of *Bacillus thuringiensis* var. *israelensis* crystal proteins in vivo and in vitro, *J. Gen. Microbiol.*, 134, 2551, 1988.

41. **Wu, D. and Chang, F. N.,** Synergism in mosquitocidal activity of 26 and 65 kDa proteins from *Bacillus thuringiensis* subsp. *israelensis* crystal, *FEBS Lett.*, 190, 232, 1985.

42. **Thomas, W. E. and Ellar, D. J.,** *Bacillus thuringiensis* var. *israelensis* crystal δ-endotoxin: effects on insect and mammalian cells *in vitro* and *in vivo*, *J. Cell Science,* 60, 181, 1983.

43. **Luthy, P. and Ebersold, H. R.,** *Bacillus thuringiensis* δ-endotoxin: Histopathology and molecular mode of action, in *Pathogenesis of Invertebrate Microbial Diseases,* Davidson, E. W., Ed., Allenheld Osmum, Totawa, NJ, 1981, 235.

44. **Dow, J. T.,** Insect midgut functions, in *Advances in Insect Physiology,* Evans, P. D. and Wigglesworth, V. B., Eds., Academic Press, London, 1986, 188.

45. **Ahmad, Z., Saleemuddin, M., and Siddiggi, M.,** Alkaline protease in the larvae of the armyworm, *Spodoptera litura, Insect Biochem.,* 6, 501, 1976.

46. **Klocke, J. and Chan, B. G.,** Effects of cotton condensed tannin on feeding and digestion in the cotton pest *Heliothis zea, J. Insect Physiol.,* 28, 911, 1982.

47. **Lecadet, M. M. and Dedonder, R.,** Enzymatic hydrolysis of the crystals of *Bacillus thuringiensis* by the proteases of *Pieris brassicae.* I. Preparation and fractionation of the lysates, *J. Invertebr. Pathol.,* 9, 310, 1967.

48. **Tojo, A. and Aizawa, K.,** Dissolution and degradation of *Bacillus thuringiensis* δ-endotoxin by gut juice protease of the silkworm, *Bombyx mori, Appl. Environ. Microbiol.,* 45, 576, 1983.

49. **Choma, C. T., Surewicz, W. K., Carey, P. R., Pozsgay, M., Raynor, T., and Kaplan, H.,** Unusual proteolysis of the protoxin and toxin from *Bacillus thuringiensis:* structural implications, *Eur. J. Biochem.,* 189, 523, 1990.

50. **Andrews, R. E., Jr., Bibilos, M. M., Bulla, L. A., Jr.,** Protease activation of the entomocidal protoxin of *Bacillus thuringiensis* subsp. *kurstaki, Appl. Environ. Microbiol.,* 50, 737, 1985.

51. **Sekar, V.,** The insecticidal crystal protein gene is expressed in vegetative cells of *Bacillus thuringiensis* var. *tenebrionis, Curr. Microbiol.,* 17, 347, 1988.

52. **Haider, M. Z., Knowles, B. H., and Ellar, D. J.,** Specificity of *Bacillus thuringiensis* var. *colmeri* insecticidal δ-endotoxin is determined by differential proteolytic processing of the protoxin by larval gut proteases, *Eur. J. Biochem.,* 156, 531, 1986.

53. **Pfannenstiel, M. A., Cray, W. C., Couche, G., and Nickerson, K. W.,** Toxicity of protease-resistant domains from the δ-endotoxin of *Bacillus thuringiensis* subsp. *israelensis* in *Culex quinquefasciatus* and *Aedes aegypti* bioassays, *Appl. Environ. Microbiol.,* 56, 162, 1990.

54. **Haider, M. Z. and Ellar, D. J.,** Characterization of the toxicity and cytopathic specificity of a cloned *Bacillus thuringiensis* crystal protein using insect cell culture, *Mol. Microbiol.,* 1, 59, 1987.

55. **Luthy, P.,** Insecticidal toxins of *Bacillus thuringiensis, FEMS Lett.,* 8, 1, 1980.

56. **Yunovitz, H. and Yawetz, A.,** Interaction between the δ-endotoxin produced by *Bacillus thuringiensis* spp. *entomocidus* and liposomes, *FEBS Lett.,* 230, 105, 1988.

57. **Haider, M. Z. and Ellar, D. J.,** Mechanism of action of *Bacillus thuringiensis* insecticidal δ-endotoxin: interaction with phospholipid vesicles, *Biochem. Biophys. Acta,* 978, 216, 1989.

58. **English, L. and Readdy, T. L.,** Delta endotoxin-induced leakage of $K^+$ and $H_2O$ from phospholipid vesicles is catalyzed by reconstituted midgut membrane, *Insect Biochem.,* in press.

59. **Hanozet, G. M., Giordana, B., and Sacchi, V. F.,** $K^+$-dependent phenylalanine uptake in membrane vesicles isolated from the midgut of *Philosamia cynthia* larvae, *Biochem. Biophys. Acta,* 596, 481, 1980.

60. **Hofmann, C., Vanderbruggen, H., Höfte, H., Van Rie, J., Jansens, S., and Van Mellaert, H.,** Specificity of *Bacillus thuringiensis* δ-endotoxins is correlated with the presence of high-affinity binding sites in the brush border membrane of target insect midguts, *Proc. Natl. Acad. Sci. U.S.A.,* 85, 7844, 1988.

61. **Hofmann, C., Luthy, P., Hutter, R., and Pliska, V.,** Binding of the δ-endotoxin from *Bacillus thuringiensis* to brush-border membrane vesicles of the cabbage butterfly *(Pieris brassicae), Eur. J. Biochem.,* 173, 85, 1988.

62. **Van Rie, J., Jansens, S., Höfte, H., Degheele, D., and Van Mellaert, H.,** Specificity of *Bacillus thuringiensis* δ-endotoxins, *Eur. J. Biochem.,* 186, 239, 1989.

63. **Van Rie, J., Jansens, S., Höfte, H., Degheele, D., and Van Mellaert, H.,** Receptors on the brush border membrane of the insect midgut as determinants of the specificity of *Bacillus thuringiensis* δ-endotoxins, *Appl. Environ. Microbiol.,* 56, 1378, 1990.

64. **Wolfersberger, M. G.,** Specificity and mode of action of *Bacillus thuringiensis* insecticidal crystal proteins toxic to lepidopteran larvae: Recent insights from studies utilizing midgut brush border membrane vesicles, in *Proc. VIth Int. Colloq. Invertebrate Pathology,* Society for Invertebrate Pathology, Adelaide, 1990, 278.

65. **Wolfersberger, M. G.,** The toxicity of two *Bacillus thuringiensis* δ-endotoxins to gypsy moth larvae is inversely related to the affinity of binding sites on midgut brush border membranes for the toxins, *Experientia,* 46, 475, 1990.

66. **Garczynski, S. G., Crim, J., and Adang, M. J.,** Identification of putative brush border membrane binding proteins specific to *Bacillus thuringiensis* δ-endotoxin by protein blot analysis, submitted.

67. **Soutar, A. K., and Wade, D. P.,** Ligand blotting, in *Protein Function: a Practical Approach,* Creighton, T. E., Ed., IRL Press, Oxford, 1989, 55.

68. **Haider, M. Z. and Ellar, D. J.,** Analysis of the molecular basis of insecticidal specificity of *Bacillus thuringiensis* crystal δ-endotoxin, *Biochem. J.,* 248, 197, 1987.

69. **Cioffi, M. and Harvey, W. R.,** Comparison of potassium transport in three structurally distinct regions of the insect midgut, *J. Exp. Biol.,* 91, 103, 1981.

70. **Moffett, D.,** Voltage-current relation and $K^+$ transport in tobacco hornworm (*Manduca sexta*) midgut, *J. Membr. Biol.,* 54, 213, 1980.

71. **Harvey, W. R. and Wolfersberger, M. G.,** Mechanism of inhibition of active potassium transport in isolated midgut of *Manduca sexta* by *Bacillus thuringiensis* endotoxin, *J. Exp. Biol.,* 106, 91, 1979.

72. **Griego, V. M., Moffett, D., and Spence, K. D.,** Inhibition of active $K^+$ transport in the tobacco hornworm (*Manduca sexta*) midgut after ingestion of *Bacillus thuringiensis* endotoxin, *J. Insect Physiol.,* 25, 283, 1979.

73. **Crawford, D. N. and Harvey, W. R.,** Barium and calcium block *Bacillus thuringiensis* subspecies *kurstaki* δ-endotoxin inhibition of potassium current across isolated midgut of larval *Manduca sexta, J. Exp. Biol.,* 137, 277, 1988.

74. **Sacchi, V. F., Parenti, P., Hanozet, G. M., Giordana, B., Luthy, P., and Wolfersberger, M. G.,** *Bacillus thuringiensis* toxin inhibits $K^+$-gradient-dependent amino acid transport across the brush border membrane of *Pieris brassicae* midgut cells, *FEBS Lett.,* 204, 213, 1986.

75. **Wolfersberger, M. G.,** Neither barium nor calcium prevents the inhibition of Bacillus thuringiensis δ-endotoxin of sodium- or potassium gradient-dependent amino acid accumulation by tobacco hornworm midgut brush border membrane vesicles, *Arch. Insect. Biochem. Physiol.,* 12, 267, 1989.

76. **Hendrickx, K., De Loof, A., and Van Mellaert, H.,** Effects of *Bacillus thuringiensis* δ-endotoxin on the permeability of brush border membrane vesicles from tobacco hornworm *(Manduca sexta)* midgut, *Comp. Biochem. Physiol.,* 95C, 241, 1990.

77. **Knowles, B. H. and Ellar, D. J.,** Colloid-osmotic lysis is a general feature of the mechanism of action of *Bacillus thuringiensis* δ-endotoxins with different insect specificities, *Biochem. Biophys. Acta,* 924, 509, 1987.

78. **Slatin, S. L., Abrams, C. K., and English, L.,** Delta-endotoxins form cation-selective channels in planar lipid bilayers, *Biochem. Biophys. Res. Commun.,* 169, 765, 1990.

79. **Parker, M. W., Tucker, A. D., Tsernoglou, D., and Pattus, F.,** Insights into membrane insertion based on studies of colicins, *Trends Biochem. Sci.,* 15, 126, 1990.

80. **English, L. and Cantley, L. C.,** Delta endotoxin is a potent inhibitor of the (Na,K)-ATPase, *J. Biol. Chem.,* 261, 1170, 1986.

81. **English, L. H. and Readdy, T. L.,** Delta endotoxin inhibits a phosphatase in midgut epithelial membranes of *Heliothis virescens, Insect Biochem.,* 19, 145, 1989.

82. **Spies, A. G. and Spence, K. D.,** Effect of sublethal *Bacillus thuringiensis* crystal endotoxin treatment on the larval midgut of a moth, *Manduca sexta:* SEM study, *Tissue Cell,* 17, 379, 1985.

83. **Goldberg, L. J. and Margalit, J.,** A bacterial spore demonstrating rapid larvicidal activity against *Anopheles sergentii, Uranotaenia unguiculata, Culex univitattus, Aedes aegypti* and *Culex pipiens, Mosquito News,* 37, 355, 1977.

84. **Herrnstadt, C., Soares, G., Wilcox, E. R., and Edwards, D. L.,** A new strain of *Bacillus thuringiensis* with activity against coleopteran insects, *Bio/Technology,* 4, 305, 1986.

85. **Shieh, T. R.,** *Bacillus thuringiensis* biological insecticide and biotechnology, in *The Impact of Chemistry on Biotechnology: Multidisciplinary Discussions,* Phillips, M. P., Shoemaker, S. P., Middlekauf, R. D., and Ottenbrite, R. M., Eds., American Chemical Society, Washington, D.C., 1988, 207.

86. **Ishiwata, S.,** On a kind of flacherie (sotto disease), *Dainihon Sanshi Keiho,* 9, 1, 1901.

87. **Berliner, E.,** Über die Schlaffsucht der Mehlmottenraupe, *Z. Gesamte Getreidewes.,* 252, 3160, 1911.

88. **Dulmage, H. T.,** Insecticidal activity of HD-1, a new isolate of *Bacillus thuringiensis* var. *alesti, J. Invertebr. Pathol.,* 15, 232, 1970.

89. **Ohba, M. and Aizawa, K.,** Serological identification of *Bacillus thuringiensis* and related bacteria isolated in Japan, *J. Invertebr. Pathol.,* 33, 387, 1978.

90. **Ohba, M. and Aizawa, K.,** Distribution of *Bacillus thuringiensis* in soils of Japan, *J. Invertebr. Pathol.,* 47, 277, 1986.

91. **Saleh, S. M., Harris, R. F., and Allen, O. N.,** Method for determining *Bacillus thuringiensis* var. *thuringiensis* Berliner in soil, *Can. J. Microbiol.,* 15, 1101, 1969.

92. **DeLucca, A. J., Simonson, J. G., and Larson, A. D.,** *Bacillus thuringiensis* distribution in soils of the United States, *Can. J. Microbiol.,* 27, 865, 1981.

93. **Travers, R. S., Martin, P. A., and Reichelderfer, C. F.,** Selective process for efficient isolation of soil *Bacillus* species, *Appl. Environ. Microbiol.,* 53, 1263, 1987.

94. **DeLucca, A. J., Palmgren, M. S., and Ceigler, A.,** *Bacillus thuringiensis* in grain elevator dusts, *Can. J. Microbiol.,* 28, 452, 1982.

95. **Dulmage, H. T. and Cooperators,** Insecticidal activity of isolates of *Bacillus thuringiensis* and their potential for pest control, in *Microbial Control of Pests and Plant Diseases, 1970—1980,* Burges, H. D., Ed., Academic Press, London, 1981, 193.

96. **McClinden, J. H., Sabourin, J. R., Clark, B. D., Gensler, D. R., Workman, W. E., and Dean, D. H.,** Cloning and expression of an insecticidal k-73 type crystal protein gene from *Bacillus thuringiensis* var. *kurstaki* into *Eschericia coli, Appl. Environ. Microbiol.,* 50, 623, 1985.

97. **Jaquet, F., Hutter, R., and Lüthy, P.,** Specificity of *Bacillus thuringiensis* δ-endotoxin, *Appl. Environ. Microbiol.,* 53, 500, 1987.

98. **Johnson, D. E.,** Specificity of cultured insect tissue cells for the bioassay of entomocidal protein of *Bacillus thuringiensis,* in *Invertebrate Cell System Applications,* 2nd ed., Mitsuhashi, J., Ed., CRC Press, Boca Raton, 1989, chap. 10.

99. **Chow, E., and Gill, S. G.,** A rapid colorimetric assay to evaluate the effects of Bacillus thuringiensis toxins on cultured insect cells, *J. Tissue Culture Methods,* 12, 39, 1989.

100. **Gringorten, J. L., Witt, D. P., Milne, R. E., Fast, P. G., Sohi, S. S., and Van Frankenhuyzen, K.,** An *in vitro* system for testing *Bacillus thuringiensis* toxins: the lawn assay, *J. Invertebr. Pathol.,* 56, 237, 1990.

101. **Rusche, J., Jellis, C. L., Farrell, K., Farley, J., Hodgon, J., Carson, H., and Witt, D.,** Expression and alteration of a *Bacillus thuringinesis* endotoxin gene in *Eschericia coli,* in *Biotechnology for Crop Protection,* Hedin, P. A., Menn, J. J., and Hollingworth, R. M., Eds., American Chemical Society, Washington, D.C., 1988, 454.

102. **Höfte, H., Van Rie, J., Jansens, S., Van Houtven, A., Vanderbruggen, H., and Vaeck, M.,** Monoclonal antibody analysis and insecticidal spectrum of three types of lepidopteran-specific insecticidal crystal proteins of *Bacillus thuringiensis, Appl. Environ. Microbiol.,* 54, 2010, 1988.

103. **MacIntosh, S. C., Stone, T. B., Sims, S. R., Hunst, P. L., Greenplate, J. T., Marrone, P. G., Perlak, F. J., Fischoff, D. A., and Fuchs, R. L.,** Specificity and efficacy of purified *Bacillus thuringiensis* proteins against agronomically important insects, *J. Invertebr. Pathol.,* 56, 258, 1990.

104. **Prefontaine, G., Fast, P., Lau, P. C. K., Hefford, M., Hanna, Z., and Brosseau, R.,** Use of oligonucleotide probes to study the relatedness of δ-endotoxin genes among *Bacillus thuringiensis* subspecies and strains, *Appl. Environ. Microbiol.,* 53, 2808, 1987.

105. **Visser, B.,** A screening method for the presence of four different crystal protein gene types in 25 *Bacillus thuringiensis* strains, *FEMS Lett.,,* 58, 121, 1989.

106. **Krywienczyck, J., Dulmage, H. T., Hall, I. M., Beegle, C. C., Arakawa, K. Y., and Fast, P. G.,** Occurrence of *kurstaki* k-1 crystal activity in *Bacillus thuringiensis* subsp. *thuringiensis* serovar (H1), *J. Invertebr. Pathol.,* 37, 62, 1981.

107. **Krywienczyck, J., Dulmage, H. T., and Fast, P. G.,** Occurrence of two serologically distinct groups with *Bacillus thuringiensis* serotype 3ab var. *kurstaki, J. Invertebr. Pathol.,* 31, 372, 1978.

108. **Lynch, M. J. and Baumann, P.,** Immunological comparisons of the crystal protein from strains of *Bacillus thuringiensis, J. Invertebr. Pathol.,* 46, 47, 1985.

109. **Smith, R. A.,** Use of crystal serology to differentiate among varieties of *Bacillus thuringiensis, J. Invertebr. Pathol.,* 50, 1, 1987.

110. **Huber-Lukac, M., Jaquet, F., Luethy, P., Huetter, R., and Braun, D. G.,** Characterization of monoclonal antibodies to a crystal protein of *Bacillus thuringiensis* subsp. *kurstaki., Infect. Immun.,* 54, 228, 1986.

111. **Adang, M. J., Idler, K. F., and Rochelaeu, T. A.,** Structural and antigenic relationships among three insecticidal crystal proteins of *Bacillus thuringiensis* subsp. *kurstaki,* in *Biotechnology in Invertebrate Pathology and Cell Culture,* Maramorosch, K., Ed., Academic Press, San Diego, 1987, 85.

112. **Gonzalez, J. M., Dulmage, H. T., and Carlton, B. C.,** Correlation between specific plasmids and δ-endotoxin production in *Bacillus thuringiensis, Plasmid,* 5, 351, 1981.

113. **Gaertner, F.,** Cellular delivery systems for insecticidal proteins: living and non-living microorganisms, in *Controlled Delivery of Crop Protection Agents,* Wilkins, R. M., Ed., Taylor and Francis, PA, 1990, 245.

114. **Heirson, A., Landen, R., Lovgren, A., Dalhammar, G., and Boman, H. G.,** Transformation of vegetative cells of *Bacillus thuringiensis* by plasmid DNA, *J. Bacteriol.,* 169, 1147, 1987.

115. **Lereclus, D., Arantes, O., Chaufaux, J., and Lecadet, M.-M.,** Transformation and expression of a cloned endotoxin gene in *Bacillus thuringiensis, FEMS Microbiol. Lett.,* 60, 211, 1989.

116. **Mahillon, J., Chungatupornchai, W., Decock, J., Dierickx, S., Michiels, F., Peferoen, M., and Joos, H.,** Transformation of *Bacillus thuringiensis* by electroporation, *FEMS Microbiol. Lett.,* 60, 205, 1989.

117. **Schurter, W., Geiser, M., and Mathé, D.,** Efficient transformation of *Bacillus thuringiensis* and *B. cereus* via electroporation: transformation of acrystalliferous strains with a cloned δ-endotoxin gene, *Mol. Gen. Genet.,* 218, 177, 1989.

118. **Baum, J. A., Coyle, D. M., Gilbert, M. P., Jany, C. S., and Gawron-Burke, C.,** Novel cloning vectors for *Bacillus thuringiensis, Appl. Environ. Microbiol.,* 56, 3420, 1990.

119. **Höfte, H., de Greve, H., Seurinck, J., Jansens, S., Mahillon, J., Ampe, C., Vanderkerckhove, J., Vanderbruggen, H., Van Montagu, M., Zabeau, M., and Vaeck, M.,** Structural and functional analysis of a cloned delta endotoxin gene of *Bacillus thuringiensis* Berliner 1715, *Eur. J. Biochem.,* 161, 273, 1986.

120. **Honee, G., van der Salm, T., and Visser, B.,** Nucleotide sequence of crystal protein gene isolation from *Bacillus thuringiensis* subspecies *entomocidus* 60.5 coding for a toxin highly active against *Spodoptera* species, *Nucleic Acids Res.,* 16, 6240, 1988.

121. **Obukowicz, M. G., Perlak, F. J., Kusano-Kretzmer, K., Mayer, E. J., Bolten, S. L., and Watrud, L. S.,** Tn5-mediated integration of the δ-endotoxin gene from *Bacillus thuringiensis* into the chromosome of root-colonizing pseudomonads, *J. Bacteriol.,* 168, 982, 1986.

122. **Dimock, M. B., Beach, R. M., and Carlson, P.,** Endophytic bacteria for the delivery of crop protection agents, in *Biotechnology for Crop Protection,* Roberts, D. W. and Granados, R. R., Eds., ACS Symposium Series 379, American Chemical Society, Washington, D.C., 1989, 88.

123. **Stock, C., McGloughlin, T. J., Klein, J. A., and Adang, M. J.,** Expression of a *Bacillus thuringiensis* crystal protein gene in *Pseudomonas cepacia* 526, *Can. J. Microbiol.,* 36, 879, 1990.

124. **Drahos, D. J., Hemming, B. C., and McPherson, S.,** Tracking recombinant organisms in the environment: B-galactosidase as a selectable non-antibiotic marker for fluorescent pseudomonads, *Bio/Technology,* 4, 439, 1986.

125. **Adang, M. J., Firoozabady, E., Klein, J., Deboer, D., Sekar, V., Kemp, J. D., Murray, E., Rocheleau, T. A., Rashka, K., Staffeld, G., Stock, C., Sutton, D., and Merlo, D.,** Expression of a *Bacillus thuringiensis* insecticidal protein gene in tobacco plants, in *Molecular Strategies for Crop Protection. UCLA Symposia on Molecular and Cellular Biology. New Series,* Arntzen, C., and Ryan, C., Eds., Alan R. Liss, New York, 1987, 345.

126. **Vaeck, M., Reynaerts, A., Hofte, H., Jansens, S., De Beuckeleer, M., Dean, C., Zabeau, M., Van Montagu, M., and Leemans, J.,** Transgenic plants protected from insect attack, *Nature,* 328, 33, 1987.

127. **Fischoff, D. A., Bowdis, K. S., Perlak, F. J., Marrone, P. G., McCormick, S. M., Niedermeyer, J. G., Dean, D. A., Kusano-Kretzmer, K., Mayer, E. J., Rochester, D. E., Rogers, S. G., and Fraley, R. T.,** Insect tolerant transgenic tomato plants, *Bio/Technology,* 5, 807, 1987.

128. **Barton, K. A., Whiteley, H. R., and Yang, N. S.,** *Bacillus thuringiensis* δ-endotoxin expressed in transgenic *Nicotiana tabacum* provides resistance to lepidopteran insects, *Plant Physiol.,* 85, 1103, 1987.

129. **Adang, M. J., Deboer, D., Endres, J., Firoozabady, E., Klein, J., Merlo, A., Merlo, D., Murray, E., Rashka, K., and Stock, C.,** Manipulation of *Bacillus thuringiensis* genes for pest insect control, in *Biotechnology, Biological Pesticides, and Novel Plant-Pest Resistance for Insect Pest Management,* Roberts, D. W., and Granados, R. R., Eds., Boyce Thompson Institute for Plant Research, Ithaca, New York, 1988, 31.

130. **Murray, E. E., Stock, C., Eberle, M., Sekar, V., Rocheleau, T. A., and Adang, M. J.,** Analysis of unstable RNA transcripts of insecticidal crystal protein genes of *Bacillus thuringiensis* in transgenic plant and electroporated protoplasts, *Plant Mol. Biol.,* in press.

131. **Adang, M. J., Rocheleau, T. A., Merlo, D. J., and Murray, E. E.,** Synthetic insecticidal crystal protein gene, European Pat. Appl., 27760, 1990.

132. **Perlak, F. J., Deaton, R. W., Armstrong, T. A., Fuchs, R. L., Sims, S. R., Greenplate, J. T., and Fischoff, D. A.,** Insect resistant cotton plants, *Bio/Technology,* 8, 939, 1990.

133. **Gordon-Kamm, W. J., Spencer, T. M., Mangano, M. L., Adams, T. R., Daines, R. J., Start, W. G., O'Brien, J. V., Chambers, S. A., Adams, W. R., Willetts, N. G., Rice, T. B., Mackey, C. J., Krueger, R. W., Kausch, A. P., and Lemaux, P. G.,** Transformation of maize cells and regeneration of fertile transgenic plants, *Plant Cell,* 2, 603, 1990.

134. **Klein, T. M., Wolf, E. D., Wu, R., and Sanford, J. C.,** High-velocity microprojectiles for delivering nucleic acids into living cells, *Nature,* 327, 70, 1987.

135. **Klein, T. M. and Maliga, P.,** Plant transformation by particle bombardment, in *Biotechnology for Biological Control of Pests and Vectors,* Maramorosch, K., Ed., CRC Press, Boca Raton, Fl, 1991, 105.

136. **Fischoff, D.,** Interactions between engineered insecticidal proteins and their hosts when expressed by transgenic plants, presented at Annu. Meeting of the Entomological Society of America, New Orleans, December 2—6, 1990.

137. **Bietlot, H., Carey, P. R., Choma, C., Kaplan, H., Lessard, T., and Pozsgay, M.,** Facile preparation and characterization of the toxin from *Bacillus thuringiensis* var. *kurstaki, Biochem. J.,* 260, 87, 1989.

138. **Chestukhina, G. G., Tyurin, S. A., Kostina, L. I., Osterman, A. L., Zalunin, I. A., Khodova, O. A., and Stepanov, V. M.,** Subdomain organization of *Bacillus thuringiensis* entomocidal proteins' N-terminal domains, *J. Protein Chem.,* 9, 501, 1990.

139. **Choma, C. T., Surewicz, W. K., Carey, P. R., and Pozsgay, M.,** Secondary structure of the entomocidal toxin from *Bacillus thuringiensis* subsp. *kurstaki* HD-73, *J. Protein Chem.*, 9, 87, 1990.

140. **Convents, D., Houssier, C., Lasters, I., and Lauwereys, M.,** The *Bacillus thuringiensis* δ-endotoxin: evidence for a two domain structure of the minimal toxic fragment, *J. Biol. Chem.*, 265, 1369, 1990.

141. **Armstrong, J. L., Rohrmann, G. F., and Beaudreau, G. S.,** Delta endotoxin of *Bacillus thuringiensis* subsp. *israelensis*, *J. Bacteriol.*, 161, 39, 1985.

142. **Knowles, B. H., Nicholls, C. N., Armstrong, G., Tester, M., and Ellar, D. J.,** Broad spectrum cytolytic toxins made by *Bacillus thuringiensis*, in Vth Int. Colloq. Invertebrate Pathology and Microbial Control, Adelaide, Australia, 1990, 283.

143. **Jany, C. S., Jelen, A., Chambers, J., Von Tersch, M., Gawron-Burke, C., and Johnson, T. B.,** Bioactivity of a novel insecticidal crystal protein from *Bacillus thuringiensis* subsp. *aizawai*, in *Annu. Meeting Entomological Society of America*, New Orleans, December 2—6, 1990.

144. **Slaney, A. C., Johnson, T. B., Donovan, W. P., and Rupar, M. J.,** *Bacillus thuringiensis* toxic to Coleoptera: comparative bioassays of strains producing different CryIII δ-endotoxins, in *Annu. Meeting Entomological Society of America*, New Orleans, 1990.

145. **Nishiisutsuji-Uwo, J., Endo, Y., and Himeno, M.,** Mode of action of *Bacillus thuringiensis* δ-endotoxin: effect on TN-368 cells, *J. Invertebr. Pathol.*, 34, 267, 1979.

146. **Knowles, B. H. and Ellar, D. J.,** Characterization and partial purification of a plasma membrane receptor for *Bacillus thuringiensis* var. *kurstaki* lepidopteran-specific δ-endotoxin, *J. Cell Sci.*, 83, 89, 1986.

147. **English, L. H. and Cantley, L. C.,** Delta endotoxin inhibits Rb$^+$ uptake, lowers cytoplasmic pH and inhibits a K$^+$-ATPase in *Manduca sexta* CHE cells, *J. Membr. Biol.*, 85, 199, 1985.

148. **Yunovitz, H., Sneh, B., Schuster, S., Oron, U., Broza, M., and Yawetz, A.,** A new sensitive method of determining the toxicity of a highly purified fraction from δ-endotoxin produced by *Bacillus thuringiensis* var. *entomocidus* on isolated larval midgut on *Spodoptera litoralis* (Lepidoptera, Noctuidae), *J. Invertebr. Pathol.*, 48, 223, 1986.

149. **Ryerse, J. S., and Beck, J. R., Jr.,** Light microscope immunolocation of *Bacillus thuringiensis kurstaki* δ-endotoxin in the midgut and malpighian tubules of the tobacco budworm, *Heliothis virescens*, *J. Invertebr. Pathol.*, 56, 86, 1990.

Chapter 2

# MOSQUITOCIDAL TOXIN GENE OF *BACILLUS THURINGIENSIS* SUBSPECIES *ISRAELENSIS*

Lee A. Bulla, Jr., Kathleen C. Raymond, and Robert M. Faust

## TABLE OF CONTENTS

# I. INTRODUCTION

*Bacillus thuringiensis* subsp. *israelensis (Bti)* produces proteinaceous parasporal crystals during sporulation. These crystals are unlike those of the lepidopteran-specific *Bt* subspecies, in that they are composed of multiple proteins ranging in size from 26 to 135 kilodaltons (kDa), rather than a major polypeptide, that of the protoxin (135 kDa). The crystals are lethal to larval and adult mosquitoes[24,32] and to blackfly larvae.[49] The larvicidal activity is extremely high, rapid, and specific for mosquitoes and blackflies.[24,49] No effect is observed on other aquatic insects, fish, or frog larvae.[21] For this reason, the World Health Organization (WHO) recommended utilization of *Bti* as an agent for biological control of mosquitoes.[3] This recommendation was approved by the Environmental Protection Agency (EPA) in the U.S. Since then, commercial sources of *Bti* have been successfully used to control mosquitoes and blackflies.

Mosquitoes and blackflies are serious pests to humans and animals. Mosquitoes transmit diseases such as filaria, elephantiasis, malaria, and yellow fever, all of which still are threats in tropical areas.[23] Blackflies act as vectors of filarial worms and blood protozoans among domestic and wild vertebrates and transmit parasites to man.[11] Synthetic chemical pesticides have been effectively used to control these pests. However, the disadvantages of chemicals are that they persist for a long time in the environment, they are hazardous to humans, and due to chemical longevity, genetically based resistance to the chemicals occurs. Consequently, the development of biological controls as an alternative means of pest control is desired.

Much research is required to develop a long lasting and viable product for field application. Direct application of *Bti* in field conditions to control mosquitoes has complications. For example, Margalit et al.[38] determined that four applications, at 10-day intervals, of *Bti* powder suspended in water, decreased larval mosquito populations, but about 48 h after each application, mosquito larvae hatched from new eggs. To further control the larval population in the absence of larval predators, spraying with *Bti* at 8- to 10-day intervals was necessary. Here, we discuss possible strategies for transferring the *Bti* toxin gene into alternate hosts that are food sources for mosquitoes and blackflies. Release of a genetically modified *Bti* biocontrol agent for mosquitoes and blackflies is feasible and could be safer and more economical than chemical or fermentation processes currently used.

# II. MOSQUITOCIDAL TOXIN

The parasporal crystals of *Bti* are composed of several distinct proteins ranging in size from 26 to 135 kDa.[17,37,39,48] The major components in these crystals, however, have an apparent molecular weight of approximately 28 and 68 kDa. There has been considerable debate as to which protein possesses dipteran insecticidal activity. This debate centers around the 26- to 28- and 65- to 68-kDa proteins. The cytolytic activity of the 28-kDa protein is well documented,[28,45,46] but there is controversy over which protein bestows larvicidal activity. Recently, it has been suggested that neither protein alone is larvicidal, but that there is synergism between the two proteins which is necessary to produce larvicidal activity.[29,59]

Thomas and Ellar[45,46] showed that solubilized *Bti* crystals caused rapid lysis of insect and mammalian cells and, through further experimental analysis, assigned the mosquitocidal activity to the 26-kDa protein. They hypothesized that toxicity is manifested by interaction of the protein with the plasma membrane causing a detergent-like rearrangement of the lipids, thereby leading to disruption of the membrane and cytolysis.[17]

Armstrong et al.[4] made an initial assumption that the toxin would be resistant to protease digestion, which is a common phenomenon among species of *Bacillus,* and purified a protease-resistant protein from *Bti*. The crystals were solubilized with alkali and the proteins were digested with trypsin and proteinase K. Using a combination of gel filtration and ion exchange chromatography, two forms of a 25-kDa protein, differing in size by two amino acids, were purified. This protein, in either form, lysed rabbit and human erythrocytes, lysed cultured mosquito cells, was insecticidal to mosquito larvae at 50 μg/ml, and was lethal to mice. When antibodies to the purified protein were tested against *Bti* proteins, only those proteins from toxic *Bti* strains cross-reacted with the antibodies. They concluded that this protein was derived from the 28-kDa peptide, which they, therefore, believe to be the *Bti*

In contrast, Lee et al.[37] assigned the mosquitocidal activity to the 65-kDa protein. Parasporal inclusion bodies from *Bti* were separated according to size through a 10 to 40% sucrose gradient. Based on protein content, the population designated as small dots was found to be most toxic to mosquito larvae. Analysis of the protein content of the small dots revealed that they were composed of the 38-kDa and 65-kDa proteins, but did not contain any detectable 28-kDa protein. Assignment of the toxic activity to the 65-kDa protein was made when further analysis showed a strong correlation between production of the 65-kDa protein and larvicidal activity.

Gel filtration chromatography of alkali-solubilized *Bti* crystals permitted separation of the 28- and 65-kDa proteins.[28] The 28-kDa protein was not mosquitocidal, but caused hemolysis of rat red blood cells. The 65-kDa protein, however, was found to be toxic to mosquito larvae (*Aedes aegypti*) with an $LC_{50}$ of 180 ng/ml and had no hemolytic activity. Purification of the toxic protein (65 kDa) from the unfractionated crystal protein mixture resulted in a sevenfold increase in specific activity, which correlated with their observation that the 65-kDa protein represented approximately one seventh of the total proteins in the crystal preparations they used.

Sriram et al.[43] used the antibotic netropsin to separate the sporulation and crystal formation events in *Bti*. At low concentrations (1 to 3 μg/ml), the antibiotic inhibited sporulation, but had no effect on crystal production or larvicidal activity ($LC_{50} = 50$ ng/ml). However, at higher concentrations of netropsin (3 to 7 μg/ml), crystal inclusions were not formed, and there was a concomitant tenfold reduction in the larvicidal activity ($LC_{50} = 500$ ng/ml). SDS polyacrylamide gel electrophoretic analysis of alkali-solubilized proteins from cells treated with varying amounts of netropsin revealed that the acrystalliferous cells contained only the low molecular weight crystal peptides, and that the 26 kDa was present in all of the cells analyzed. From this observation, they concluded that the larger molecular weight proteins are responsible for crystal production and that the 28-kDa protein is the larvicidal component in the cells.

A synergism between the 26- and 65-kDa proteins also has been suggested.[59] *Bti* crystals were alkali-solubilized overnight at 4°C and then chromatographed, resulting in separation of three major protein components of 26, 65, and 130 kDa. The 26-kDa protein was inactive against mosquito larvae at concentrations of 6.4 μg/ml. The 65-kDa protein also was inactive in the assay system at low concentrations, and only slightly active at higher concentrations. What little activity it had at concentrations of 1.6 μg/ml was attributed to contamination by the 26-kDa protein. Maximum larvicidal activity against *Ae. aegypti* was observed only when both of the proteins were present simultaneously.

A similar conclusion was reached by Ibarra and Federici.[29] An inclusion body from *Bti*, predominantly composed of the 65-kDa protein, was isolated and used in a larvicidal assay. This inclusion body was less toxic to mosquito larvae (*Ae. aegypti*) than were the native *Bti* crystals in their assay system. They also found that the toxicity of this inclusion body was directly correlated to the extent of contamination with the 28-kDa protein.

Held et al.[26] removed the 28-kDa cytolytic protein from solubilized *Bti* crystals by affinity chromatography using a monoclonal antibody directed against the 28-kDa protein. Bioassays were performed on third instar *Ae. aegypti* larvae to determine $LC_{50}$ values. Total solubilized crystals had an $LC_{50}$ value of 0.64 µg/ml. The crystal protein fraction depleted of the 28-kDa protein was nonhemolytic and retained nearly full toxicity to mosquito larvae ($LC_{50}$ = 0.75 µg/ml). The purified 28-kDa protein was hemolytic and relatively nontoxic to mosquito larvae ($LC_{50}$ = 21.7 µg/ml). They concluded that because the 65-kDa protein is the predominant protein in the flow-through fraction, devoid of the 28-kDa cytolytic protein, that this particular protein is the mosquitocidal toxin.

We believe that the 65-kDa protein is solely responsible for larvicidal activity.[28,37] However, only by cloning each of the genes will a definitive assignment of the toxin activity to either the 28- or 65-kDa proteins be possible.

## III. CLONING OF THE MOSQUITOCIDAL GENE

*B. thuringiensis* subsp. *israelensis,* serotype 14, has been analyzed by agarose gel electrophoresis. It contains eight plasmids ranging in size from 3.3 to 135 megadaltons (MDa) and one linear piece of DNA of approximately 10 MDa.[25] Plasmid-curing studies implicated the 75-MDa plasmid in crystal production, which is synonymous with larvicidal activity. Plasmid transfer experiments involving acrystalliferous (cry⁻) Bti strains further implicated the 75-MDa plasmid as the one encoding the larvicidal activity. Other plasmids were transferred into the Cry⁻ strain, but only transfer of the 75-MDa plasmid converted the transcipient strain to crystal and toxin production.

Ward et al.[55] cloned HindIII fragments from this 75-MDa plasmid into the vector pUC13. Two clones, p1P173 and p1P174, both containing a 9.7-kilobase pair (kbp) HindIII fragment, produced a 26-kDa protein in an *E. coli in vitro* transcription-translation system. The protein produced by these clones was precipitable by antibody prepared against the 26-kDa protein from native *Bti* crystals. *E. coli* cells harboring p1P174 were toxic to *Ae. aegypti* larvae, and protein extracts caused cytolysis of *Ae. albopictus* cells. The amount of protein required for cells to be lethal to *Ae. aegypti* larvae suggests that this clone does not encode the protoxin but, rather, the cytolytic activity. Moreover, that cytolytic activity is observed substantiates this notion.

Using virtually the same approach, Waalwijk et al.[53] also cloned a 9.7-kbp HindIII fragment into pBR322. This clone, p425, produced a 28-kDa protein in an *in vitro* transcription-translation system. Amino acid composition of this protein agreed well with that determined for the 28-kDa protein from the native *Bti* crystals. Insect toxicity was not examined, however. This clone most likely is identical to that previously obtained by Ward et al.,[55] and, therefore, we believe that it contains the gene encoding the cytolytic activity and not the protoxin activity.

Because expression of the protoxin gene is sporulation specific in native *Bti*, Sekar and Carlton[41] employed a *Bacillus* cloning system, rather than an *E. coli* one, to enhance production of the cloned gene product. DNA from the 75-MDa plasmid of *Bti* was partially digested with XbaI and ligated into the *B. cereus*-cloning vector, pBC16, carrying a tetracycline resistance marker. Polyethyleneglycol-induced protoplasts of *B. megaterium* VT1660 were transformed with the resultant DNA, and tetracycline-resistant transformants were selected. The transformants were allowed to sporulate and lyse; the lysates were used directly for bioassays against *Ae. aegypti* larvae. A toxic clone, VB131, was produced that contained a 6.3-kbp insert composed of three XbaI fragments (2.7, 1.8, and 1.8 kbp). This clone produced phase-refractile bodies during sporulation that were toxic to *Ae. aegypti* larvae. Alkali-solubilized inclusion bodies from this clone produced a precipitin band in a double immunodiffusion assay using antiserum prepared against solubilized *Bti* crystals. No polyacrylamide gel analysis was provided, and, therefore, it is not known what polypeptide species was produced from this clone.

Recently, Thorne et al.[47] described the cloning of a fragment from the 75-MDa plasmid of *Bti* that encodes larvicidal activity. The restriction map of this insert bears no resemblance to either of the clones previously described.[41,53] Furthermore, the clone has no hemoloytic activity on human red blood cells.[60] Thorne et al.[47] have sequenced the region of DNA encoding the gene product responsible for larvicidal activity. Two large open reading frames (ORFs) were identified in the 3750 nucleotides sequenced. The first ORF could code for a protein of 72 kDa and the second ORF, which extends beyond the sequenced region, could encode a protein of at least 26 kDa. When gene expression was carried out *in vitro* with an *E. coli* transcription-translation system, polypeptides up to 72 kDa were generated; however, when expressed in *B. subtilis*, a 58-kDa protein accumulated. The 5′-flanking sequence containing the promoter and ribosome binding site is virtually identical to that of *B. thuringiensis* subsp. *kurstaki* (toxic to lepidopteran species), which these investigators[47] also have cloned and sequenced. The coding regions of the two different subspecies exhibited little DNA homology, although the deduced amino acid sequences were strikingly similar.

Both Waalwijk et al.[53] and Thorne et al.[47] have sequenced the DNA responsible for encoding what they believe to be the mosquito toxin, which is known to be expressed only during sporulation in *Bti*. Interestingly, the promoter region of the gene encoded in the clone of Thorne et al.[47] appears to be sporulation-specific, as evidenced by its increased production in *B. subtilis* during sporulation and a DNA sequence homologous to the promoter of the cloned toxin gene from *B.t. kurstaki*. The sequence of the promoter encoding the 28-kDa protein, which Waalwijk et al.[53] claim to be larvicidal, however, is different from that described for other *Bacillus* genes expressed during sporulation.

From the deduced amino acid sequences of the two genes,[47,53] the hydropathy and secondary structure of the molecules have been predicted.[54] The predicted hydropathy of the 28-kDa protein bears no resemblance to the predicted hydropathy of either the 68-kDa protein from *Bti* or the 130-kDa protoxin from *B.t. thuringiensis*.[54] Curiously, the 68-kDa and 130-kDa proteins are strikingly similar. Likewise, a comparison of the predicted secondary structure of the three molecules lends additional evidence to the conclusion that the 68-kDa and 130-kDa proteins are quite similar.[54] These comparisons, showing a predicted similarity between the lepidopteran and dipteran killing toxins, along with the fact that the 68-kDa protein is not cytolytic to human red blood cells,[60] indicate that, in fact, the 68-kDa protein is the *Bti* toxin, as previous work indicated.[28,37]

Still, the debate as to which protein in the *Bti* crystal complex is larvicidal continues and now includes the 130-kDa protein. Visser et al.[52] solubilized *Bti* crystals and separated three proteins of 230, 130, and 28 kDa, by sucrose gradient centrifugation. Mosquitocidal activity coincided with the 230- and 130-kDa peaks, whereas hemolytic activity was found only in the peak corresponding to the 28-kDa protein. The protein with the greatest mosquitocidal activity was the 130-kDa protein. They also observed that solubilization of the proteins in NaOH pH 12.0 vs. pH 9.5, favored production of a 65-kDa protein with a concomitant reduction in the amounts of the 230- and 130-kDa proteins. Unfortunately, they never performed bioassays with the 65-kDa protein, nor did they determine whether the 230- or 130-kDa proteins are protoxins containing a 65-kDa toxic moiety. Their data do strongly suggest that *Bti* makes a 130-kDa protoxin and a 65-kDa toxin in a fashion similar to the lepidopteran specific *B. thuringiensis* subspecies.

Three groups of researchers using different restriction enzymes and cloning vectors, have cloned the same gene from *Bti* that encodes a 130-kDa protein.[2,5,56] In all cases, the gene was cloned from the 72-MDa plasmid of *Bti*. The 130-kDa gene product is mosquitocidal and has no hemolytic activity. A 68-kDa protein also is produced, again indicating that the 68-kDa protein probably is the toxin generated from the 130-kDa protoxin.

Recently, Bourgouin et al.[5] cloned a gene from *Bti* that produces a 125-kDa protein that is mosquitocidal. The insecticidal activity of the 125- and 130-kDa proteins were compared. The 125-kDa protein kills larvae of *Ae. aegypti*, *Culex pipiens*, and *Anopheles stephensi*, whereas the 130-kDa protein has no effect on *Culex pipiens* larvae, but is effective against the other two. Restriction endonuclease mapping of the two genes indicates that the 3'-halves are similar, but the 5'-halves are not. Also, the gene encoding the 125-kDa protein is flanked by inverted repeat sequences. This report is the first of repeat sequences in *Bti*.

The predicted similarities between the *Bti* 68-kDa protein and *Bt* subsp. *thuringiensis* 130-kDa protein[54] suggest that there is a conserved structure/function relationship, or common mechanism of toxicity, between the two *Bt* subspecies. With the recent cloning of a *Bti* gene producing a 130-kDa protein,[6] which may in fact be a protoxin containing a 68-kDa toxin, the similarities between lepidopteran- and dipteran-specific *Bt* subspecies are even more striking. More experimentation, however, is necessary to further understand the mechanism of toxic action and to determine what is responsible for the varied spectra of insect host ranges among the different *Bt* subspecies.

## IV. FOOD SOURCES OF MOSQUITOES AND BLACKFLIES

Mosquito larvae live either in stagnant or running water. They also are found at the edges of clear, running streams, especially ones with well-developed algal populations. The larvae of a few species (*Ae. australis* and *A. detritas*) actually live in pools containing seawater.[23]

Mosquito larvae are filter feeders; large particles are excluded whereas small particles enter the mouth and are further segregated before reaching the pharyngeal lumen.[23] They feed on particulate matter ranging in size from that of microscopic bacteria to that of clearly visible particles.[27] Depending upon the mosquito species, feeding tendencies differ. *Anopheles* species feed close to the surface of water, whereas *Culex* species feed slightly below the surface as well as on the very bottom.[23]

Blackfly larval habitat, however, is restricted to running water.[30] Two modes of feeding are utilized by blackflies: browsing and filter feeding. Particulate matter ingested by blackfly larvae ranges in size from 0.5 to 300 μm in length and 0.5 to 120 μm in width.[9]

The Gram-negative bacterium *Escherichia coli,* present in domestic sewage, ranges from 2 to 5 μm in length during exponential growth,[12] which is within the size range of particulate matter capable of passing through the internal organs of larvae. The Gram-positive bacterium *Bacillus subtilis* is a normal inhabitant of soil. Both bacteria are present in storage waters such as lakes and ponds that receive drainage.[20] During periods of snow melt and runoff when blackfly larvae are active, *E. coli* and *B. subtilis* are abundant.

Fredeen[20] has tested various bacteria found in natural blackfly larval habitats to determine their potential as larval food sources. He fed Simulidae larvae the alga *Chlamydamonas,* the Gram-positive bacterium *B. subtilis,* and Gram-negative bacteria *Aerobacter aerogenes* and *E. coli.* Under laboratory conditions, *Simulium venustum, S. verecundum,* and *S. vittatum* larvae developed from first instar larvae to pupae when they were provided only bacteria as a food source. Some of the pupae developed to adults, indicating that bacteria alone are an adequate larval food source. However, more reached adulthood when the alga was added as a food source.

Anderson and Dicke[1] dissected blackfly larvae to determine the gut microflora. They found that all species of larvae analyzed contained diatoms and algae. One in particular was the blue-green bacterium *Cyanobacter.* This organism is a photosynthetic bacterium and is widely distributed not only in fresh water but also in marine habitats.[19,44,57] The size of *Cyanobacter* species, as well as other blue-green bacteria, render them a potential food source for both blackfly and mosquito larvae.

## V. MOSQUITO TOXIN GENE IN LARVAL FOOD SOURCES

In our opinion, the ultimate insecticide for mosquito and blackfly control involves incorporating the gene encoding entomicidal activity into these insects' natural food source.

The blue-green bacteria (cyanobacteria) are logical candidates because genetic manipulation of these organisms is possible and because they are a source of larval food. Introduction of the dipteran toxin gene from *Bti* into these bacteria would provide a form of pest control having persistence in nature and would alleviate multiple applications otherwise necessary with direct application of *Bti* and chemical insecticides.

The cyanobacteria blue-green algae are photosynthetic prokaryotes that carry out oxygenic photosynthesis similar to higher plants. Many species of cyanobacteria contain endogenous plasmids.[35,36,42,51] As yet, no function has been determined for these plasmids that range in size from 1.8 MDa to 74 MDa. As many as five plasmids are present in various strains of *A. nidulans* and other species of *Anacystis*.

A number of researchers have used these endogenous plasmids to construct hybrid plasmids capable of transformation and replication in *E. coli* and *A. nidulans*. These plasmids contain selective markers for antibiotic resistance and have at least one unique restriction site for cloning.[22,33,51] The major problem with these hybrid plasmids is that they are relatively large, thereby limiting the size of a potential cloned insert.

Similar hybrid plasmids have been constructed that are capable of transforming the cyanobacterium *Agmenellum quadruplicatum* (PR-6). Buzby et al.[7] used the smallest plasmid endogenous to PR-6 (3.0 MDa) and combined it with various *E. coli* plasmids containing selective markers. To increase transformation efficiency they found that it was necessary to eliminate all AvaI restriction sites in the plasmid so that the exogenous DNA would not be restricted by the PR-6 restriction system (AvaI is an isoschizomer of the AquI restriction endonuclease of PR-6). Buzby et al.[7] observed expression of the *E. coli lacZ* gene from a plasmid vector in PR-6. Expression of the gene product in PR-6 was comparable to the level observed in *E. coli*. These investigators now are studying gene fusions with various PR-6 promoters in hope of finding foreign gene products formed in PR-6 under the control of the *lac* promoter. Genetic engineering of this kind would be of potential use when considering such systems for introduction of the toxin gene into cyanobacteria.

Wolk et al.[58] constructed a plasmid capable of transforming the cyanobacter *Anabaena*. Construction of the plasmid was accomplished by combining DNA from a plasmid endogenous to the cyanobacter *Nostoc* to DNA from pBR322, a plasmid of *E. coli*. To increase transformation efficiency, it was necessary to delete that portion of the plasmid which could be restricted in *Anabaena*. The resultant plasmids were capable of transference between *E. coli* and *Anabaena* by a conjugation-like process.

Most studies have indicated that transformation of cyanobacteria with *E. coli* plasmids is not possible unless combined with DNA from plasmids endogenous to cyanobacteria. However, recently, Dzelzkalns and Bogorad[16] stably transformed the cyanobacterium *Synechocystis* sp. PCC 6803 (6803), using a pretreatment involving ultraviolet irradiation. In *E. coli* and other prokaryotes, irradiation induces a repair process known as the SOS response. Apparently, irradiating the 6803 cells alleviated the host-controlled restriction system and low frequency recombination occurred, allowing foreign DNA to integrate into the host chromosome.

Daniell et al.[13] reported the transformation of *A. nidulans* 6301 by the *E. coli* plasmid pBR322. Transformed cells were selected by ampicillin resistance, and β-lactamase activity was detected in these cells. Transformation was performed with both intact cells and permeaplasts.

We believe that, in time, cyanobacter transformation techniques will be greatly improved as more knowledge is gained about the system. Indeed, the recent finding that cyanobacteria can be transformed with heterologous DNA without the presence of endogenous DNA in the vector is noteworthy.[13,16] This fact now makes it much more feasible to introduce a foreign gene into cyanobacteria, alleviating recloning into shuttle vectors and thereby saving much time and effort.

The gene encoding mosquito larvicidal activity from *Bacillus sphaericus* 1593M has been cloned and expressed in the cyanobacterium *Anacystis nidulans* R2.[15] The gene was cloned into the vector pHV33[40] for expression in *E. coli* and *B. subtilis,* and into the vector pUC303[33] for expression in *A. nidulans* R2. Bioassays against second instar larvae of *Culex pipiens* were performed to determine larvicidal activity expressed as 100% mortality per μg of protein per ml in 48 h. The results indicated that expression of the larvicidal activity in *A. nidulans* R2 is comparable to that in *E. coli* (1 μg/ml). These results bring us a step closer to developing a cyanobacterial based biopesticide although larvicidal activity is much reduced as compared to *B. sphaericus* 1593M $[1(10^{-3})$ μg/ml]. This problem may be resolved by putting the expression of the *B. sphaericus* gene under the control of a cyanobacterial promoter.

Because transformation vectors and procedures are available for various bacteria and cyanobacteria as well, several of which are known to be larval food sources, there are a number of potential systems that can be utilized in developing biological agents to control mosquitoes and blackflies. A biopesticide would be much more economical and safer for the environment than chemical pesticides now widely in use.[18]

## VI. SAFETY

*Bti* itself has no effect on aquatic organisms such as water mites, shrimps, oysters, and other organisms.[14] The only affected species are the dipteran insects: mosquitoes and black-flies. No effects from the *Bti* toxin have been observed in the food chain after exposure to the bacterium although alkali-solubilized parasporal crystals of *Bti* are hemolytic to erythrocytes from many species, including humans. This general cytolytic or hemolytic activity has been assigned to the 28-kDa protein in the native parasporal crystals.[28,45,46] Genetically engineering an organism to encode only the larvicidal activity (68 kDa), which is not cytolytic,[28,60] would alleviate any safety problems concerning cytolytic activity. Introducing the *Bti* entomicidal toxin gene into a larval food source capable of replicating naturally also would preclude repeated applications of the control agent. Furthermore, it would eliminate the costly fermentation process and chemical synthesis of pesticides now in use. If a bio-control agent as described herein were developed and used, the animal and human population, as well as the environment, would be exposed to fewer chemical irritants and pollutants.

## REFERENCES

1. **Anderson, J. R. and Dicke, R. J.,** Ecology of the immature stages of some Wisconsin blackflies (Diptera: Simuliidae), *Ann. Entomol. Soc. Am. J.,* 3, 386, 1960.
2. **Angsuthanasombat, C., Chungjatupornchai, W., Kertbundit, S., Luxananil, P., Settasatian, C., Wilairat, P., and Panyim, S.,** Cloning and expression of 130-kd mosquito-larvicidal d-endotoxin gene of *Bacillus thuringiensis* var. *israelensis* in *Escherichia coli, Mol. Gen. Genet.,* 208, 384, 1987.
3. **Arata, A. A., Chapman, H. C., Cupello, J. M., Davidson, E. W., Laird, M., Margalit, J., and Roberts, D. W.,** Status of biocontrol in medical entomology, *Nature (London),* 276, 669, 1978.
4. **Armstrong, J. L., Rohrmann, G. F., and Beaudreau, G. S.,** Delta endotoxin of *Bacillus thuringiensis* subsp. *israelensis, J. Bacteriol.,* 161, 1, 39, 1985.
5. **Bourgouin, C., Klier, A., and Rapoport, G.,** Characterization of the genes encoding the haemolytic toxin and the mosquitocidal delta-endotoxin of *Bacillus thuringiensis israelensis, Mol. Gen. Genet.,* 205, 390, 1986.
6. **Bourgouin, C., Delécluse, A., Ribier, J., Klier, A., and Rapoport, G.,** A *Bacillus thuringiensis* subsp. *israelensis* gene encoding a 125-kilodalton larvicidal polypeptide is associated with inverted repeat sequences, *J. Bacteriol.,* 170, 3575, 1988.

7. **Buzby, J. S., Porter, R. D., and Stevens, S. E., Jr.**, Plasmid transformation in *Agmenellum quadruplicatum* PR-6: construction of biphasic plasmids and characterization of their transformation properties, *J. Bacteriol.*, 154, 3, 1446, 1983.

8. **Buzby, J. S., Porter, R. D., Stevens, S. E., Jr.**, Expression of the *Escherichia coli lacz* gene on a plasmid vector in a cyanobacterium, *Science*, 230, 805, 1985.

9. **Chance, M. M.**, The functional morphology of the mouthparts of blackfly larvae (Diptera: Simuliidae), *Quaest. Entomol.*, 6, 245, 1970.

10. **Chou, P. Y. and Fasman, G. D.**, Prediction of the secondary structure of proteins from their amino acid sequence, *Adv. Enzymol.*, 47, 45, 1978.

11. **Crosskey, R. W.**, Simuliid taxonomy—the contemporary scene, in *Black Flies*, Laird, M., Ed., Academic Press, London, 1981, 3.

12. **Cullum, J. and Vicente, M.**, Cell growth and length distribution in *Escherichia coli*, *J. Bacteriol.*, 134, 330, 1978.

13. **Daniell, H., Sarojini, G., and McFadden, B. A.**, Transformation of the cyanobacterium, *Anacystis nidulans* 6301 with the *Escherichia coli* plasmid pBR322, *Proc. Natl. Acad. Sci. U.S.A.*, 83, 2546, 1986.

14. **Davidson, E. W.**, Bacteria for the control of arthropod vectors of human and animal disease, in *Microbial and Viral Pesticides*, Kurstak, E., Ed., Marcel Dekker, New York, 1982, 289.

15. **de Marsac, N. T., de la Torre, F., and Szulmajster, J.**, Expression of the larvicidal gene of *Bacillus sphaericus* 1593M in the cyanobacterium *Anacystis nidulans* R2, *Mol. Gen. Genet.*, 20, 396, 1987.

16. **Dzelzkalns, V. A. and Bogorad, L.**, Stable transformation of the Cyanobacterium *Synechocystis* sp. PCC 6803 induced by UV irradiation, *J. Bacteriol.*, 165, 964, 1986.

17. **Ellar, D. J., Thomas, W. E., Knowles, B. H., Ward, S., Todd, J., Drobniewski, F., Lewis, J., Sawyer, T., Last, D., and Nichols, C.**, Biochemistry, genetics, and mode of action of *Bacillus thuringiensis* d-endotoxins, in *Molecular Biology of Microbial Differentiation*, Hoch, J. A. and Setlow, P., Eds., American Society for Microbiology, Washington, D.C., 1985, 230.

18. **Faust, R. M. and Bulla, L. A., Jr.**, Bacteria and their toxins as insecticides, in *Microbial and Viral Pesticides*, Kurstak, E., Eds., Marcel Dekker, New York, 1982, 75.

19. **Fogg, B. E.**, Physiology and ecology of marine blue-green algae, in *The biology of blue-green algae*, Carr, N. G. and Witton, B. A., Eds., University of California Press, Berkeley and Los Angeles, 1973, 368.

20. **Fredeen, F. J. H.**, Bacteria as food for blackfly larvae (Diptera:Simuliidae) in laboratory cultures and in natural streams, *Can. J. Zool.*, 42, 527, 1964.

21. **Garcia, R. and Goldberg, L. J.**, University of California Mosquito Control Research, Annu. Rep. 1977, p. 29, 1978.

22. **Gendel, S., Straus, N., Pulleyblank, D., and Williams, J.**, Shuttle cloning vectors for the cyanobacterium *Anacystis nidulans*, *J. Bacteriol.*, 156, 148, 1983.

23. **Gillett, J. D.**, in *Mosquitoes*, Clay, R., Ed., The Chaucer Press, Bungay, U.K., 1971.

24. **Goldberg, L. J., and Margalit, J.**, A bacterial spore demonstrating rapid larvicidal activity against *Anopheles serget II, Uranotaenia unquiculata, Culex univitattus, Aedes aegypti*, and *Culex pipiens*, *Mosq. News*, 37, 355, 1977.

25. **Gonzalez, J. M., Jr. and Carlton, B. C.**, A large transmissible plasmid is required for crystal toxin production in *Bacillus thuringiensis* variety *israelensis*, *Plasmid*, 11, 28, 1984.

26. **Held, G. A., Huang, Y.-S., and Kawanishi, C. Y.**, Effect of removal of the cytolytic factor of *Bacillus thuringiensis* subsp. *israelensis* on mosquito toxicity, *Biochem. Biophys. Res. Commun.*, 141, 937, 1986.

27. **Horsfall, W. R.**, in *Mosquitoes*, The Ronald Press Company, New York, 1955.

28. **Hurley, J. M., Lee, S. G., Andrews, R. E., Jr., Klowden, M. J., and Bulla, L. A., Jr.**, Separation of the cytolytic and mosquiticidal proteins of *Bacillus thuringiensis* subsp. *israelensis*, *Biochem. Biophys. Res. Commun.*, 126, 961, 1985.

29. **Ibarra, J. E. and Federici, B. A.**, Isolation of a relatively nontoxic 65-kilodalton protein inclusion from the parasporal body of *Bacillus thuringiensis* subsp. *israelensis*, *J. Bacteriol.*, 165, 527, 1986.

30. **Jamnback, H.**, The origins of blackfly control programmes in *Black Flies*, Laird, M., Ed., Academic Press, London, 1981, 71.

31. **Kim, K.-H., Ohba, M., and Aizawa, K.**, Purification of the toxic protein from *Bacillus thuringiensis* serotype 10 isolate demonstrating a preferential larvicidal activity to the mosquito, *J. Invertebr. Pathol.*, 44, 214, 1984.

32. **Klowden, M. J., Held, G. A., Bulla, L. A., Jr.**, Toxicity of *Bacillus thuringiensis* subsp. *israelensis* to adult *Aedes aegypti* mosquitoes, *Appl. Environ. Microbiol.*, 46, 312, 1983.

33. **Kuhlemeier, C. J., Thomas, A. A. M., van der Ende, A., van Leen, R. W., Borrias, W. E., van den Hondel, C. A. M. J. J., and van Arkel, G. A.**, A host-vector system for gene cloning in the cyanobacterium *Anacystis nidulans* R2, *Plasmid*, 10, 156, 1983.

34. **Kyte, J. and Doolittle, R. F.**, A simple method for displaying the hydropathic character of a protein, *J. Mol. Biol.*, 157, 105, 1982.

35. **Lau, R. and Doolittle, W.**, Covalently closed circular DNAs in closely related unicellular cyanobacteria, *J. Bacteriol.*, 137, 648, 1979.
36. **Lau, R. H., Sapienza, C., and Doolittle, W. F.**, Cyanobacterial plasmids: their widespread occurrence, and the existence of regions of homology between plasmids in the same and different species, *Mol. Gen. Genet.*, 178, 203, 1980.
37. **Lee, S. G., Eckblad, W., and Bulla, L. A., Jr.**, Diversity of protein inclusion bodies and identification of mosquiticidal protein in *Bacillus thuringiensis* subsp. *israelensis*, *Biochem. Biophys. Res. Commun.*, 126, 953, 1985.
38. **Margalit, J., Zomer, E., Erel, Z., and Barak, Z.**, Development and application of *Bacillus thuringiensis* var. *israelensis* serotype H14 as an effective biological control agent against mosquitoes in Israel, *Biotechnology*, 1, 74, 1983.
39. **Pfannenstiel, M. A., Ross, E. J., Kramer, V. C., and Nickerson, K. W.**, Toxicity and composition of protease-inhibited *Bacillus thuringiensis* var. *israelensis* crystals, *FEMS Microbiol. Lett.*, 21, 39, 1984.
40. **Primrose, S. B. and Ehrlich, S. D.**, Isolation and plasmid deletion mutants and study of their instability, *Plasmid*, 6, 193, 1981.
41. **Sekar, V. and Carlton, B. C.**, Molecular cloning of the delta-endotoxin gene of *Bacillus thuringiensis* var. *israelensis*, *Gene*, 33, 151, 1985.
42. **Simon, R. D.**, Survey of extrachromosomal DNA found in the filamentous cyanobacteria, *J. Bacteriol.*, 136, 414, 1978.
43. **Sriram, R., Kandar, H., and Jayaraman, K.**, Identification of the peptides of the crystals of *Bacillus thuringiensis* var. *israelensis* involved in the mosquito larvicidal activity, *Biochem. Biophys. Res. Commun.*, 132, 19, 1985.
44. **Stanier, R. Y. and Cohen-Bazire, G. C.**, Phototrophic prokaryotes: the cyanobacteria, *Annu. Rev. Microbiol.*, 31, 225, 1977.
45. **Thomas, W. E. and Ellar, D. J.**, *Bacillus thuringiensis* var. *israelensis* crystal d-endotoxin effects on insect and mammalian cells *in vitro* and *in vivo*, *J. Cell. Sci.*, 60, 181, 1983.
46. **Thomas, W. E. and Ellar, D. J.**, Mechanism of action of *Bacillus thuringiensis* var. *israelensis* insecticidal d-endotoxin, *FEBS Lett.*, 154, 362, 1983.
47. **Thorne, L., Garduno, F., Thompson, T., Decker, D., Zounes, M., Wild, M., Walfield, A. M., and Pollock, T. J.**, Structural similarity between the lepidoptera- and diptera-specific insecticidal endotoxin genes of *Bacillus thuringiensis* subsp. *"kurstaki"* and *"israelensis"*, *J. Bacteriol.*, 166, 801, 1986.
48. **Tyrell, D. J., Bulla, L. A., Jr., Andrews, R. E., Jr., Kramer, K. J., Davidson, L. I., and Nordin, P.**, Comparative biochemistry of entomocidal parasporal crystals of selected *Bacillus thuringiensis* strains, *J. Bacteriol.*, 145, 1052, 1981.
49. **Undeen, A. H. and Nagel, W. L.**, The effect of *Bacillus thuringiensis* ONR-60A strain (Goldberg) on *Simulium* larvae in the laboratory, *Mosq. News*, 38, 524, 1978.
50. **van den Hondel, C., Keegstra, W., Borrias, W., and van Arkel, G.**, Homology of plasmids in strains of unicellular cyanobacteria, *Plasmid*, 2, 323, 1979.
51. **van den Hondel, C. A. M. J. J., Verbeek, S., van der Ende, A., Weisbeek, P. J., Borrias, W. E., and van Arkel, G. A.**, Introduction of transposon Tn 901 into a plasmid of *Anacystis nidulans:* Preparation for cloning cyanobacteria, *Proc. Natl. Acad. Sci. U.S.A.*, 77, 1570, 1980.
52. **Visser, B., van Workum, M., Dullemans, A., and Waalwijk, C.**, The mosquitocidal activity of *Bacillus thuringiensis* var. *israelensis* is associated with $M_r$ 230,000 and 130,000 crystal proteins, *FEMS Microbiol. Lett.*, 30, 211, 1986.
53. **Waalwijk, C., Dullemans, A. M., van Workum, M. E. S. and Visser, B.**, Molecular cloning and the nucleotide sequence of the $M_r$ 28,000 crystal protein gene of *Bacillus thuringiensis* subsp. *israelensis*, *Nucleic Acids Res.*, 13, 8207, 1985.
54. **Wabiko, H., Raymond, K. C. and Bulla, L. A., Jr.**, *Bacillus thuringiensis* entomocidal protoxin gene sequence and gene product analysis, *DNA*, 5, 305, 1986.
55. **Ward, E. S., Ellar, D. J., and Todd, J. A.**, Cloning and expression in *Escherichia coli* of the insecticidal d-endotoxin gene of *Bacillus thuringiensis* var. *israelensis*, *FEBS Lett.*, 175, 377, 1984.
56. **Ward, E. S. and Ellar, D. J.**, Cloning and expression of two homologous genes of *Bacillus thuringiensis* subsp. *israelensis* which encode 130-kilodalton mosquitocidal proteins, *J. Bacteriol.*, 170, 727, 1988.
57. **Whitton, B. A.**, Freshwater plankton, in *The Biology of Blue-Green Algae*, Carr, N. G. and Whitton, B. A., Eds., University of California Press, Berkeley and Los Angeles, 1973, 353.
58. **Wolk, C. P., Vonshak, A., Kehoe, P., and Elhai, J.**, Construction of shuttle vectors capable of conjugative transfer from *Escherichia coli* to nitrogen-fixing filamentous cyanobacteria, *Proc. Natl. Acad. Sci. U.S.A.*, 81, 1561, 1984.
59. **Wu, D. and Chang, F. N.**, Synergism in mosquiticidal activity of 26 and 65 kDa proteins from *Bacillus thuringiensis* subsp. *israelensis* crystal, *FEBS Lett.*, 190, 232, 1985.
60. **Walfield, Alan**, personal communication.

Chapter 3

# THE *BACILLUS SPHAERICUS* TOXINS AND THEIR POTENTIAL FOR BIOTECHNOLOGICAL DEVELOPMENT

**Colin Berry, John Hindley, and Coreen Oei**

## TABLE OF CONTENTS

# I. DISCOVERY AND PROPERTIES OF *BACILLUS SPHAERICUS*

*Bacillus sphaericus* is an aerobic, Gram-positive, spore-forming bacterium which is widespread in soil and aquatic environments. Some strains produce a toxin which is lethal when ingested by filter-feeding mosquito larvae. In 1965, Kellen et al.[1] described the isolation of a larvicidal *B. sphaericus* strain (strain K) from moribund larvae in the U.S. A more pathogenic strain (strain SSII-1) was later isolated by Singer[2] from infected larvae collected in India. Subsequently, more highly toxic strains were isolated from around the world, among which are strain 1593 from Indonesia,[3] strain 2297 from Sri Lanka,[4] and strain 2362 from Nigeria.[5] The potency and specificity of the *B. sphaericus* toxins make them promising agents for biological control of mosquito pests.

*B. sphaericus* strains have unusual nutritional requirements, requiring only acetate or a few other simple compounds as carbon sources.[6,7] This has permitted the development of selective media for the isolation of *B. sphaericus* strains.[7,8] In addition, Yousten et al.[9] have designed a medium for the isolation of *B. sphaericus* (utilizing arginine as the sole carbon and nitrogen source and containing the antibiotic streptomycin), which favor the selective isolation of larvicidal strains from the nonpathogenic strains. A more problematic feature of *B. sphaericus* is its inability to metabolize carbohydrates.[10-12] This means that cheap media, based on readily available carbohydrates such as molasses and starch, are unsuitable for the production of *B. sphaericus*, necessitating the use of more expensive proteinaceous media. This may limit the production of large-scale cultures of the bacteria in developing countries.

# II. CHARACTERIZATION OF *BACILLUS SPHAERICUS*

The species *Bacillus sphaericus* is a heterogeneous group of bacteria which share a set of phenotypic features, but which are difficult to distinguish from each other by simple biochemical tests.[13] Studies to differentiate *B. sphaericus* strains were undertaken by Krych et al.,[14,15] who examined DNA homologies between 62 strains. Five DNA homology groups were identified of which one (group II) was further subdivided into two subgroups. All seven pathogenic strains tested fell into one of these subgroups (group IIA).

Two methods have been found useful for the differentiation of pathogenic from non-pathogenic *B. sphaericus:* phage-typing[16,17] and H-serotyping.[16] These microbiological methods have allowed the insecticidal strains to be classified into at least four major groups. Many highly toxic strains are found in phage group 3, serotype H5a5b (e.g., 1593, 2362) and phage group 4 and serotype H25 (e.g., 2297). The less insecticidal strain Kellen K falls in phage group 1, serotype H1a while strain SSII-1 is classified under phage group 2, serotype H2. In 1988, de Barjac et al.[18] identified a new set of highly insecticidal *B. sphaericus* isolates from Ghana (of which strain IAB59 is representative) which belonged to serotype H6 (phage type not determined), although other previously identified strains of serotype H6, were not found to be insecticidal. In general, there is a good correlation between insecticidal activity and the phage type and H-serotype of *B. sphaericus* strains.

# III. THE MOSQUITOCIDAL TOXINS OF *BACILLUS SPHAERICUS*

## A. COMPARISON OF THE TOXINS OF THE HIGH AND LOW TOXICITY *BACILLUS SPHAERICUS* STRAINS

The insecticidal activity of *B. sphaericus* isolates has been attributed to a toxic component found in the bacteria rather than an invasive, replicative process of the bacteria in the larval gut. Davidson et al.[19] observed that chloroform-treated isolates of *B. sphaericus* SSII-1 retained toxicity. When the cells were centrifuged, toxic activity was retained in the cell pellet, and not in the supernatant, and toxicity was lost when the cells were heated. These findings were confirmed by Myers and Yousten,[20] who found that chloroform or UV treatments did not cause any decrease in the larvicidal activity of the *B. sphaericus* SSII-1 cells although there was a $10^4$- to $10^5$-fold decrease in viable cell numbers. In the same study, it was also reported that all stages of the bacterial cell cycle exhibited toxicity, which indicated that the activity was not related to sporulation in this strain. Electron microscopic examination of the cells and spores revealed no parasporal body or special inclusion body that could explain the toxic activity of the bacteria.

In 1979, a comparative study of the toxins produced by *B. sphaericus* strains SSII-1 and 1593 was performed by Myers et al.[21] In contrast to SSII-1, toxicity in strain 1593 increased dramatically as sporulation occurred. Moreover, oligosporogenic mutants of 1593 failed to increase in toxicity and were also very much less toxic than the sporulating parent strain. These observations, together with the finding that spores of this strain exhibited high insecticidal activity, suggested very strongly that in strain 1593, toxin production was a sporulation-associated event. Strain 2297, another highly insecticidal strain, showed about a 1000-fold toxicity increase during sporulation.[22] The development of spores in strain 2297 and other highly insecticidal strains (1593, 2013-4, 1691) was accompanied by the formation of crystal-like polyhedral bodies.[22-24] These parasporal crystals were not found in the low toxicity strains or the nontoxic strains,[23] suggesting that the toxin could be located in these parasporal inclusions. Charles et al.[25] examined the ultrastructure and protein profiles of asporogenous mutants of strains 2297 and 1593 and confirmed that synthesis of crystal in *B. sphaericus* was a sporulation event. Mutants which were blocked at an early stage in sporulation did not produce crystals and were poorly toxic to mosquito larvae. However, mutants blocked later retained the ability to produce crystals and did not differ in toxicity to the wild type strains.

Comparison of the larvicidal toxins from *B. sphaericus* strains SSII-1 and 1593 revealed several important differences.[21] The toxin of strain SSII-1 was found to be relatively unstable, as its activity was destroyed by heat and reduced by refrigeration, a freeze-thaw cycle, or two methods of cell breakage.[20] The relative instability of the toxin made fractionation or purification difficult. In contrast, the toxin from strain 1593 was more stable than that of SSII-1. The toxin was unaffected by many of the treatments which inactivated the SSII-1 toxin, including processes such as lyophilization, urea extraction, sonic disruption, or French pressure cell treatment.[21] The results summarized above led several investigators[21,26,27] to suggest that different toxins might be active in highly toxic strains to those in strains such as SSII-1. This conclusion is supported by the finding that DNA probes derived from cloned toxin genes from highly toxic strains failed to hybridize to DNA extracted from the less toxic strains SSII-1, 1889, 2173 and Kellen K.[28-31]

## B. ISOLATION AND CHARACTERIZATION OF THE TOXINS

One of the first attempts to isolate the protein toxin from *B. sphaericus* was undertaken by Tinelli and Bourgouin.[32] A toxic fraction was isolated from an extract of *B. sphaericus* spores by ion-exchange chromatography. When the fraction was subjected to electrophoresis (without SDS) in a polyacrylamide gel, three proteins were identified. Upon further analysis of the three proteins, it was concluded that at least one protein or a heteroprotein with a molecular weight of about 55 kDa was responsible for toxicity.

Partial purification of the toxin from the cytoplasm of sporulating *B. sphaericus* 1593 cells was also undertaken by Davidson,[33] utilizing various methods such as ammonium sulfate precipitation, ion-exchange, and gel filtration chromatography. The activity of the toxin of approximately 100 kDa, was destroyed by incubation with the enzymes pronase or subtilisin or by heating to 80°C for 30 min. Davidson[34] solubilized a toxin of 54 kDa from spores of *B. sphaericus* 1593 in 0.05 *M* NaOH. This treatment caused dissolution of the parasporal crystals found in highly insecticidal strains,[22-24] and the resulting solution, when neutralized, was toxic to mosquito larvae. This supported the suggestion that these crystal inclusions could be directly responsible for larvicidal activity. The toxin, which was further purified by precipitation from the extract in the presence of 1 $M$ acetate buffer (pH 4), was found to be resistant to subtilisin but sensitive to pronase treatment. The similarity in the molecular weight and the protease sensitivity pattern suggested that it might be the same toxin as that reported by Tinelli and Bourgouin.[32] In a later study, Payne and Davidson[35] separated the crystals from the spore-crystal complex of strain 1593 by passage through a French pressure cell and subsequent centrifugation through a NaBr gradient. The highest larvicidal activity was found in the crystal-enriched fractions. Bourgouin et al.[36] also reported the isolation of larvicidal toxins from *B. sphaericus* strain 2297. The toxic fractions contained two major polypeptides of 40 and 56 kDa, which were continuously synthesized during sporulation, and were absent during the vegetative stage of bacterial growth.

Baumann et al.[37] purified crystals from spore-crystal complexes of *B. sphaericus* strain 2362 by a method similar to that of Payne and Davidson.[35] The layer from the NaBr gradient containing most of the crystals was used for all subsequent protein purifications. The intact crystal preparations gave protein bands of 43, 57, 63, 98, 110, and 125 kDa when they were solubilized with NaOH at pH 9, 10 or 11. When the pH was raised to 12, all the high molecular weight proteins were eliminated, and only the 43- and 63-kDa proteins could be detected on SDS-PAGE. These two proteins were further purified by various chromatographic techniques. The $LC_{50}$ value (the concentration of protein needed to kill 50% of test larvae) against the mosquito *Culex pipiens*, was 6 ng/ml for the crystals, while the $LC_{50}$ value for the 43-kDa protein was 35 ng/ml. The 63-kDa protein had no larvicidal activity, even up to a concentration of 100 μg/ml. Further analysis of the two proteins revealed that they were quite distinct, as shown by their amino acid compositions, their lack of immunological cross-reactivity, their opposite charges at neutral pH and their susceptibility to midgut protease digestion. When the crystal toxins were subjected to electrophoresis and immunoblotting with antibodies raised against the purified 43- or 63-kDa proteins, it was found that the high molecular weight proteins (98 to 125 kDa) contained antigenic determinants of both the 43- and 63-kDa proteins. Based on these observations, it was suggested that the high molecular weight crystal proteins were precursors of the lower molecular weight proteins. The high molecular weight proteins and the 63-kDa protein were rapidly broken down in the midgut of *Culex* larvae and a 40-kDa protein, antigenically related to the 43-kDa protein, remained after 4 h of digestion. Further experiments showed that there was a high degree of similarity between the crystal proteins of strains 1593, 1691, 2297, and 2362. No cross-reactivity was found between antisera raised against either the 63- or 43-kDa proteins from *B. sphaericus*, with the solubilized insecticidal toxins from *Bacillus thuringiensis israelensis* or *B. thuringiensis kurstaki*.

Further studies were performed to investigate whether the high molecular weight crystal proteins of 125 and 110 kDa, were the precursors of the 43- and 63-kDa peptides.[38] The 110-kDa protein was found to be toxic to *Culex pipiens* larvae, but had no effect on cultured cells of *Culex quinquefasciatus*, while the 43-kDa protein was toxic to both. The 43-kDa toxin was not toxic to cultured cells from the fall armyworm (*Spodoptera frugiperda*) or bovine kidney. Following studies relating the kinetics of growth to synthesis of crystal proteins, proteases and heat-resistant spores, by strains 2362 and 2297, Broadwell and Baumann[38] suggested that the earliest detectable crystal protein (125 kDa) was a precursor of the 110-kDa protein, which they proposed was subsequently cleaved to produce the 63- and 43-kDa proteins. Other studies[39-42] showed a further activation of the 43-kDa protein to a 40-kDa toxin by larval gut enzymes, resulting in a 54-fold increase in toxicity to cultured cells (see below).

Other investigators have also purified larvicidal toxins from *B. sphaericus*. Larvicidal proteins from strain 1593 were purified using an immunoaffinity method by Narasu and

Gopinathan.[43] Coat proteins from the spores were separated by preparative gel electrophoresis under nondenaturing conditions. Proteins were isolated from the gel and assayed for toxicity to *C. pipiens quinquefasciatus* larvae. Antibodies were raised against the major toxic band and were used to make an immunoaffinity column for the purification of larvicidal proteins. Protein eluted from the column produced a single band on nondenaturing polyacrylamide gels but was shown to contain four peptides of apparent molecular weight 42.6, 44.1, 50.7, and 51.3 kDa, respectively, under denaturing conditions. The eluate was estimated to have approximately a 12% carbohydrate content and it was suggested that differential glycosylation might account for the four bands on SDS-PAGE. The $LC_{50}$ value for the eluate was 8.3 ng protein per milliliter against *C. pipiens*. Sgarrella and Szulmajster[44] reported the isolation of toxins from solubilized cell walls and membranes of *B. sphaericus* strain 1593M, by purification through Sephadex G-100 and DEAE Sephacel chromatography. They found that toxic activity resided in a protein of apparent molecular weight 38 kDa, which appeared to form higher molecular weight aggregates when the pH was raised from 7.5 to 8.5.

## C. ENZYMATIC ACTIVATION OF THE 43-kDa *BACILLUS SPHAERICUS* TOXIN

In their early studies, Baumann et al.[37] observed the conversion of the 43-kDa protein to a 40-kDa protein, both *in vivo* in the guts of *Culex pipiens* larvae and *in vitro* on treatment with gut extracts from these larvae. Davidson[45] showed that *in vitro* treatment of both toxic extracts from *B. sphaericus* 1593 and purified 43-kDa toxin from strain 2362, with *Culex quinquefasciatus* larval gut homogenates increased the cytotoxicity to cultured *C. quinquefasciatus* cells. Further studies[39-42] confirmed that the activation of the 43-kDa protein involved a reduction in molecular weight of 2 to 4 kDa, and Broadwell and Baumann[40] showed that this was due in part to the removal of six amino acids (Asp-Phe-Ile-Asp-Ser-Phe) from the N-terminus of the purified 43-kDa protein. In the latter study,[40] the reduction in size of the protein was found to correlate to a 54-fold increase in cytotoxicity to *C. quinquefasciatus* cells. A similar decrease in molecular weight and increase in toxicity was observed when the 43-kDa toxin was treated with the proteases trypsin or α-chymotrypsin, or with gut extracts from larvae of the species *C. pipiens, C. quinquefasciatus, Aedes aegypti*, or *Anopheles gambiae*.[39-42] It has also been shown that the activated forms of the toxin produced by these treatments, are not cytotoxic to cultured cells of *Aedes dorsalis, Anopheles gambiae, Anopheles stephensi, Spodoptera frugiperda*, or rat CSH2 cells.[39,40] These results indicate that larvicidal specificity of the 43-kDa toxin is not due to activation of the toxin by specific proteases found only in susceptible species, but is likely to be due to differences in the target cells between sensitive and nonsensitive species. It has been shown that when the 43-kDa toxin is incubated with cultured *C. quinquefasciatus* cells in the presence of *N*-acetyl-D-glucosamine, *N*-acetyl-D-galactosamine or wheat germ agglutinin, both cytotoxicity[40,46,47] and binding of the toxin to the cell surface[46,47] were reduced. This has led investigators to suggest that there may be a glycoprotein receptor for the toxin.

A two-step pathway of activation of the *B. sphaericus* toxin was postulated by Broadwell and Baumann.[40] The first step was thought to be associated with sporulation and to involve the processing of 125-kDa protoxin to a 43-kDa toxin. The second step was thought to be associated with the larvae and to involve a further activation of the toxin by proteolytic cleavage of the 43-kDa protein to the 40-kDa toxin.

Recently, Hindley and Berry[48] cloned and sequenced a gene which encoded a 41.9-kDa protein from *B. sphaericus* 1593. From its deduced amino acid sequence, this protein was identified as that described by Baumann et al.[37] as the 43-kDa toxin. The finding of a single open reading frame encoding this protein, with defined translational initiation and termination codons, indicated that it was not derived from a higher molecular weight precursor. Baumann et al.[49] later confirmed that the 41.9- and 51.4-kDa proteins (previously described as the

43- and 63-kDa proteins) were the products of two different genes. The 41.9-kDa protein produced in *E. coli* was not toxic to mosquito larvae but a mixture of the 41.9- and 51.4-kDa proteins was toxic.

# IV. TARGET RANGE, PATHOLOGY, AND SAFETY

## A. TARGET RANGE

The larvicidal toxins of *B. sphaericus* have a relatively narrow target spectrum, affecting mainly species from the major mosquito genera (reviewed by Davidson[50]). The toxicity of *B. sphaericus* is influenced by many factors and this has led to difficulties in the comparison and quantitation of the efficacy of toxin towards various mosquito species. Some factors which affect the biological activity of the bacteria, are the growth media, preparation methods, and formulations, and the availability of other food sources to the larvae in mosquito breeding areas.[51-54] Results of toxicity bioassays were influenced by the age of the larvae used. The two most commonly noted patterns of response to *B. sphaericus* when tested against four species of mosquito larvae, were either decreasing susceptibility with increasing larval instar[55-57] or high susceptibility of first instars with lower, approximately equal susceptibility of the older larvae.[27,58]

It has been found that, in general, the *B. sphaericus* strains show greatest activity towards the Culicine and Anopheline species, while the *Aedes* species are relatively resistant towards the bacteria, *Ae. aegypti* being very resistant to the toxin. Mulla et al.[57] evaluated the efficacy of several *B. sphaericus* strains on five species of mosquitoes and found that all strains tested were equally and highly toxic to *C. quinquefasciatus* (LC$_{50}$ of 0.02 to 0.04 mg/ml) and *Culex tarsalis* (LC$_{50}$ of 0.02 to 0.06 mg/ml). They had intermediate levels of activity against *Anopheles freeborni* (LC$_{50}$ of 0.65 mg/ml) and *Anopheles quadrimaculatus* (LC$_{50}$ of 0.53 to 0.65 mg/ml) but had insignificant levels of activity towards *Ae. aegypti* (LC$_{50}$ of >40 mg/ml). The same trend was also reported by Lacey and Singer,[59] using strains 2013-4 and 1023-6. Lacey et al.[60] bioassayed strain 2297 against eight species of mosquitoes from three subfamilies and found that the most susceptible species were in the genus *Culex* and the least susceptible were the *Aedes* species and *Toxorhynchites r. rutilus*. Cheong and Yap[61] bioassayed strain 1593 against *Mansonia uniformis* larvae and found them to be less susceptible than *C. quinquefasciatus* but more susceptible than *Ae. aegypti*.

Although the toxicity of *B. sphaericus* to mosquito larvae has been well documented, there has been little information on the effect of the toxin on adult mosquitoes. It has been shown that *B. thuringiensis israelensis* is lethal to adult mosquitoes.[62,63] Similar studies by Stray et al.[64] showed that adult *C. quinquefasciatus* were killed by the soluble *B. sphaericus* toxin when it was introduced by enema into the midgut of the insect. Oral administration of the toxin to adult *Culex quinquefasciatus* was not fatal and the toxin also had no effect on adult *Ae. aegypti*. In a study to examine the nematicidal effects of *B. sphaericus,* Bone and Tinelli[65] incubated the spore extracts of strain 1593 on the eggs of the nematode, *Trichostrongylus colubriformis*. The spore extracts were lethal to the eggs of *T. colubriformis* and may represent a new source of anthelmintic activity. Blackfly larvae (Simuliidae) have been found to be resistant to all *B. sphaericus* strains, although they are susceptible to some toxins of *B. thuringiensis*.[53]

Recently, Thiery and de Barjac[66] assayed more than 180 *B. sphaericus* strains belonging to six H-serotypes, for activity against three mosquito species (*C. pipiens, An. stephensi,* and *Ae. aegypti*). Serotype H5a5b strains were generally toxic to all three mosquito species, strains belonging to serotypes H6 and H25 were toxic to *C. pipiens* and *An. stephensi* while strains of serotypes 26a26b and H2a2b were much less toxic and often only to *C. pipiens*. The fact that *B. sphaericus* strains can have different LC$_{50}$ values against two or three mosquito species means that the activity of the strain can be expressed by activity ratios (the ratio of the $LC_{50}$ of the strain on one mosquito species, to the $LC_{50}$ of the same strain on another species). Characterization of *B. sphaericus* toxicity using activity ratios in addition to the $LC_{50}$ values against specific mosquito species would allow the larvicidal power to be more efficiently assessed. For example, application of strains from serotype H5a5b would be effective for the control of *C. pipiens* alone but simultaneous control of *C. pipiens* and *An. stephensi* would be more efficacious when treated with a strain from serotype H25 for which the activity ratio for these two species is approximately 1. Another advantage of using activity ratios when evaluating toxicity of *B. sphaericus* strains is their reproducibility. Data expressed in this way allow direct comparison of bioassays performed in different laboratories, regardless of the variations in bioassay conditions.

## B. PATHOLOGY

Histological examination of moribound *Culex* larvae which had been fed with *B. sphaericus* strain Kellen K showed that all bacterial cells were confined to the larval gut.[1] Bacteremia did not occur until near or after the death of the larvae. These results were also seen in mosquito larvae infected with *B. sphaericus* strain SSII-1,[19,67] where bacteria were confined to the peritrophic membrane of the gut. Although there was bacterial multiplication in the gut, larval mortality was shown to occur without multiplication of the bacteria, as chloroform-treated *B. sphaericus* were also toxic to larvae. Electron microscopic studies revealed that *B. sphaericus* cells were rapidly degraded in the midgut and gross histological changes occurred in the larval midgut cells. Swelling of the midgut was accompanied by basal separation of the midgut cells. There was also an increase in the size and number of the lysosomes per cell. Ultrastructural studies on the midgut of the mosquito larvae intoxicated with *B. sphaericus* strain 2297 spore complex, revealed that there was rapid dissolution of the parasporal inclusions within the midgut lumen.[68] Alterations to the midgut cells caused by the action of the toxins differed according to the mosquito species tested, but were all localized within the gastric cecum and posterior stomach.

The activated form of the *B. sphaericus* toxin has been shown to be cytolytic to cultured cells of *C. quinquefasciatus*.[38,39,45] When treated with the toxin, these cultured cells became swollen, rounded and phase-dark before lysing.[45] Electron microscopy of toxin treated cells of *C. quinquefasciatus* revealed that ultrastructural changes to the mitochondria, cell membrane, endoplasmic reticulum, and Golgi apparatus occurred minutes after intoxication.[69] Binding of the *B. sphaericus* toxins to susceptible and nonsusceptible cultured insect cells was studied using fluorescent labeled toxin and antibody-secondary antibody techniques.[46] The toxin bound very strongly to *C. quinquefasciatus* cells but less strongly to cells of insensitive insects. Evidence was also presented, showing that internalization of the toxin may occur by receptor-mediated endocytosis and that the toxin could form pores in the cell membrane. *In vivo* binding of the toxins to the midgut cells of mosquito larvae, showed that strong binding was restricted to the gastric cecum and posterior midgut of *C. quinquefasciatus,* while there was no binding to the midgut cells of two resistant *Aedes* species.[70] In *Anopheles* species, the fluorescent toxin bound weakly to cells throughout the midgut.[47] Internalization of the toxin occurred in *C. pipiens* but not in *Anopheles* species. Binding of the toxin to midgut cells of highly susceptible *Culex* species may be very specific and of high affinity, while binding to *Anopheles* cells may be nonspecific and/or of low affinity.

## C. SAFETY

The safety of *B. sphaericus* to nontarget organisms, invertebrates and vertebrates, is an important consideration for its use as a large-scale biological insecticide. Many of the strains that are active against mosquito larvae have been shown to have a high degree of specificity. *B. sphaericus* SSII-1 was shown to be nonpathogenic to the commercially important honey bee (*Apis mellifera*) and other nontarget insects.[71] The following organisms were tested for susceptibility to larvicidal *B. sphaericus* and were found to be unaffected by exposure to the bacteria: the mayfly *Callibaetis pacificus;* ostracods; conchostracans; diving beetle (larvae and adult); tadpoles; *Moina* spp.; *Ceriodaphnia* spp.; and the larvivorous fish, *Gambusia affinis affinis* and *Aphyosemion gardneri.*[53,57,72]

Strain 1593-4 was injected subcutaneously, intraperitoneally and intracerebrally into rats and mice in maximum challenge tests.[73] Rats, mice and guinea pigs were also fed the bacteria to determine their toxicity to mammals. No death or clinical illness resulted from these tests. However, high doses of *B. sphaericus* produced mild lesions in the brains of rats and severe lesions were found in the eyes of rabbits with intraocular injections. These lesions were due partly to the high concentrations of foreign substances present. Shadduck et al.[74] injected viable or autoclaved preparations of several isolates of *B. sphaericus* into mice, rats, and rabbits via the following routes: intracerebral, intraocular, subcutaneous, and intraperitoneal. Their results confirmed those of the WHO tests; there was no death or clinical illness. Shadduck et al.[74] therefore concluded that all the *B. sphaericus* strains tested were avirulent for the mammals tested and concluded that the larvicidal strains are highly unlikely to be hazardous to man.

## V. FIELD TRIALS USING *BACILLUS SPHAERICUS*

The use of *B. sphaericus* preparations for mosquito control in the field has produced a range of results concerning persistence and efficacy. The various strains of *B. sphaericus,* doses of spores, and formulations used, in addition to the different target larvae against which toxicity was assessed, make the results from different trials difficult to compare. Most investigators conclude however, that *B. sphaericus* may prove to be a useful and economically viable means for mosquito control, especially in regions where larval resistance to conventional pesticides has developed.

The major problem experienced with the use of *B. sphaericus* in the field has been the rapid sedimentation of spores, out of the larval feeding zone in the upper layers of the aquatic habitat.[72,75,76] In most studies this has resulted in a duration of control of 1 to 3 weeks. Nicolas et al.[77] found no evidence of recycling of *B. sphaericus* in mud sediments, and Karch and Charles[76] found loss of toxicity of the sedimented spores and also noted the disappearance of the parasporal crystals associated with toxicity, after 4 weeks in water.

Some formulations may exhibit slow sedimentation, for example the fluid concentrate of *B. sphaericus* 2362 (BSP1, $2 \times 10^{10}$ spores per g, produced by Solvay, Brussels, Belgium) used by Nicolas et al.,[77] which gave complete control of *Culex quinquefasciatus* larvae for 5 to 6 weeks after a single treatment. It should be noted however, that the application of 10 g/m² used in this investigation, was in the order of 50 to 500 times the dose used in many of the other studies.[72,75,78] Clearly, further trials are necessary to find the most suitable formulations of *B. sphaericus* for persistence in the field.

Another finding from field trials is the ability of *B. sphaericus* to recycle in larval cadavers.[75,77] Although this could be potentially useful for maintaining the bacterium in the environment, Nicolas et al.[77] found the cadavers sedimented within 48 h and thus were removed from larval feeding zones. Although cells and spores of *B. sphaericus* may remain viable after settling into the pond bed,[50,76,77] sediments would have to be agitated or stirred in order to return the bacteria to the feeding zones.

## VI. PRODUCTION AND FORMULATION OF THE *BACILLUS SPHAERICUS* TOXINS

### A. GROWTH CONDITIONS FOR TOXIN PRODUCTION

Once a promising strain of *B. sphaericus* has been isolated, it is important to optimize growth conditions for maximum toxin yield and for ease of production on a large scale.

*B. sphaericus* 2362 has been identified as a useful strain for development as it is more toxic to *Culex pipiens* than strain 1593[66,79] and is able to grow at higher temperatures without inhibition of toxin production.[79] Yousten has investigated the effects of various conditions on growth, sporulation and toxin synthesis for *B. sphaericus* 2362.[79] Under normal growth conditions, the pH of the medium rises due to bacterial production of ammonia from protein substrates. In his experiments, Yousten demonstrated that toxin yield could be increased by holding the pH of the medium close to 7.0 by the addition of sulfuric acid.

Other studies aimed at possible commercial development of *B. sphaericus* strain 2362[80,81] have examined seven industrial protein hydrolysates as the basis of media for toxin production. Most of the proteins were waste products from industrial processes, all of which had been subjected to autolysis or treatment with proteases before being used for media preparation. It was found that preparations containing proteins with short average peptide chain lengths gave better toxin yields. In addition, the presence of 0.5% (w/v) glycerol in the media was found to increase toxin yields, while in simple media, L-arginine appeared to inhibit production.

It will be necessary to perform studies similar to those outlined above for the development of other strains of *B. sphaericus*, as conditions for optimum toxin yield may vary from strain to strain, according to the number and nature of the toxins being produced. It has been suggested that a barrier to the production of *B. sphaericus* in the third world may be its requirement for proteinaceous substances which may also be useful as food for humans and animals.[79] The use of industrial protein waste products as the basis for media may help to circumvent this problem. Obeta and Okafor[82] have studied growth, sporulation, and toxicity of *B. sphaericus* in media, the components of which were readily available for local production in Nigeria. Their media, based on dried cow's blood, and in some cases, supplemented with extracts from various leguminous seeds, gave acceptable levels of toxicity and good yields of spores for local use. The use of simple media and methods for production of bioinsecticides, which can be performed close to the sites of their eventual use, may reduce costs and so help to make them more available for use in developing countries.

## B. FORMULATION FOR FIELD APPLICATION

There are many possible formulations of *B. sphaericus* which could be developed for application in the field. In addition, a large range of products formulated from recombinant organisms in which *B. sphaericus* toxin genes have been cloned may also be developed, for example, aquatic microorganisms or nonviable cells in which toxins have been overexpressed. Preparations containing a single strain of *B. sphaericus* may be produced for use in rotation with other pesticides. Alternatively, mixed formulations of strains of *B. sphaericus* and *B. thuringiensis israelensis* could be developed to control a wider range of target mosquitoes. Both of these strategies would reduce the risk that resistance to a single toxin may develop.

The mosquitocidal preparations chosen should have a long shelf life and provide control of target larvae over a reasonable period of time. As mentioned previously, *B. sphaericus* applied in the field gives good mosquito control, but settling of spores[72,75,76] leads to a relatively short control duration under normal field conditions (up to 6 weeks[77]). Formulations of *B. sphaericus* with well-dispersed (rather than clumped) spores may show slower settling rates. Insecticidal powders may be produced from cultures by a lactose-acetone precipitation method.[83] However, powders produced by spray drying are more easily suspended, give higher larval mortalities, and have longer shelf lives, and so may prove to be more useful preparations for storage and application.[78,84] It may also be possible to prepare *B. sphaericus* spores in such a way as to increase their buoyancy, or to produce time release formulations of spores for greater duration of control. Lacey et al.[85] described reduction of larval populations of *Culex quinquefasciatus* in test containers, for over 8 weeks using slow release floating pellets made up of 30% *B. sphaericus* 1593 primary powder, 30% powdered sugar (as the releasing agent) and 40% polypropylene foam (for buoyancy). Pelleted formulations have the advantage that their greater bulk allows them to penetrate foliage covering aquatic habitats more easily. It has been shown in studies on *B. sphaericus* strains SSII-1 and 1593[86,87] that UV radiation, while not affecting toxicity, could produce a rapid loss of spore viability. This in turn reduces the ability of *B. sphaericus* to persist in the environment. For this reason, it was suggested[87] that formulations of *B. sphaericus* designed to float in larval feeding zones should contain screening agents to block the germicidal effects of UV light. These are some of the challenges for chemical engineering in the development of bioinsecticides for field application.

## VII. CLONING OF *BACILLUS SPHAERICUS* TOXIN GENES

### A. CLONING IN *ESCHERICHIA COLI*

*Escherichia coli* has been used by several groups as a cloning host in the first steps to the identification and characterization of toxin genes from *B. sphaericus*. It is an important step, as it allows individual proteins involved in toxicity to be studied away from the background of other *B. sphaericus* proteins, an unknown number of which may also be toxic.

The first cloning of toxin genes from *B. sphaericus* was reported by Ganesan et al. in 1983.[88] In this work, a 3.7-kb fragment from a partial *Sau*3AI digest of total *B. sphaericus* 1593M DNA was found to encode larvicidal activity in *E. coli*, comparable to that found with *B. sphaericus*. A later report by the same group[89] however recounted an unexplained decrease in the activity of the clones without apparent sequence deletions.

The genes encoding two proteins of molecular weight 41.9 and 51.4 kDa have been cloned from highly toxic strains of *B. sphaericus* by several groups.[29,48,89-92] There is sequence homology between the two proteins[49] and results using *E. coli* produced toxins indicate that both proteins may be necessary for toxicity.[29,93] No sequence homology has been found between either the 51.4- or 41.9-kDa proteins and the insecticidal toxins isolated from *Bacillus thuringiensis* strains. The genes encoding the 41.9- and 51.4-kDa proteins from various *B. sphaericus* strains show a high degree of sequence conservation[49,92] with strains 1593, 2362, and 2317.3 having identical coding regions. Strains 2297 and IAB59 have, respectively, 3 and 5 amino acid substitutions in the 51.4-kDa protein, and 5 and 1 substitutions in the 41.9-kDa protein compared with the above three strains. The genes for these two proteins are to date the best characterized and have received most attention in terms of their development for use in the field.

In earlier studies, Broadwell and Baumann[38] found a protein of 110 kDa from *B. sphaericus* 2362 to be toxic to *Culex pipiens* larvae. Recently, Bowditch et al.[94] have cloned a gene from strain 2362 encoding a protein of 125 kDa which appears to be the precursor of the 110-kDa toxin (having a putative leader sequence at the N-terminus). The 125-kDa protein itself, seems to be a surface layer protein of *B. sphaericus* and is nontoxic. *E. coli* carrying the cloned gene for the 125-kDa protein, accumulate a recombinant product of 125 kDa as well as peptides of 110 to 113 kDa, but exhibit no larvicidal toxicity. In addition to the gene for the 125-kDa protein, a second, related open reading frame was cloned and sequenced. No evidence was found for the production of protein from this reading frame when cloned in either *E. coli*, *B. subtilis* DB104 or *B. sphaericus* SSII-1.

Other groups[95,96] have reported the cloning of mosquitocidal toxins from strains of *B. sphaericus*. The sequences of these clones have not been reported and their relationship to the toxin genes described above is unclear.

## B. CLONING IN OTHER ORGANISMS

The cloning of *B. sphaericus* toxin genes into other organisms may be desirable both to study the toxins and their expression and to develop new hosts for better control of mosquitoes in the field.

Toxin genes from *B. sphaericus* have been expressed in *Bacillus subtilis*[89,93,94,96,97] and *Bacillus sphaericus* SSII-1[94] to study their expression. Levels of expression and times of toxin production appear to be similar to those of the original host strains.

In field studies using *B. sphaericus,* the spores with which the toxins are associated, have been found to settle out of the mosquito larval feeding zones, reducing the duration of control.[72,75-77] It may be possible to increase the effectiveness of *B. sphaericus* toxins in the field by cloning into alternative hosts which might persist in the larval feeding zones for longer. With this in mind, a toxin clone from *B. sphaericus* 1593M, isolated by de Marsac et al.[89] has been cloned and expressed in *Anacystis nidulans* R2 (also called *Synechococcus* PCC 7942), a unicellular cyanobacterium which grows in the upper layers of aquatic habitats. In further work, cyanobacteria belonging to several genera have been isolated from mosquito breeding sites in France.[98,99] The genetic characterization of these bacteria, along with studies of their use by mosquito larvae as food sources, is being undertaken with a view to the possible production of toxic recombinants for use in the field.

There is much potential for further development of the *B. sphaericus* toxins expressed in new hosts. Other aquatic organisms might prove useful to overcome the spore settling problem encountered with the natural host. Another solution to the problem might be the use of asporogenic strains of *Bacilli* with the toxin genes placed under the control of a *Bacillus* vegetative promoter.

The designing of recombinant organisms for use in the field presents many challenges. Recombinants with long term persistence in mosquito habitats may be desirable from the point of view of cost and the potential difficulty involved in repeated application, but it is necessary to consider the problems which might arise from the exploitation of such organisms. The production of a new toxic organism which persists in the field may be undesirable for several reasons. The constant presence of toxin in mosquito habitats, at levels which might fall below the lethal dose, may promote the development of resistance among target insects. For proper integrated pest management, a pesticide which does not remain in the habitat is better for use in rotation with other larvicidal agents (e.g., other bioinsecticides, insect growth regulators, or chemical insecticides). In addition, purely commercial pressures favor the production of a toxin formulation which has to be used repeatedly, rather than one which might remain active in the environment indefinitely. Furthermore, the production and field application of live recombinant organisms is viewed with some suspicion by the public at large and will be subject to rigorous scrutiny by regulatory agencies. As yet, guidelines for production, testing, and use of recombinant microorganisms have not been finalized either in the U.S. or in the European Community.[100] These factors lead to greater difficulty in gaining acceptance for a product and the associated increase in costs may discourage commercial development. One possible way of avoiding the problems of overpersistence might be to use an application formulated from dead bacteria. Such a process as the MCAP™ formulation developed by Mycogen Corp. (San Diego) for encapsulation of *Pseudomonas fluorescens* carrying *Bacillus thuringiensis* toxins,[101,102] might prove effective in this respect. This is an example of a strategy using a novel host for overproduction of toxin, with subsequent field application of nonviable cells. The overexpression of *B. sphaericus* toxins in *E. coli*, bacilli, pseudomonads, or other organisms may allow similar development of these proteins.

## VIII. INCREASING THE RANGE OF TARGET INSECTS

The cloning of mosquitocidal toxin genes offers the opportunity to produce recombinant organisms with combined toxicities derived from more than one type of toxin. The production of multiple toxins in the same bacteria would make the development of toxin resistance in target insects less likely. Several strategies may be employed for this purpose. Fridlender et al.[81] have succeeded in cloning a toxin gene from *Bacillus thuringiensis israelensis* into *B. sphaericus* 2362 using a *Bacillus subtilis* plasmid. Their resulting clones appear to have increased activity against *Aedes* larvae, while activity against *Culex* larvae appears to be of the same level as the *B. sphaericus* 2362 and the *B. thuringiensis israelensis* strains from which the toxin genes were derived. The recombinant *B. sphaericus* has obvious potential for use in the field against a wide range of mosquito targets.

The genes for the 41.9- and 51.4-kDa proteins from *B. sphaericus* 1593M have been cloned into both a nontoxic crystal minus strain (strain 4Q2-81) and a toxic strain (4Q2-72) of *B. thuringiensis israelensis*.[103] The transferred genes appear to be expressed as in *B. sphaericus*, with strain 4Q2-81 showing activity against *Aedes aegypti*, *Anopheles stephensi*, and *Culex pipiens* similar to that of *B. sphaericus*. In contrast to the results of Fridlender et al.[81] however, Bourgouin et al.[103] found no evidence of an additive effect of the toxins of *B. sphaericus* and *B. thuringiensis israelensis* when expressed in the same cell.

Further experiments will show whether it is possible to combine the activities of more than one *B. sphaericus* toxin for increased effectiveness against larvae. Bowditch et al.[94] have cloned the possibly cryptic gene 80 from *B. sphaericus* 2362 into the lower toxicity *B. sphaericus* strain SSII-1, but expression of this gene was not observed. However, other combinations of *B. sphaericus* and *B. thuringiensis israelensis* toxins are possible and may yield recombinants with greater toxicity than the parent strains.

It should be remembered when developing new host organisms containing *Bacillus* toxin genes that both *B. sphaericus* and *B. thuringiensis israelensis* are known to produce more than one type of mosquitocidal protein. It may therefore be necessary to clone multiple genes in order to reproduce the activity of the original host. Further study of the interactions between the known mosquitocidal toxins is necessary in order to determine whether combinations of toxins can be made, which do not interact antagonistically.

## IX. POTENTIAL FOR THE DEVELOPMENT OF GENETICALLY ENGINEERED TOXINS

With the isolation of *B. sphaericus* toxin genes and their cloning into suitable strains of *E. coli,* the possibility of introducing mutations into the genes arises. Investigations relating protein sequence to toxicity, along with mechanistic studies on the mode of action of the *B. sphaericus* toxins, are important in identifying amino acids, likely to produce altered toxicity upon mutation. The separate expression of the genes for the 51.4- and 41.9- kDa proteins will allow investigation of the roles of both proteins in larvicidal activity. When expressed in *E. coli,* both proteins appear to be required for toxicity.[29,93,104] In contrast, de la Torre et al.[93] have suggested that it is necessary to express only the 41.9-kDa protein in *B. subtilis* in order to produce larvicidal activity and have proposed an enzymatic role for the 51.4-kDa protein in activating the 41.9-kDa toxin. Deletion studies being performed in the authors' laboratory, may help to delineate sequences within the two proteins which are important for toxicity, thus allowing us to concentrate further studies on these regions. In other work[92] the sequences of the 41.9- and 51.4-kDa toxins from five highly toxic *B. sphaericus* strains have been compared. A high degree of sequence conservation was noted, with three strains (1593, 2362, and 2317.3) having identical sequences. Other highly active strains have differences in amino acid sequences, which may account for some of their differences in toxicity to target insects. Thiery and de Barjac[66] have assayed the toxicities of many strains of *B. sphaericus* to three species of target mosquitoes and confirmed that different levels of toxicity and different target spectra exist for *B. sphaericus* strains. Similarly, Yap et al.[105] have shown that strain 2297 (group H25), in contrast to strains 1593 and 2362 (group H5a,5b), is nontoxic to larvae of the mosquito *Mansonia uniformis*. It should be noted that not all differences in larvicidal activity can be attributed to sequence changes within the 51.4- and 41.9-kDa proteins. The amino acid sequences for these two proteins from strains 1593 and 2362 are identical, yet these strains have different toxicities.[66] It is possible that strain variations in toxicity are due to combinations of different *B. sphaericus* toxins, which remain to be identified. It seems likely, however, that amino acid sequence changes between related toxins could be responsible for some of the strain differences in activity. Identification of variable amino acids may be important in designing mutagenesis experiments to alter toxicity, and thereby improve the mosquitocidal activity of the proteins.

## X. CONCLUSIONS

The insecticidal toxins of *B. sphaericus* are both potent and highly specific in their activity against mosquito larvae. Mosquitoes are major pest insects and the vectors of many diseases (including malaria, Bancroftian and Brugian filariasis, dengue fever, Japanese encephalitis, and yellow fever). In order for the *B. sphaericus* toxins to be used as effective control agents with widespread application, further research is required. An appropriate formulation which can be produced and applied on a local basis in developing countries would be an advantage. The cloning of *B. sphaericus* toxins into other organisms may provide a useful vector in which the toxin can be applied in the field and further insight into the mechanism of action of the toxins may help in the production of new control agents. With increasing resistance of mosquitoes to conventional pesticides, alternative means of control must be developed. In this respect the mosquitocidal toxins of *B. sphaericus* and *B. thuringiensis israelensis* are likely to be important in future vector control programs.

## ACKNOWLEDGMENT

The authors would like to thank Dr. K. Pardy for her help in the preparation of this manuscript.

## REFERENCES

1. **Kellen, W. R., Clark, T. B., Lindegren, J. E., Ho, B. C., Rogoff, M. H., and Singer, S.,** *Bacillus sphaericus* Neide as a pathogen of mosquitoes, *J. Invertebr. Pathol.,* 7, 442, 1965.
2. **Singer, S.,** Insecticidal activity of recent bacterial isolates and their toxins against mosquito larvae, *Nature,* 244, 110, 1973.
3. **Singer, S. and Murphy, D. J.,** New insecticidal strains of *Bacillus sphaericus* useful against *Anopheles albimanus* larvae, Abstr. 76th Annu. Meet., American Society of Microbiology, 1976, 181.
4. **Wickremesinghe, R. S. B. and Mendis, C. L.,** *Bacillus sphaericus* spore from Sri Lanka demonstrating rapid larvicidal activity on *Culex quinquefasciatus*, *Mosq. News,* 40, 387, 1980.
5. **Weiser, J.,** A mosquito-virulent *Bacillus sphaericus* in adult *Simulium damnosum* from northern Nigeria, *Zentralbl. Mikrobiol.,* 139, 57, 1984.
6. **White, P. J. and Lotay, H. K.,** Minimal nutritional requirements of *Bacillus sphaericus* NCTC 9602 and 26 other strains of this species: the majority grow and sporulate with acetate as sole major source of carbon, *J. Gen. Microbiol.,* 118, 13, 1980.
7. **Massie, J., Roberts, G., and White, P. J.,** Selective isolation of *Bacillus sphaericus* from soil by use of acetate as the only major source of carbon, *Appl. Environ. Microbiol.,* 49, 1478, 1985.

8. **Travers, R. S., Martin, P. A. W., and Reichelderfer, C. F.**, Selective process for efficient isolation of soil *Bacillus* spp., *Appl. Environ. Microbiol.*, 53, 1263, 1987.

9. **Yousten, A. A., Fretz, S. B., and Jelley, S. A.**, Selective medium for mosquito-pathogenic strains of *Bacillus sphaericus*, *Appl. Environ. Microbiol.*, 49, 1532, 1985.

10. **Baumann, L., Okamoto, K., Unterman, B., Lynch, M., and Baumann, P.**, Phenotypic characterization of *Bacillus thuringiensis* and *B. cereus*, *J. Invertebr. Pathol.*, 44, 329, 1984.

11. **de Barjac, H., Véron, M., and Cosmao-Dumanoir, V.**, Charactérisation biochimique et sérologique de souches de *Bacillus sphaericus* pathogènes ou non pour les moustiques, *Ann. Microbiol. (Inst. Pasteur)*, 131B, 191, 1980.

12. **Russell, B. L., Jelley, S. A., and Yousten, A. A.**, Carbohydrate metabolism in the mosquito pathogen *Bacillus sphaericus* 2362, *Appl. Environ. Microbiol.*, 55, 294, 1989.

13. **de Barjac, H.**, Insect pathogens in the genus Bacillus, in *The aerobic endospore-forming bacteria*, Berkeley, R. C. W. and Goodfellow, H., Eds., Academic Press, New York, 1981, 241.

14. **Krych, V. K., Johnson, J. L., Hendrick, J. C., and Yousten, A. A.**, Taxonomy and identification of mosquito-pathogenic strains of *Bacillus sphaericus*, in *Progress in Invertebrate Pathology*, Proc. Int. Colloq. Invertebr. Pathol., Weiser, J., Ed., Prague, 1978, 99.

15. **Krych, V. K., Johnson, J. L., and Yousten, A. A.**, Deoxyribonucleic acid homologies among strains of *Bacillus sphaericus*, *Int. J. Syst. Bacteriol.*, 30, 476, 1980.

16. **Yousten, A. A., de Barjac, H., Hedrick, J., Cosmao Dumanoir, V., Myers, P.**, Comparison between bacteriophage typing and serotyping for the differentiation of *Bacillus sphaericus* strains, *Ann. Microbiol. (Inst. Pasteur)*, 131B, 297, 1980.

17. **Yousten, A. A.**, Bacteriophage typing of mosquito pathogenic strains of *Bacillus sphaericus*, *J. Invertebr. Pathol.*, 43, 124, 1984.

18. **de Barjac, H., Thiery, I., Cosmao-Dumanoir, V., Frachon, E., Laurent, P., Charles, J.-F., Hamon, S., and Ofori, J.**, Another *Bacillus sphaericus* serotype harbouring strains very toxic to mosquito larvae: serotype H6, *Ann. Microbiol. (Inst. Pasteur)*, 139, 363, 1988.

19. **Davidson, E. W., Singer, S., and Briggs, J. D.**, Pathogenesis of *Bacillus sphaericus* strain SSII-1 infections in *Culex pipiens quinquefasciatus* (= *C. pipiens fatigans*) larvae, *J. Invertebr. Pathol.*, 25, 179, 1975.

20. **Myers, P. and Yousten, A. A.**, Toxic activity of *Bacillus sphaericus* SSII-1 for mosquito larvae, *Infect. Immun.*, 19, 1047, 1978.

21. **Myers, P., Yousten, A. A., and Davidson, E. W.**, Comparative studies of the mosquito-larval toxin of *Bacillus sphaericus* SSII-1 and 1593, *Can. J. Microbiol.*, 25, 1227, 1979.

22. **Yousten, A. A. and Davidson, E. W.**, Ultrastructural analysis of spores and parasporal crystals formed by *Bacillus sphaericus* 2297, *Appl. Environ. Microbiol.*, 44, 1449, 1982.

23. **Davidson, E. W. and Myers, P.**, Parasporal inclusions in *Bacillus sphaericus*, *FEMS Microbiol. Letts.*, 10, 261, 1981.

24. **Kalfon, A., Larget-Thiéry, I., Charles, J.-F., and de Barjac, H.**, Growth, sporulation and larvicidal activity of *Bacillus sphaericus*, *Eur. J. Appl. Microbiol. Biotechnol.*, 18, 168, 1983.

25. **Charles, J.-F., Kalfon, A., Bourgouin, C., and de Barjac, H.**, *Bacillus sphaericus* asporogenous mutants: morphology, protein pattern and larvicidal activity, *Ann. Microbiol. (Inst. Pasteur)*, 139, 243, 1988.

26. **Davidson, E. W., Myers, P., and Yousten, A. A.**, Cellular location of insecticidal activity in *Bacillus sphaericus* strain 1593, in Rep. Int. Congr. Invertebr. Pathol., Prague, 1978, 47.

27. **Wraight, S. P., Molloy, D. P., and Singer, S.**, Studies on the culicine mosquito host range of *Bacillus sphaericus* and *Bacillus thuringiensis* var. *israelensis* with notes on the effects of temperature and instar on bacterial efficacy, *J. Invertebr. Pathol.*, 49, 291, 1987.

28. **Louis, J., Jayaraman, K., and Szulmajster, J.**, Biocide gene(s) and biocidal activity in different strains of *Bacillus sphaericus*. Expression of the gene(s) in *E. coli* maxicells, *Mol. Gen. Genet.*, 195, 23, 1984.

29. **Baumann, P., Baumann, L., Bowditch, R. D., and Broadwell, A. H.**, Cloning of the gene for the larvicidal toxin of *Bacillus sphaericus* 2362: Evidence for a family of related sequences, *J. Bacteriol.*, 169, 4061, 1987.

30. **Berry, C.**, A Study of the Insecticidal Toxin Genes of *Bacillus sphaericus* Strains 1593, 2362 and 2297, Ph.D. thesis, University of Bristol, Bristol, England, 1988.

31. **Jackson-Yap, J., Hindley, J., and Berry, C.**, unpublished data, 1989.

32. **Tinelli, R. and Bourgouin, C.**, Larvicidal toxin from *Bacillus sphaericus* spores, *FEBS Letts.*, 142, 155, 1982.

33. **Davidson, E. W.**, Purification and properties of soluble cytoplasmic toxin from the mosquito pathogen *Bacillus sphaericus* strain 1593, *J. Invertebr. Pathol.*, 39, 6, 1982.

34. **Davidson, E. W.**, Alkaline extraction of toxin from spores of the mosquito pathogen, *Bacillus sphaericus* strain 1593, *Can. J. Microbiol.*, 29, 271, 1983.

35. **Payne, J. M. and Davidson, E. W.**, Insecticidal activity of the crystalline parasporal inclusions and other components of the *Bacillus sphaericus* 1593 spore complex, *J. Invertebr. Pathol.*, 43, 383, 1984.

36. **Bourgouin, C., Charles, J.-F., Kalfon, A. R., and de Barjac, H.,** *Bacillus sphaericus* 2297. Purification and biogenesis of parasporal inclusions, toxic for mosquito larvae, in *Bacterial Protein Toxins,* Alouf, J. E., Fehrenbach, F. J., Freer, J. F., and Jeljaszeweir, J., Eds., Academic Press, London, 1984, 389.

37. **Baumann, P., Unterman, B. M., Baumann, L., Broadwell, A. H., Abbene, S. J., and Bowditch, R. D.,** Purification of the larvicidal toxin of *Bacillus sphaericus* and evidence for high-molecular-weight precursors, *J. Bacteriol.,* 163, 738, 1985.

38. **Broadwell, A. H. and Baumann, P.,** Sporulation-associated activation of *Bacillus sphaericus* larvicide, *Appl. Environ. Microbiol.,* 52, 758, 1986.

39. **Davidson, E. W., Bieber, A. L., Meyer, M., and Shellabarger, C.,** Enzymatic activation of the *Bacillus sphaericus* mosquito larvicidal toxin, *J. Invertebr. Pathol.,* 50, 40, 1987.

40. **Broadwell, A. H. and Baumann, P.,** Proteolysis in the gut of mosquito larvae results in further activation of the *Bacillus sphaericus* toxin, *Appl. Environ. Microbiol.,* 53, 1333, 1987.

41. **Aly, C., Mulla, M. S., and Federici, B. A.,** Ingestion, dissolution, and proteolysis of the *Bacillus sphaericus* toxin by mosquito larvae, in *Fundamental and Applied Aspects of Invertebrate Pathology,* Foundation of the 4th Int. Colloq. Invertebrate Pathology, Waginingen, Sampson, R. A., Vlak, J. M., and Peters, D., Eds., 1986, 549.

42. **Aly, C., Mulla, M. S., and Federici, B. A.,** Ingestion, dissolution and proteolysis of the *Bacillus sphaericus* toxin by mosquito larvae, *J. Invertebr. Pathol.,* 53, 12, 1989.

43. **Narasu, M. L. and Gopinathan, K. P.,** Purification of larvicidal protein from *Bacillus sphaericus* 1593, *Biochem. Biophys. Res. Commun.,* 141, 756, 1986.

44. **Sgarrella, F., and Szulmajster, J.,** Purification and characterization of the larvicidal toxin of *Bacillus sphaericus* 1593M, *Biochem. Biophys. Res. Commun.,* 143, 901, 1987.

45. **Davidson, E. W.,** Effects of *Bacillus sphaericus* 1593 and 2362 spore/crystal toxin on cultured mosquito cells, *J. Invertebr. Pathol.,* 47, 21, 1986.

46. **Davidson, E. W., Shellabarger, C., Meyer, M., and Bieber, A. L.,** Binding of the *Bacillus sphaericus* mosquito larvicidal toxin to cultured insect cells, *Can. J. Microbiol.,* 33, 982, 1987.

47. **Davidson, E. W.,** Variation in binding of *Bacillus sphaericus* toxin and wheat germ agglutinin to larval midgut cells of six species of mosquitoes, *J. Invertebr. Pathol.,* 53, 251, 1989.

48. **Hindley, J. and Berry, C.,** Identification, cloning and sequence analysis of the *Bacillus sphaericus* 1593 41.9 kD larvicidal toxin gene, *Mol. Microbiol.,* 1, 187, 1987.

49. **Baumann, L., Broadwell, A. H., and Baumann, P.,** Sequence analysis of the mosquitocidal toxin genes encoding 51.4- and 41.9-kilodalton proteins from *Bacillus sphaericus* 2362 and 2297, *J. Bacteriol.,* 170, 2045, 1988.

50. **Davidson, E. W.,** Bacteria for the control of arthropod vectors of human and animal disease, in *Microbial and Viral Pesticides,* Kurstak, E., Ed., Marcel Dekker, New York, 1982, 289.

51. **Goldberg, L. J., Ford, I., Tanabe, A. M., and Watkins, H. M. S.,** Effectiveness of *Bacillus sphaericus* var. *fusiformis* (SSII-1) as a potential mosquito larval control agent: The role of variations in natural microbial flora in the larval environment, *Mosq. News,* 37, 465, 1977.

52. **Ramoska, W. A. and Pacey, C.,** Food availability and period of exposure as factors of *Bacillus sphaericus* efficacy on mosquito larvae, *J. Econ. Entomol.,* 72, 523, 1979.

53. **Singer, S.,** Potential of *Bacillus sphaericus* and related spore-forming bacteria for pest control, in *Microbial Control of Insects, Mites and Plant Diseases,* Burges, H. D., Ed., Academic Press, London, 1981, 283.

54. **Lacey, L. A., Urbina, M. J., and Heitzman, C. M.,** Sustained release formulations of *Bacillus sphaericus* and *Bacillus thuringiensis* (H-14) for control of container-breeding *Culex quinquefasciatus, Mosq. News,* 44, 26, 1984.

55. **Wraight, S. P., Molloy, D., and Jamnback, H.,** Efficacy of *Bacillus sphaericus* strain 1593 against the four instars of laboratory reared and field collected *Culex pipiens pipiens* and laboratory reared *Culex salinarius, Can. Entomol.,* 113, 379, 1981.

56. **Wraight, S. P., Molloy, D., Jamnback, H., and McCoy, P.,** Effects of temperature and instar on the efficacy of *Bacillus thuringiensis* var. *israelensis* and *Bacillus sphaericus* strain 1593 against *Aedes stimulans* larvae, *J. Invertebr. Pathol.,* 38, 78, 1981.

57. **Mulla, M. S., Darwazeh, H. A., Davidson, E. W., Dulmage, H. T., and Singer, S.,** Larvicidal activity and field efficacy of *Bacillus sphaericus* strains against mosquito larvae and their safety to nontarget organisms, *Mosq. News,* 44, 336, 1984.

58. **Balaraman, K.,** Comparative studies on the virulence of three strains of *Bacillus sphaericus* Meyer and Niede against mosquito larvae, *Indian J. Med. Res.,* 72, 55, 1980.

59. **Lacey, L. A. and Singer, S.,** Larvicidal activity of new isolates of *Bacillus sphaericus* and *Bacillus thuringiensis* (H-14) against anopheline and culicine mosquitoes, *Mosq. News,* 42, 537, 1982.

60. **Lacey, L. A., Lacey, C. M., Peacock, B., and Thiery, I.,** Mosquito host range and field activity of *Bacillus sphaericus* isolate 2297 (serotype 25), *J. Am. Mosq. Control Assoc.,* 4, 51, 1988.

61. **Cheong, W. C. and Yap, H. H.,** Bioassays of *Bacillus sphaericus* (strain 1593) against mosquitoes of public health importance in Malaysia, *Southeast Asian J. Trop. Med. Public Health,* 16, 54, 1985.

62. **Klowden, M. J., Held, G. A., and Bulla, L. A., Jr.,** Toxicity of *Bacillus thuringiensis* subsp. *israelensis* to adult *Aedes aegypti* mosquitoes, *Appl. Environ. Microbiol.,* 46, 312, 1983.

63. **Klowden, M. J. and Bulla, L. A., Jr.,** Oral toxicity of *Bacillus thuringiensis* subsp. *israelensis* to adult mosquitoes, *Appl. Environ. Microbiol.,* 48, 665, 1984.

64. **Stray, J. E., Klowden, M. J., and Hurlbert, R. E.,** Toxicity of *Bacillus sphaericus* crystal toxin to adult mosquitoes, *Appl. Environ. Microbiol.,* 54, 2320, 1988.

65. **Bone, L. W. and Tinelli, R.,** *Trichostrongylus colubriformis:* larvicidal activity of toxic extracts from *Bacillus sphaericus* (strain 1593) spores, *Exp. Parasitol.,* 64, 514, 1987.

66. **Thiery, I. and de Barjac, H.,** Selection of the most potent *Bacillus sphaericus* strains based on activity ratios determined on three mosquito species, *Appl. Microbiol. Biotechnol.,* 31, 577, 1989.

67. **Davidson, E. W.,** Ultrastructure of midgut events in the pathogenesis of *Bacillus sphaericus* strain SSII-1 infections of *Culex pipiens quinquefasciatus* larvae, *Can. J. Microbiol.,* 25, 178, 1979.

68. **Charles, J.-F.,** Ultrastructural midgut events in Culicidae larvae fed with *Bacillus sphaericus* 2297 spore/complex, *Ann. Microbiol. (Inst. Pasteur),* 138, 471, 1987.

69. **Davidson, E. W. and Titus, M.,** Ultrastructural effects of the *Bacillus sphaericus* mosquito larvicidal toxin on cultured mosquito cells, *J. Invertebr. Pathol.,* 50, 213, 1987.

70. **Davidson, E. W.,** Binding of the *Bacillus sphaericus* (Eubacteriales: Bacillaceae) toxin to midgut cells of mosquito (Diptera: Culicidae) larvae: relationship to host range, *J. Med. Entomol.,* 25, 151, 1988.

71. **Davidson, E. W., Morton, H. L., Moffett, J. O., and Singer, S.,** Effect of *Bacillus sphaericus* strain SSII-1 on honey bees, *Apis mellifera, J. Invertebr. Pathol.,* 29, 344, 1977.

72. **Mulligan, F. S., III, Schaefer, C. H., and Miura, T.,** Laboratory and field evaluation of *Bacillus sphaericus* as a mosquito control agent, *J. Econ. Entomol.,* 70, 774, 1978.

73. **WHO,** Biological control agent data sheet, *Bacillus sphaericus,* strain 1593-4, unpubl. document, VBC/BCDS/79.09, 8 pp., 1979.

74. **Shadduck, J. A., Singer, S., and Lause, S.,** Lack of mammalian pathogenicity of entomocidal isolates of *Bacillus sphaericus, Environ. Entomol.,* 9, 403, 1980.

75. **Davidson, E. W., Urbina, M., Payne, J., Mulla, M. S., Darwazeh, H., Dulmage, H. T., and Correa, J. A.,** Fate of *Bacillus sphaericus* 1593 and 2362 spores used as larvicides in the aquatic environment, *Appl. Environ. Microbiol.,* 47, 125, 1984.

76. **Karch, S. and Charles, J.-F.,** Toxicity, viability and ultrastructure of *Bacillus sphaericus* 2362 spore/crystal complex used in the field, *Ann. Microbiol. (Inst. Pasteur.),* 138, 485, 1987.

77. **Nicolas, L., Dossou-Yovo, J., and Hougard, J.-M.,** Persistence and recycling of *Bacillus sphaericus* 2362 spores in *Culex quinquefasciatus* breeding sites in West Africa, *Appl. Microbiol. Biotechnol.,* 25, 341, 1987.

78. **Mulla, M. S., Darwazeh, H. A., Davidson, E. W., and Dulmage, H. T.,** Efficacy and persistence of the microbial agent *Bacillus sphaericus* against mosquito larvae in organically enriched habitats, *Mosq. News,* 44, 166, 1984.

79. **Yousten, A. A.,** *Bacillus sphaericus:* Microbiological factors related to its potential as a mosquito larvicide, *Adv. Biotechnol. Proc.,* 3, 315, 1984.

80. **Klein, D., Yanai, P., Hofstein, R., Fridlender, B., and Braun, S.,** Production of *Bacillus sphaericus* larvicide on industrial peptones, *Appl. Microbiol. Biotechnol.,* 30, 580, 1989.

81. **Fridlender, B., Keren-Zur, M., Hofstein, R., Bar, E., Sandler, N., Keynan, A., and Braun, S.,** The development of *Bacillus thuringiensis* and *Bacillus sphaericus* as biocontrol agents: From research to industrial production, in *Memórias do Instituto Oswaldo Cruz,* Vol. 84, 1° Simpósio Nacional de Controle Biológico de Pragas e Vetores, Rio de Janeiro, Brazil, 1989.

82. **Obeta, J. A. N. and Okafor, N.,** Production of *Bacillus sphaericus* strain 1593 primary powder on media made from locally obtainable Nigerian agricultural products, *Can. J. Microbiol.,* 29, 704, 1983.

83. **Dulmage, H. T., Correa, J., and Martinez, A. J.,** Coprecipitation with lactose as a means of recovering the spore-crystal complex of *Bacillus thuringiensis, J. Invertebr. Pathol.,* 15, 15, 1970.

84. **Lacey, L. A.,** Production and formulation of *Bacillus sphaericus, Mosq. News,* 44, 153, 1984.

85. **Lacey, L. A., Urbina, M. J., and Heitzman, C. M.,** Sustained release formulations of *Bacillus sphaericus* and *Bacillus thuringiensis* (H-14) for control of container-breeding *Culex quinquefasciatus, Mosq. News,* 44, 26, 1984.

86. **Mulligan, F. S., III, Schaefer, C. H., and Wilder, W. H.,** Efficacy and persistence of *Bacillus sphaericus* and *B. thuringiensis* H-14 against mosquitoes under laboratory and field conditions, *J. Econ. Entomol.,* 73, 684, 1980.

87. **Burke, W. F., Jr., McDonald, K. O., and Davidson, E. W.,** Effect of UV light on spore viability and mosquito larvicidal activity of *Bacillus sphaericus* 1593, *Appl. Environ. Microbiol.,* 46, 954, 1983.

88. **Ganesan, S., Kamdar, H., Jayaraman, K., and Szulmajster, J.,** Cloning and expression in *Escherichia coli* of a DNA fragment from *Bacillus sphaericus* coding for biocidal activity against mosquito larvae, *Mol. Gen. Genet.,* 189, 181, 1983.

89. **de Marsac, N. T., de la Torre, F., and Szulmajster, J.,** Expression of the larvicidal gene of *Bacillus sphaericus* 1593M in the cyanobacterium *Anacystis nidulans* R2, *Mol. Gen. Genet.,* 209, 396, 1987.

90. **Berry, C. and Hindley, J.,** *Bacillus sphaericus* strain 2362: identification and nucleotide sequence of the 41.9 kDa toxin gene, *Nucleic Acids Res.,* 15, 5891, 1987.

91. **Hindley, J. and Berry, C.,** *Bacillus sphaericus* strain 2297: nucleotide sequence of 41.9 kDa toxin gene, *Nucleic Acids Res.,* 16, 4168, 1988.

92. **Berry, C., Jackson-Yap, J., Oei, C., and Hindley, J.,** Nucleotide sequence of two toxin genes from *Bacillus sphaericus* IAB59: sequence comparisons between five highly toxinogenic strains, *Nucleic Acids Res.,* 17, 7516, 1989.

93. **de la Torre, F., Bennardo, T., Sebo, P., and Szulmajster, J.,** On the respective roles of the two proteins encoded by the *Bacillus sphaericus* 1593M toxin genes expressed in *Escherichia coli* and *Bacillus subtilis, Biochem. Biophys. Res. Commun.,* 164, 1417, 1989.

94. **Bowditch, R. D., Baumann, P., and Yousten, A. A.,** Cloning and sequencing of the gene encoding a 125-kilodalton surface-layer protein from *Bacillus sphaericus* 2362 and of a related cryptic gene, *J. Bacteriol.,* 171, 4178, 1989.

95. **Lee, H.-H., Kim, S. Y., Lim, P. O., and Lee, H. S.,** Cloning and expression of *Bacillus sphaericus* mosquitocidal crystalline protein gene in *E. coli, Han'guk J. Gen. Eng.,* 2, 19, 1987.

96. **Souza, A. E., Rajan, V., and Jayaraman, K.,** Cloning and expression in *Escherichia coli* of two DNA fragments from *Bacillus sphaericus* encoding mosquito-larvicidal activity, *J. Biotechnol.,* 7, 71, 1988.

97. **Baumann, L. and Baumann, P.,** Expression in *Bacillus subtilis* of the 51- and 42-kilodalton mosquitocidal toxin genes of *Bacillus sphaericus, Appl. Environ. Microbiol.,* 55, 252, 1989.

98. **Nicolas, L., Thiery, I., Rippka, R., Houmard, J., and Tandeau de Marsac, N.,** Characterization of cyanobacteria isolated from mosquito breeding sites and their potential use as vector control agents, in Abstr. 12th Annu. Meet. Soc. Invertebr. Pathol., University of Maryland, College Park, MD, August 20 to 24, 1989, 83.

99. **Thiery, I., Nicolas, L., Rippka, R., and Tandeau de Marsac, N.,** Isolation and identification of cyanobacteria from mosquito breeding sites and their properties as potential agents for mosquito control, in Abstr. 4th Regional Meet. SOVE Eur. Region, Inst. Plant Protection, Novi Sad, Yugoslavia, August 20th to 24th, 1989, 40.

100. **Godown, R. D.,** Environmental release: the battles and the war on two fronts, *Bio/technol.,* 7, 1096, 1989.

101. **Finlayson, M. and Gaertner, F.,** Alternative hosts for *Bacillus thuringiensis* delta-endotoxin, *J. Cell. Biochem.,* 13A, 152, 1989.

102. **Gelernter, W. D. and Quick, T. C.,** The MCap™ delivery system: a novel approach for enhancing efficacy and foliar persistence of biological toxins in Soc. for Invertebr. Pathology 22nd Annu. Meet., Maryland, August 20 to 24, 1989.

103. **Bourgouin, C., Delécluse, A., de la Torre, F., and Szulmajster, J.,** Transfer of the toxin protein genes of *Bacillus sphaericus* into *Bacillus thuringiensis* subsp. *israelensis* and their expression, *Appl. Environ. Microbiol.,* 56, 340, 1990.

104. **Oei, C., Hindley, J., and Berry, C.,** An analysis of the genes encoding the 51.4- and 41.9-kDa toxins of *Bacillus sphaericus* 2297 by deletion mutagenesis: the construction of fusion proteins, *FEMS Microbiol. Lett.,* 72, 265, 1990.

105. **Yap, H. H., Ng, Y. M., Foo, A. E. S., and Tan, H. T.,** Bioassays of *Bacillus sphaericus* (strain 1593, 2297 and 2362) against *Mansonia* and other mosquitoes of public health importance in Malaysia, *Malaysian Appl. Biol.,* 17, 9, 1988.

Chapter 4

# INSECT RESISTANCE TO *BACILLUS THURINGIENSIS*

**Terry B. Stone, Steven R. Sims, Susan C. MacIntosh, Roy L. Fuchs, and Pamela G. Marrone**

## TABLE OF CONTENTS

# I. INTRODUCTION

The ability of insects to overcome and adapt to the stresses of their environment has resulted in their being the most abundant group of animals on the earth today. This success has also resulted in their surmounting most methods developed for controlling their harmful effects to agriculture, man and his environment. These have included chemical, biological, and cultural strategies. Recent additions to the arsenal of control tactics are the products of biotechnology. This review focuses on the potential for insects to develop resistance to an important recourse for bioengineered insecticides — *Bacillus thuringiensis.* Detailed is the history of insect resistance to *B. thuringiensis,* genetic and physiological mechanisms of resistance, and lastly, proposed strategies to delay or prevent resistance to *B. thuringiensis*-based products from occurring.

# II. HISTORICAL BACKGROUND

## A. INSECT RESISTANCE TO CHEMICAL INSECTICIDES

Insect resistance to chemical insecticides is well documented[1] and will not be reviewed here. Over 500 species of insects are now resistant to chemical insecticides, spanning the organophosphates, carbamates, organochlorines, pyrethroids, juvenoids, and benzoylureas. These include the neurotoxins, chitin synthesis inhibitors, and juvenile hormone mimics.

In the past, regular prophylactic treatment, without concern for the development of resistance, was the norm. When a compound was lost due to resistance, another was there to replace it. However, at present, the introduction of new chemistry is not keeping pace with the development of resistance and the removal of registered pesticides from the marketplace for environmental reasons. As a result, the choices for insect control are becoming increasingly limited. Therefore, the practice of integrated pest management (IPM) and insect resistance management strategies has become not only desirable, but necessary for the preservation of existing pesticides. The ultimate consequence of increased resistance is an increased cost of pest control.

With the increasing scarcity of chemical pesticides due to environmental, health, and resistance concerns, interest is turning to microbial and other biological means for insect control. Biopesticides offer many advantages, such as narrow host range, safety to the environment and little activity to nontarget organisms. An increase in usage of these biopesticides, either naturally occurring or bioengineered, will increase selection pressure on target insect populations, necessitating a critical evaluation of the potential for resistance development.

## B. INSECT RESISTANCE TO MICROBIAL PATHOGENS

Although there are several reports of resistance development and shifts in susceptibility in laboratory selection experiments, there are no documented cases of field failure due to resistance to viral, fungal, or protozoan pathogens. There are many studies documenting the variation in susceptibility of insect populations to viruses, as well as laboratory studies selecting for resistance development. For a complete review of resistance to viral insecticides, see Briese.[2] Much less is known about the fungi and protozoa, but there are reports of intraspecific differences in susceptibility to fungal pathogens, such as *Ascosphaera apis,* the causal agent of chalkbrood disease of honeybees.[3,4] Pasteur reportedly selected a strain of silkworm resistant to the protozoan pathogen, *Nosema bombycis.*[5]

# III. INSECT RESISTANCE TO *BACILLUS THURINGIENSIS*

Products based on *Bacillus thuringiensis* are the most widely used biological insecticides. They are registered for the control of many economically important lepidopteran, coleopteran, and dipteran pests. Historically, they have been applied on limited acreage in comparison to many chemical insecticides because of a narrow spectrum of activity and short field longevity. Recent advancements in formulation, biotechnology, and the discovery of *B. thuringiensis* strains with greater potency and a broader spectrum of activity have resulted in a second generation of products with increased efficacy. Starch and polymer encapsulation systems may extend field longevity from about 48 h to longer than 7 days through the incorporation of UV protectants.[6] Formation of *B. thuringiensis* transconjugants have broadened insect spectrum.[7] Genetically modified plants and plant colonizing bacteria with insect control proteins have the potential to increase efficacy, prolong field persistence, and reduce application costs.[8-10] Not all of these products have been commercialized, but with their competitive pricing and low environmental and human health risk, they are expected to capture a substantially increased share of the insecticide market. Increased use of *B. thuringiensis* products with enhanced field stability may also increase the potential for the development of insect resistance.

## A. FIELD SUSCEPTIBILITY

Reports of field resistance or reduced efficacy to *B. thuringiensis* are few and restricted to situations where the insecticide was used intensively. In surveying the susceptibility of several laboratory and field populations of *Ephestia cautella* Hubner and *Plodia interpunctella* Hubner to Dipel®. (Abbott Laboratories, North Chicago, IL) a commercial formulation of *B. thuringiensis* subsp. *kurstaki* strain HD-1, Kinsinger and McGaughey[11] demonstrated up to a tenfold and 42-fold difference, respectively, between strains. The slopes of the dose-mortality lines were not significantly different, indicating that actual differences in susceptibility existed and not just a reduction in variability due to the elimination of susceptible individuals. Subsequent evaluation of grain storage bins over a five-state area revealed a moderate reduction in susceptibility to Dipel® of *P. interpunctella* populations in wheat and corn storage bins.[12] However, in the laboratory, comparison of the susceptibility of populations taken from these bins, 1 to 5 months after treatment, demonstrated that the strains were more resistant than those from untreated bins. McGaughey[12] concluded that the reduction in efficacy was related to the development of resistance.

It is important to note that resistance development among grain storage pests represents a unique situation in comparison to field crop insects. Grain storage pests are continually exposed to *B. thuringiensis* which retains its potency for a prolonged period of time without exposure to UV light. In addition, there is little immigration of susceptible individuals. Consequently, several successive generations are exposed, providing an enhanced opportunity for resistance to develop.

Reduced efficacy of *B. thuringiensis* among field crop insects has been reported with *Plutella xylostella* L. in Hawaii by Tabashnik.[13] Populations of *P. xylostella* on watercress have shown a decrease in susceptibility to *B. thuringiensis* subsp. *kurstaki* of 25- to 33-fold after 2 years of continual treatment. Tabashnik[14] also reported an 11-fold difference in susceptibility to Dipel® among diamondback moth populations evaluated from the Hawaiian islands of Oahu (two strains) and Maui (one strain). There is additional evidence of decreased *P. xylostella* susceptibility in both the Philippines and Thailand, where *B. thuringiensis* is used extensively, because of widespread resistance to chemical insecticides.[15,16]

Surveys of the intraspecific differences in susceptibility of several insect species have shown that variation exists among populations. Briese[2] summarized the findings of studies performed with *Bombyx mori* L. and *Musca domestica* L. Differences in $LC_{50}$ values were observed, but because the slopes of the dose-mortality lines were not provided, actual resistance among these populations could not be accurately determined. Stone et al.[17] demonstrated 4 to 15-fold differences in susceptibility to Dipel® and purified *B. thuringinesis* subsp. *kurstaki* strain HD-73 endotoxin (63-kDa active fragment), among 12 *Heliothis virescens* F. and *H. zea* Boddie populations collected from the southern U.S. and the Virgin Islands. Greater variation in susceptibility was observed among the *H. zea* populations than those of *H. virescens*. *H. zea* $LC_{50}$ values ranged from 11-fold for Dipel® and 16-fold for HD-73 endotoxin. $LC_{50}$ values for the *H. virescens* populations were fourfold for Dipel® and sixfold for HD-73 endotoxin. Though similarities exist between the two species, *H. zea* is inherently tenfold less susceptible to *B. thuringiensis* than *H. virescens*.[18] This natural tolerance may explain the greater range in response of *H. zea* to the two materials, but it is not necessarily an indication that the rate or potential of resistance will be greater for either species.

## B. LABORATORY SELECTION

Selection for resistance in the laboratory to *B. thuringiensis* has produced varying results depending on the bacterial products involved, i.e., β-exotoxin or δ-endotoxin, the insect species, and selection procedure. Selection to the β-exotoxin has typically required several generations and a homogenous population for low levels of resistance to occur. These studies are summarized by Briese[2] and Georghiou.[19]

Only low levels of resistance to *B. thuringiensis* subsp. *israelensis* have been achieved in laboratory selection on Diptera. Goldman[20] demonstrated twofold resistance in a field collected strain of *Aedes aegypti* L. from Brazil after 15 generations of selection pressure at the $LC_{50}$. This difference was statistically significant; however, the slopes of the dose-mortality lines were not. Resistance did not develop in either a field collected strain from Sri Lanka or a laboratory strain.

Georghiou[19] summarized several laboratory resistance selection studies with *B. thuringiensis* subsp. *israelensis* on field collected *Culex quinquefasciatus* Say. Georghiou et al.[21,22] and Georghiou and Vazquez-Garcia[23] collected *C. quinquefasciatus* from four sites (I-IV). Strains I, III, and IV were selected at the $LC_{90}$ for 60, 12, and 11 generations, respectively. Strain II was selected for 36 generations with gradually increasing pressure to simulate field conditions. Maximum resistance development at the $LC_{95}$ for the four strains was 16.5-fold (I), 4.4-fold (II), 4.1-fold (III), and 5.9-fold (IV). Resistance of strain I was found by Vazquez-Garcia to plateau by the F40 then gradually decline to about 12-fold by the F60. When the strain was reared in the absence of selection for three generations, resistance declined by about 50%. Additionally, selected larvae developed more slowly and adults laid fewer eggs than the unselected parents, indicating that resistance was developed at a biologic cost to the insects.

Recently, Dai and Gill[24] demonstrated enhanced resistance of *C. quinquefasciatus* to *B. thuringiensis* subsp. *israelensis* with selection on purified 72-kDa endotoxin protein. A strain of *C. quinquefasciatus* selected for 20 generations on a mixture of *B. thuringiensis* subsp. *israelensis* endotoxin proteins and spores was subcultured on purified 72-kDa endotoxin. Subculturing resulted in a threefold increase in resistance to the endotoxin:spore mix and a 70-fold difference in susceptibility to the 72-kDa endotoxin. The slopes of the dose-mortality lines before and after subculturing on purified endotoxin were not significantly different.

Several unsuccessful attempts to select for *B. thuringiensis* resistance among lepidoptera have been reported. These also have been summarized by Briese[2] and Georghiou.[19] These unsuccessful selection experiments indicate that resistance development may not be routine even in the laboratory. The remainder of this discussion will focus on successful selection experiments because of the insight that can be gained into potential mechanisms available to insects for resistance development.

The first case of laboratory resistance selection development was reported by McGaughey.[25] *P. interpunctella* larvae from a population not effectively controlled by Dipel®, were used for the study. After establishment in the laboratory for 20 generations, larvae were fed Dipel®-treated diet at an expected $LC_{70-80}$. Survival of the F1 was 19%. By the F4, survival had increased to 82%. The $LC_{50}$ of the F2 was 27-fold that of unselected insects and by the F15 it increased to 97-fold. The slopes of the dose-mortality lines were not significantly different. Resistance was stable after seven generations without selection pressure, indicating no reduction in fitness as a result of selection.

McGaughey and Beeman[26] additionally selected for resistance to Dipel® in a laboratory colony of *E. cautella* in culture for 10 years and five *P. interpunctella* colonies taken from midwestern grain storage bins. The *E. cautella* colony developed 7.5-fold resistance after 21 generations of selection. Resistance developed in all *P. interpunctella* colonies selected, but was found to increase at different rates. The greatest level of resistance developed was 250-fold at the F36. The other *P. interpunctella* colonies developed, 63-fold (F36), 73-fold (F42), 33-fold (F31) and 15-fold (F39) resistance. The different rates of *P. interpunctella* selection were attributed to the colonies representing different biotypes. The presence of spores and several types of insecticidal proteins in formulations of Dipel®[27] may have also influenced the disparate rates of selection, as each of the colonies developed resistance to the factors at different rates. The low level of *E. cautella* resistance was thought to reflect a decrease in genetic heterogeneity and therefore, a loss of resistance genes resulting from the extensive length of time the colony had been in culture.[26] It might also indicate inherent differences in potential for resistance development in different insect species.

Stone et al.[28] reported the first case of laboratory resistance to a field crop insect. *H. virescens* were selected to a root colonizing bacterium, *Pseudomonas fluorescens* (Ps112-12a),[29,30] genetically engineered to express a 130-kDa protoxin of *B. thuringiensis* subsp. *kurstaki* strain HD-1, the *crylA(b)* gene product. Ps112-12a was incorporated in the diet at the $LC_{60-80}$. Threefold resistance had developed by the F3. Resistance continued to increase to the F7 where it stabilized through the F14 at about 24-fold. Slopes of the dose-mortality lines were not significantly different. Subsequent subculturing on Dipel® for 4 generations potentiated Ps112-12a resistance to 65-fold presumably as a result of the several other insecticidal proteins or spores present in the formulation.[27] Resistance to Dipel® also increased from fourfold prior to subculturing to about 50-fold and to purified *B. thuringiensis* subsp. *kurstaki* strain HD-73 endotoxin by fivefold. Dipel® is known to contain the cry1A(c) protein characteristic of HD-73.[27]

Finally, Whalon[31] has selected *Leptinotarsa decimlineata* Say larvae for resistance to *B. thuringiensis* subsp. *san diego*. The selected population was initiated from field collected insects that had survived several applications of the commercial *B. thuringiensis* subsp. *san diego* formulation, M-One® (Mycogen Corp., San Diego, Calif.). Selection was continued in the laboratory and at the F10 resistance was 20-fold.

These studies provide evidence that resistance to *B. thuringiensis* could develop in the field. However, Georghiou[19] states that laboratory selection experiments are ''worst case'' scenarios because they do not take into consideration biological factors such as immigration, refugia, and incomplete coverage. In addition, they should be viewed with skepticism since the laboratory population represents only a small fraction of the field population. Consequently, resistance in the laboratory cannot predict that resistance will develop in the field. Laboratory resistance selected lines should be used as models for understanding the potential development, the mechanisms of resistance, and in developing strategies for delaying or preventing the occurrence of resistance to *B. thuringiensis*.

## C. CROSS RESISTANCE

Insect colonies resistant to *B. thuringiensis* toxins have not shown cross resistance to chemical insecticides.[32-36] This is not surprising given the difference in mechanism of action of chemical insecticides and *B. thuringiensis* toxins and the subsequent mechanism(s) of resistance.

The cross resistance of Dipel® resistant *P. interpunctella* to other *B. thuringiensis* isolates was determined by McGaughey and Johnson.[37] Fifty-seven isolates with known toxicity to

*P. interpunctella* were evaluated. Twenty-one isolates from subsp. *kenyae, entomocidus, tolworthi,* and *darmstadiensis,* were equally toxic to susceptible and resistant *P. interpunctella.* Thirty-six isolates representing subspecies *thuringiensis, kurstaki, galleriae, aizawai,* and *tolworthi* were found to be less toxic to Dipel® resistant *P. interpunctella.* The greatest degree of cross resistance was demonstrated to isolates of subsp. *kurstaki,* the subspecies of which Dipel® is comprised. There was no evidence of cross resistance to the β-exotoxin. The authors concluded that *P. interpunctella* resistance was specific toward the HD-1 type spore:crystal complex. More specifically, Höfte and Whiteley[27] reported the extent of cross resistance was to the *cryIA(a-c)* toxins present in Dipel® and not to the other crystal protein types.

## IV. GENETICS OF RESISTANCE

Only two insects have been subjects of resistance genetics studies to date. McGaughey[25] and McGaughey and Beeman[26] examined the genetics of *P. interpunctella* resistance to Dipel®, and Sims and Stone[38] analyzed the genetics of *H. virescens* resistance to a genetically engineered *Pseudomonas fluorescens* (Ps112-12a) expressing *B. thuringiensis* subsp. *kurstaki* strain HD-1 protoxin. Stability of resistance varied among the *P. interpunctella* strains when they were removed from selection pressure. At least one strain showed a significant decline in resistance between generations 11 to 28 postselection.[26] This may have been due to susceptible alleles remaining in the population although the possibility of deleterious effects of resistance on reproduction and development cannot be discounted.

F1 hybrid progeny produced by crossing the five selected *P. interpunctella* strains with the parental susceptible colony showed that resistance is recessive but variable in expression of recessiveness.[26] Attempts to determine the number of genes involved in resistance were inconclusive because presumed heterozygous and homozygous genotypes in the backcross could not be clearly differentiated.

Sims and Stone[38] found that resistance to Ps112-12a in a laboratory *H. virescens* strain was partially dominant and controlled by several genetic factors. Resistance was unstable, declining significantly in a subculture of the F19 resistant line after five generations of nonselection.

Existing studies on the genetics of *B. thuringiensis* resistance have been hampered by several factors. First, all studies to date have relied on the resistance attained partially or completely under laboratory selection. The genetic control of the resistance thus obtained may not correspond to the control established under field selection because field selection operates on a much larger sample of total population variability. If a parallel is drawn to examples of insect resistance to chemical pesticides, we might predict that *B. thuringiensis* resistance in the field is more likely to be controlled by a Mendelian-single gene system compared to multiple loci in the laboratory.[39]

A second problem with studies of *B. thuringiensis* resistance genetics is the insufficient level of resistance difference between resistant and susceptible lines. For example, despite the >300-fold difference found between resistant and susceptible *P. interpunctella* strains,[26] the shallow slopes of the resistant, susceptible, and F1 dosage-mortality lines resulted in significant overlap. This made genetic inferences based on the traditional backcrossing technique difficult.

Finally, our knowledge of the genetics of most *B. thuringiensis* target insects is inadequate for linkage analysis. Among the Lepidoptera, few visible genetic markers exist and it is unlikely that these will be associated with resistance. Fortunately, two relatively new techniques offer promise for greatly increasing the number of genetic markers available for linkage analysis of *B. thuringiensis* resistance. Electrophoresis of isozymes in pest insects may reveal the presence of diagnostic alleles that can be made homozygous and used for linkage mapping. Restriction fragment length polymorphisms employ restriction enzymes to cleave chromosomes into fragments of distinct lengths, the distribution of which can be scored using standard procedures. Polymorphisms in the lengths of individual fragments can also be used as markers for linkage mapping. Considerable progress has been made using these two techniques to develop a complete linkage map for *H. virescens*.[40] Evaluation of both *B. thuringiensis* and chemical resistant strains of *H. virescens* using linkage analysis offers the potential of greatly increasing our understanding of resistance genetics in this species.

# V. MODE OF ACTION AND RESISTANCE MECHANISMS

## A. MODE OF ACTION OF *BACILLUS THURINGIENSIS*

The susceptibility of lepidopteran larvae to *B. thuringiensis* subsp. *kurstaki* crystal protein is well documented. The ultrastructural effects have been studied by both light and electron microscopy.[41-43] Histopathological studies of insects treated with gut juice activated endotoxin illustrate that the *B. thuringiensis* subsp. *kurstaki* protein attacks the epithelium of the midgut lumen.[41] The endotoxin causes swelling of the microvilli which line the midgut lumen, eventually destroying the cells and exposing the hemolymph to infection by spores or vegetative cells inducing septicemia.[42] Other ultrastructural effects involve vesiculation of the endoplasmic reticulum, ribosomal loss, disintegration of the mitochondria, and swelling of the goblet cells.[42,43] It appears that the target of the endotoxin is the cell membrane and not directly on the intracellular organelles.[41,44] Similar histopathological effects are seen in tissue culture cells.[45-47] Leakage of the gut contents into the hemolymph is the ultimate cause of larval death, which may take days. Gut paralysis and cessation of feeding occur minutes after ingestion of *B. thuringiensis* endotoxin.[48,49] The termination of feeding is, of course, the important effect in terms of the agronomic value of *B. thuringiensis*.

Proteolytic processing of the full-length *B. thuringiensis* subsp. *kurstaki* protoxin (130 to 140 kDa) is necessary for insecticidal activity.[50-52] The alkaline environment of the lepidopteran midgut dissolves the *B. thuringiensis* subsp. *kurstaki* protoxin and gut juice proteases enzymatically cleave it to a trypsin stable protein (55 to 70 kDa). The smaller activated protein crosses the peritrophic membrane to the site of action on the midgut epithelium. Insect specificity may be determined by the type of proteolytic cleavage that occurs in the insect gut as with *B. thuringiensis* subsp. *colmeri,* which has dual activities against lepidopteran and dipteran cell lines.[53,54] When the protoxin from *B. thuringiensis* subsp. *colmeri* is mixed with lepidopteran gut juice, a lepidopteran active toxin is created, lacking activity against dipteran cell lines.[53] Alternatively, dipteran gut juice treatment of *B. thuringiensis* subsp. *colmeri* produces a protein with only dipteran activity. Analysis of the endotoxins produced by the two different gut proteases show that the dipteran activated protein is 2.5 kDa lower in molecular weight than that of the lepidopteran-activated protein. The two proteins with differing specificities may have very different conformational structures even though the amino acid sequence is identical except for an extra 20 amino acids on the lepidopteran active endotoxin. Other *B. thuringiensis* proteins, such as *B. thuringiensis* subsp. *aizawai,* have been identified that have dual specificity but activation by either lepidopteran or dipteran gut proteases produces an endotoxin active against both classes of insects.[14]

Recent studies have identified receptor binding as the primary interaction of the *B. thuringiensis* protein with the lepidopteran midgut epithelium.[55-57] Hofmann et al.[56] isolated brush border membrane vesicles (BBMV) from *Pieris brassicae* and identified two binding site populations, of low and high affinity, for *B. thuringiensis* subsp. *thuringiensis* (cryIB). A later report[58] linked insect specificity to binding by comparing the affinity of two *B. thuringiensis* proteins, *B. thuringiensis* subsp. *thuringiensis* (cryIB) and *B. thuringiensis* subsp. *berliner* (cryIA(b)) with differing sensitivities against BBMV isolated from *P. brassicae* and *Manduca sexta*. *P. brassicae* was sensitive to both *B. thuringiensis* proteins which bound to high affinity receptors on the BBMV of this insect. *M. sexta* was only sensitive to *B. thuringiensis* subsp. *berliner*, which bound with high affinity to midgut cells. *B. thuringiensis* subsp. *thuringiensis* was not toxic towards *M. sexta* and did not bind to isolated BBMV. Finally, Van Rie et al.[59] investigated the binding of three *B. thuringiensis* proteins, subsps. *kurstaki* (cryIA(c)), *berliner* (cryIA(b)) and *aizawai* (cryIA(a)) to BBMV of *H. virescens*. Although all three *B. thuringiensis* proteins exhibited toxicity toward *H. virescens*, the affinities of these three proteins for midgut vesicles were not significantly different. Instead, the concentration of binding sites correlated with the observed toxicity. It is important to note that an 80-fold range of $LC_{50}$s was observed but only a fivefold difference was found in binding site concentration. The complexity of the protein-membrane interaction was further illustrated by the hypothesis of a three-binding site model supported by competition studies. Van Rie[59] speculated that the cryIA(c) protein bound to all three binding site populations, cryIA(b) protein to two sites, and cryIA(a) protein to only one site. These results clearly link midgut binding of the *B. thuringiensis* protein to toxicity, but more research is required before the level of toxicity of a particular *B. thuringiensis* endotoxin can be accurately predicted from its binding parameters.

Secondary effects of *B. thuringiensis* subsp. *kurstaki* protein on the midgut of lepidopteran larvae have been studied in depth by Wolfersberger and colleagues.[60] BBMV isolated from the midgut of *P. brassicae* accumulate amino acids; alanine, phenylalanine, histidine, glutamic acid and lysine, in the presence of an inward $K^+$ gradient.[61] The addition of *B. thuringiensis* subsp. *kurstaki* protein inhibits the ability of the BBMV to transport amino acids.[60] The endotoxin increases the $K^+$ permeability of the membrane but does not interfere with the $K^+$-amino acid symport. The increase in $K^+$ permeability of the midgut membrane disrupts the ion and fluid balance resulting in the swelling and destruction of the microvilli. Ellar et al.[62] theorized that *B. thuringiensis* subsp. *kurstaki* protein actually forms a $K^+$ pore in the midgut epithelium membrane.

In summary, the data support a distinct mode of action: once the *B. thuringiensis* protoxin is ingested, it is first proteolyzed to the activated protein, then crosses the peritrophic membrane and binds to high affinity receptors present on the midgut epithelium. The gut becomes paralyzed and the larva stops feeding. Once bound, the protein probably inserts into the membrane, which causes an opening or pore to form, thus disrupting the inward $K^+$ gradient and inhibiting the transport of amino acids. The microvilli swell and eventually are destroyed, allowing for the gut contents to enter the hemocoel, causing larval death.

## B. MECHANISM OF RESISTANCE

Study of mechanisms of resistance have centered on the effects of chemical insecticides. The onset of resistance is dependent on a multitude of interdependent factors that include, genetic, biochemical, physiological, and ecological elements. Mechanisms of resistance can be grouped into three major areas: biochemical, physiological, and behavioral.[63]

Biochemical mechanisms[63] encompass (1) the increased ability of an insect to metabolize or detoxify an insecticide, known as metabolic resistance;[64] (2) the decreased sensitivity of a target site to the effects of an insecticide, or target site insensitivity;[65] and (3) the decreased ability of an insect to adsorb insecticides. Physiological mechanisms are closely related to biochemical mechanisms, especially insecticide adsorption.[63] Typically physiological mechanisms include reducing insecticide penetration, increasing insecticide segregation and elimination.[65] Behavioral mechanisms are difficult to study and quantitate but may be crucial in the development of resistance.[66,67]

Most chemical insecticides are neurotoxins with contact activity, a very different mode-of-action than the gut activity of *B. thuringiensis* proteins. Due to the complexity of the mode-of-action of *B. thuringiensis* insect control proteins, the potential mechanisms of resistance are difficult to predict. The mechanism of resistance to *B. thuringiensis* could occur at any one or more steps; activation or cleavage of the protoxin, passage through the peritrophic membrane, binding to the midgut epithelium, or insertion and pore formation in the midgut membrane. In addition, behavioral responses will undoubtedly play a major role in resistance development in a species specific manner.

Resistance to *B. thuringiensis* has been identified only recently with a limited number of insects. Studies are in process to ascertain the mode of resistance for each of these insects. Van Rie et al.[68] demonstrated a decrease in the affinity of the *B. thuringiensis* protein for receptors found on the BBMV isolated from the Dipel® resistant *P. interpunctella*. The concentration of binding sites remained constant but the binding affinity decreased 70-fold. This biochemical mechanism would be classified as target site insensitivity. As mechanisms of resistance are recognized, appropriate strategies to prevent or delay the onset of resistance to *B. thuringiensis* products can be developed.

# VI. STRATEGIES TO PREVENT OR DELAY THE DEVELOPMENT OF INSECT RESISTANCE TO *B. THURINGIENSIS* PRODUCTS

Numerous strategies have been proposed for the integration of the new, more effective *B. thuringiensis* products with existing pest management practices. Although the actual potential for development of resistance in the field is unknown, it is prudent to develop strategies to prevent or delay its occurrence as new *B. thuringiensis* products are developed. These products should be used in an IPM mode and not as a stand alone approach to control insect damage. Advantages and disadvantages of several potential strategies are discussed with respect to efficacy, implementation, and commercial acceptability. Since significant insect resistance to *B. thuringiensis* products has not developed under field conditions, these strategies have not been tested directly. Selection of appropriate strategies are nonetheless important, because of the investment required for the development of these products and their environmental advantages. Although subsequent discussions will focus on insect tolerant plants, some of these strategies and their implementation are equally applicable to other *B. thuringiensis* products. Two short-term and two long-term strategies will be discussed.

A particularly attractive strategy is to use chemicals and/or biorational insecticides, such as insect sex pheromones or insect growth regulators, in combination or rotation with insect tolerant genetically modified plants. This strategy would limit the exposure of target pests to the plants and would enable the elimination of insects that have developed resistance. A similar strategy has proven successful in combining genetic plant resistance and chemicals.[69] A reduction in the use of chemical or biorational insecticide should result when used in conjunction or rotation with insect resistant genetically modified plants. This approach is a natural fit with existing agronomic practices and IPM approaches. Since insecticides will have to be used to control insects not sensitive to *B. thuringiensis* proteins, limited application of insecticides to genetically modified plants would also require minimal costs. This strategy provides the flexibility to adopt the most effective insecticides available and to address the insect/crop specificity of the insecticide. Although this is an attractive approach, field research is needed to optimize the implementation, to identify the preferred insecticides for each crop and to optimize the rates and timing used in conjunction or rotation with genetically modified plants.

Another strategy that could be implemented at the time of introduction of insect resistant plants is the use of mixed seed varieties.[70-72] This approach essentially dilutes the number of insecticide expressing plants in a field with nonexpressing plants. In so doing, the susceptible pests will have refuges for feeding so that any resistance genes that may evolve will be diluted out by the susceptible insect population. Some damage will occur to the crop, but depending on the ratio of *B. thuringiensis*-expressing and -nonexpressing plants, the damage may not be of economic importance. Specific implementation of this strategy, in terms of acceptable ratios, will be crop dependent and is consistent with current IPM approaches. Likewise, it can be used in conjunction with any one or more of the other strategies discussed. Farmer acceptance is critical to the success of this strategy since the susceptible plants may show significant insect damage. Field data is required to determine the applicable ratios, the effectiveness and the commercial potential in terms of the effect on crop yield.

Long-term approaches include the tissue-specific expression of the *B. thuringiensis* protein in plants and the use of multiple insecticidal genes. Expression of the *B. thuringiensis* protein in specific tissues and not in others could be effective only when expression in restricted tissues is sufficient for control of the target insect(s). Lack of expression of the *B. thuringiensis* protein in some tissues may reduce selection pressure and delay or prevent the development of insect resistance. While this approach may be effective in crops where a single insect with a specific tissue preference is targeted, its utility where multiple insects with different tissue specificities are targeted for control is less attractive. For example, *H. zea* may be controlled by expressing the *B. thuringiensis* protein solely in the tomato fruit. Whether other important lepidopteran insects, such as *Keiferia lycopersicella*, *Spodoptera exigua*, and *M. sexta* would be effectively controlled by fruit-specific expression is not clear. In addition, promoters are not available that direct *B. thuringiensis* gene expression exclusively in many of the specific target tissues. Although research on tissue-specific expression is ongoing, successful implementation of this strategy will require significant additional research and time to implement. Evaluation of this approach awaits that development.

The preferred approach for delaying development of insect resistance, based on host plant resistance, is the introduction of multiple insecticidal genes with independent modes of action.[72,73] Implementation of this type of approach has also proven successful for chemical insecticides.[74] Two gene products with unique modes of actions could be simultaneously introduced into plants, i.e., multiple resistance. Alternatively, independent plants each with a unique insect resistance gene product could be used in rotation. Unfortunately, no available proteins other than the *B. thuringiensis* proteins have been proven particularly effective for insect control. Protease inhibitors[75] or plant lectins[76] could be useful in conjunction with the plants expressing the *B. thuringiensis* proteins. Recently, MacIntosh et al.[77] showed that the expression of a fusion protein containing a protease inhibitor and *B. thuringiensis* led to a significant potentiation in insecticidal activity of the *B. thuringiensis* protein. The impact of this approach on the development of insect resistance remains to be determined. Much research is now focused on identification of novel gene products with unique modes of action. As these are identified, the approach of combining these with the *B. thuringiensis* gene product may be the preferred approach to prevent or delay the development of resistance of insects to *B. thuringiensis* proteins.

Efforts are underway to identify *B. thuringiensis* proteins that maintain insecticidal activity against insects that are resistant to other *B. thuringiensis* proteins. McGaughey and Johnson[37] identified *B. thuringiensis* strains that maintain all or most of their insecticidal activity against the *P. interpunctella* colony that developed resistance to Dipel®. Data[68] suggest that the resistant *P. interpunctella* line developed resistance by modifying the receptor to which *B. thuringiensis* binds. Based on this information, the receptor binding assay could be used to identify specific *B. thuringiensis* proteins which bind to different receptors. These specific *B. thuringiensis* proteins and the gene(s) encoding these proteins may prove valuable to use in combination or rotation with the *B. thuringiensis* genes currently used to delay or prevent the development of resistance.

# VII. CONCLUSION

*B. thuringiensis* products with enhanced activity and prolonged field longevity have resulted from recent scientific advances. These improvements increase the potential for insect resistance to develop. Information on the biology and biochemistry of insect resistance to *B. thuringiensis* has been generated from laboratory selection studies. Considerable effort is now being directed toward elucidating the genetic and physiological mechanisms by which resistance can develop. Based on the information, strategies are being developed to implement these products with existing IPM practices to delay or prevent the occurrence of resistance under field conditions. Implementation of the approved strategies should maximize the agronomic benefit of *B. thuringiensis*-based products.

# REFERENCES

1. **Georghiou, G. P. and Lagunes, A.,** *The Occurrence of Resistance to Pesticides: Cases of Resistance Reported Worldwide through 1988,* Food and Agricultural Organization, Rome, 1988, 325 pp.
2. **Briese, D. T.,** Resistance of insect species to microbial pathogens, in *Pathogenesis of Invertebrate Microbial Diseases,* Davidson, E., Ed., Allenheld Osmun, Totowa, N.J., 1981, 511.
3. **Moeller, F. E. and Williams, P. H.,** Chalkbrood research at Madison, Wisconsin, *Am. Bee J.,* 116, 484, 1976.
4. **DeJong, D.,** Experimental enhancement of chalkbrood infections, *Bee World,* 57, 114, 1976.
5. **Steinhaus, E. A.,** *Principles of Insect Pathology,* McGraw-Hill, New York, 1949, 757.
6. **Dunkle, R. L. and Shasha, B. S.,** Response of starch-encapsulated *Bacillus thuringiensis* containing ultraviolet (UV) screens to sunlight, *Environ. Entomol.,* 18, 1035, 1989.
7. **Carlton, B.,** Genetic improvements of *Bacillus thuringiensis* genes for pest control, in *Biotechnology, Biological Pesticides, and Novel Plant-Pest Resistance for Insect Pest Management,* Roberts, D. W. and Granados, R. R., Eds., Boyce Thompson Institute, Ithaca, N.Y., 1988, 175.
8. **Fischhoff, D. A., Bowdish, K. S., Perlak, F. J., Marrone, P. G., McCormick, S. M., Niedermeyer, J. G., Dean, D. A., Kusano-Kretzmer, K., Mayer, E. J., Rochester, D. E., Rogers, S. G., and Fraley, R. T.,** Insect tolerant transgenic tomato plants, *Bio/technology,* 5, 807, 1987.
9. **Barton, K. A., Whiteley, H. R., Yang, N.,** *Bacillus thuringiensis* delta-endotoxin expressed in transgenic *Nicotiana tabacum* provides resistance to lepidopteran insects, *Plant Physiol.,* 85, 1103, 1987.
10. **Vaeck, M., Reynaerts, A., Höfte, H., Jansens, S., DeBeuckeleer, M., Dean, C., Zabeau, M., Van Montagu, M., and Leemans, J.,** Transgenic plants protected from insect attack, *Nature,* 328, 33, 1987.
11. **Kinsinger, R. A. and McGaughey, W. H.,** Susceptibility of populations of Indianmeal moth and almond moth to *Bacillus thuringiensis, J. Econ. Entomol.,* 72, 346, 1979.
12. **McGaughey, W. H.,** Evaluation of *Bacillus thuringiensis* for controlling Indianmeal moths (Lepidoptera:Pyralidae) in farm grain bins and elevator silos, *J. Econ. Entomol.,* 78, 1089, 1985.
13. **Tabashnik, B. E., Cushing, N. L., Finson, N., and Johnson, M. W.,** Field development of resistance to *Bacillus thuringiensis* in Diamondback moth (Lepidoptera:Plutellidae), *J. Econ. Entomol.,* 83, 1671, 1990.
14. **Tabashnik, B. E.,** unpublished data, 1989.
15. **Kirsch, K. and Schmutterer, H.,** Low efficacy of a *Bacillus thuringiensis* (Berl.) formulation in controlling the diamondback moth, *Plutella xylostella* (L.), in the Philippines, *J. Appl. Entomol.,* 105, 249, 1988.
16. **Miyata, T. and Sinchaisiri, N.,** Toxicological experiments: insecticide resistance patterns, in Report of Meeting of Joint Research Project on Insect Toxicological Studies on Resistance to Insecticides and Integrated Control of the Diamondback Moth, Bangkok, March 9 to 12, 1988, 86.
17. **Stone, T. B., Sims, S. R., MacIntosh, S. C., Marrone, P. G., Armbruster, B. A., and Fuchs, R. L.,** Insect resistance to *Bacillus thuringiensis* delta-endotoxins, presented at Int. Symp. on Molecular Insect Science, Tucson, Arizona, October 22 to 27, 1989.
18. **MacIntosh, S. C., Stone, T. B., Sims, S. R., Hunst, P. L., Greenplate, J. T., Marrone, P. G., Perlak, F. J., Fischhoff, D. A., and Fuchs, R. L.,** Specificity and efficacy of purified *Bacillus thuringiensis* proteins against agronomically important insects, *J. Invertebr. Pathol.,* 56, 258, 1990.
19. **Georghiou, G. P.,** Implications of potential resistance to biopesticides, in *Biotechnology, Biological Pesticides, and Novel Plant-Pest Resistance for Insect Pest Management,* Roberts, D. W. and Granados, R. R., Eds., Boyce Thompson Institute, Ithaca, N.Y., 1988, 175.

20. **Goldman, I. F., Arnold, J., and Carlton, B. C.,** Selection of resistance to *Bacillus thuringiensis* subsp. *israelensis,* in field and laboratory populations of *Aedes aegypti, J. Invertebr. Pathol.,* 47, 317, 1986.

21. **Georghiou, G. P.,** Insecticide resistance in mosquitoes: research on new chemicals and techniques for management, *Annu. Rep., Mosquito Control Research,* University of California, 1983, 132.

22. **Georghiou, G. P.,** Insecticide resistance in mosquitoes: research on new chemicals and techniques for management, *Annu. Rep., Mosquito Control Research,* University of California, 1984, 140.

23. **Georghiou, G. P. and Vazquez-Garcia, M.,** Assessing the potential for development of resistance to *Bacillus thuringiensis* subsp. *israelensis* toxin (*Bti*) by mosquitoes, *Annu. Rep., Mosquito Control Research,* University of California, 1982, 117.

24. **Dai, S. M. and Gill, S. S.,** Development of resistance to the 72 kD toxin of *Bacillus thuringiensis israelensis* in *Culex quinquefasciatus,* presented at Int. Symp. on Molecular Insect Science, Tucson, Ariz., October 22 to 27, 1989.

25. **McGaughey, W. H.,** Insect resistance to the biological insecticide *Bacillus thuringiensis, Science,* 229, 193, 1985.

26. **McGaughey, W. H. and Beeman, R. W.,** Resistance to *Bacillus thuringiensis* in colonies of the Indianmeal moth and almond moth (Lep:Pyralidae), *J. Econ. Entomol.,* 81, 28, 1988.

27. **Höfte, H. and Whiteley, H. R.,** Insecticidal crystal proteins of *Bacillus thuringiensis, Microbiol. Rev.,* 53, 242, 1989.

28. **Stone, T. B., Sims, S. R., and Marrone, P. G.,** Selection of tobacco budworm to a genetically engineered Pseudomonas fluorescens containing the delta-endotoxin of *Bacillus thuringiensis* subsp. *kurstaki, J. Invertebr. Pathol.,* 53, 228, 1989.

29. **Watrud, L. S., Perlak, F. J., Tran, M.-T., Kusano, K., Mayer, E. J., Miller-Wideman, M. A., Obukowicz, M. G., Nelson, D. R., Kreitenger, J. P., and Kaufman, R. J.,** Cloning of the *Bacillus thuringiensis* subsp. *kurstaki* delta-endotoxin gene into *Pseudomonas fluorescens:* molecular biology and ecology of an engineered microbial pesticide, in *Engineered Organisms in the Environment,* Halverson, H. O., Pramer, D., and Rogul, M., Eds., American Society for Microbiology, Washington, D.C., 1985, 40.

30. **Obukowicz, M. G., Perlak, F. J., Kusano-Kretzmer, K., Mayer, E. J., and Watrud, L. S.,** Integration of the delta-endotoxin gene of *Bacillus thuringiensis* into the chromosome of root colonizing strains of pseudomonads using Tn5, *Gene,* 45, 327, 1986.

31. **Whalon, M. E.,** Biotechnology and the development of diagnostic tools for insect management, presented at Annu. Entomological Society of America Natl. Conf., San Antonio, TX, December 10 to 14, 1989.

32. **Hall, I. M. and Arakawa, K. Y.,** The susceptibility of the housefly, *Musca domestica* Linnaeus, to *Bacillus thuringiensis* var. *thuringiensis* Berliner, *J. Insect. Pathol.,* 1, 351, 1959.

33. **Galichet, P. F.,** Sensitivity to the soluble heat-stable toxin of *Bacillus thuringiensis* of strains of *Musca domestica* tolerant to chemical insecticides, *J. Invertebr. Pathol.,* 9, 261, 1967.

34. **Sun, C. N., Georghiou, G. P., and Weiss, K.,** Toxicity of *Bacillus thuringiensis* var. *israelensis* to mosquito larvae variously resistant to conventional insecticides, *Mosq. News,* 40, 614, 1980.

35. **Sun, C. N., Wu, T. K., Chen, J. S., and Lee, W. T.,** Insecticide resistance in diamondback moth, in Diamondback Moth Management, Proc. 1st Int. Workshop, Asian Vegetable Research and Development Center, Shanhua, Taiwan, 1986, 359.

36. **Carlberg, G. and Lindstrom, R.,** Testing fly resistance to thuringiensin produced by *Bacillus thuringiensis,* ser. H-1, *J. Invertebr. Pathol.,* 49, 194, 1987.

37. **McGaughey, W. H. and Johnson, D. E.,** Toxicity of different serotypes of *Bacillus thuringiensis* to resistant and susceptible Indianmeal moths (Lep:Pyralidae), *J. Econ. Entomol.,* 73, 228, 1987.

38. **Sims, S. R. and Stone, T. B.,** unpublished data, 1989.

39. **Roush, R. T.,** Approaches for resistance management: Pyramiding and multilines, presented at Entomological Society of America Natl. Conf., Louisville, KY, December 4 to 8, 1988.

40. **Heckel, D. G., Bryson, P. K., and Brown, T. M.,** unpublished data, 1989.

41. **Lüthy, P., Jaquet, F., Huber-Lukac, H. E., and Huber-Lukac, M.,** Physiology of the delta-endotoxin of *Bacillus thuringiensis* including the ultrastructure and histopathological studies, in *Basic Biology of Microbial Larvicides of Vectors of Human Diseases,* Michal, F., Ed., UNDP/World Bank/WHO, Geneva, 1982, 29.

42. **Percy, J. and Fast, P. G.,** *Bacillus thuringiensis* crystal toxin: ultrastructural studies of its effect on silkworm midgut cells, *J. Invertebr. Pathol.,* 41, 86, 1983.

43. **deLello, E., Hanton, W. K., Bishoff, S. T., and Misch, D. W.,** Histopathological effects of *Bacillus thuringiensis* on the midgut of Tobacco hornworm larvae (*Manduca sexta*): low doses compared with fasting, *J. Invertebr. Pathol.,* 43, 169, 1984.

44. **Orion, U., Sokolover, M., Yawetz, A., Broza, M., Sneh, B., and Honigman, A.,** Ultrastructural changes in the larval midgut epithelium of *Spodoptera littoralis* following ingestion of a delta-endotoxin of *Bacillus thuringiensis* var. *entomocidus, J. Invertebr. Pathol.,* 45, 353, 1985.

45. **Nishiitsutsuji-Uwo, J. and Endo, Y.,** Mode of action of *Bacillus thuringiensis* delta-endotoxin: effect on TN-368 cells, *J. Invertebr. Pathol.,* 34, 267, 1979.

46. **Ebersold, H. R., Lüthy, P. and Huber, H. E.,** Membrane damaging effect of the delta-endotoxin of *Bacillus thuringiensis, Experientia,* 36, 495, 1980.

47. **Nishiitsutsuji-Uwo, J. and Endo, Y.,** Mode of action of *Bacillus thuringiensis* delta-endotoxin: general characteristics of intoxicated *Bombyx larvae, J. Invertebr. Pathol.,* 35, 219, 1980.

48. **Dulmage, H. T., Graham, H. M., and Martinez, E.,** Interactions between the Tobacco budworm, *Heliothis virescens,* and the delta-endotoxin produced by the HD-1 isolate of *Bacillus thuringiensis* var. *kurstaki:* relationship between length of exposure to the toxin and survival, *J. Invertebr. Pathol.,* 32, 40, 1978.

49. **Salma, H. S. and Sharaby, A.,** 1985. Histopathological changes in *Heliothis armigera* infected with *Bacillus thuringiensis* as detected by electron microscopy, *Insect Sci. Appl.,* 6, 503, 1985.

50. **Fast, P. G.,** *Bacillus thuringiensis* parasporal toxin: aspects of chemistry and mode-of-action, *Toxicon,* Suppl. 3, 123, 1983.

51. **Fast, P. G.,** Bacteria: the crystal toxin of *Bacillus thuringiensis,* in *Microbial Control of Pests and Plant Diseases, 1970—1980,* Burges, H. D., Ed., Academic Press, London, 1981, 223.

52. **Huber, H. E. and Lüthy, P.,** *Bacillus thuringiensis* delta-endotoxin composition and activation, in *Pathogenesis of Invertebrate Microbial Diseases,* Davidson, E. W., Ed., Allenheld Osmun, Totowa, N.J., 1981, 209.

53. **Ellar, D. J., Knowles, B. H., Haider, M. Z., and Drobniewski, F. A.,** Investigation of the specificity, cytotoxic mechanisms and relatedness of *Bacillus thuringiensis* insecticidal δ-endotoxins from different pathotypes, in *Bacterial Protein Toxins,* Falmagne, P., Fehrenbach, F. J., Jeljaszewics, J., and Thelestam, M., Eds., Gustav Fischer, New York, 1986, 41.

54. **Haider, M. Z., Knowles, B. H., and Ellar, D. J.,** Specificity of *Bacillus thuringiensis* var. *colmeri* insecticidal δ-endotoxin is determined by differential proteolytic processing of the protoxin by larval gut proteases, *Eur. J. Biochem.,* 156, 531, 1986.

55. **Wolfersberger, M. G., Hofmann, C., and Lüthy, P.,** Interaction of *Bacillus thuringiensis* delta-endotoxin with membrane vesicles isolated from lepidopteran larval midgut, in *Bacterial Protein Toxins,* Falmagne, P., Fehrenbach, F. J., Jeljaszewics, J., and Thelestam, M., Eds., Gustav Fischer, New York, 1986, 237.

56. **Hofmann, C., Lüthy, P., Hütter, R. and Pliska, V.,** Binding of the delta-endotoxin from *Bacillus thuringiensis* to brush-border membrane vesicles of the cabbage butterfly *(Pieris brassicae), Eur. J. Biochem.,* 173, 85, 1988.

57. **Knowles, B. H. and Ellar, D. J.,** Characterization and partial purification of a plasma membrane receptor for *Bacillus thuringiensis* var. *kurstaki* lepidopteran-specific delta-endotoxin, *J. Cell Sci.,* 83, 89, 1986.

58. **Hofmann, C., Vanderbruggen, H., Höfte,H., Van Rie, J., Jansens, S., and Van Mellaert, H.,** Specificity of *Bacillus thuringiensis* d-endotoxins is correlated with the presence of high-affinity binding sites in the brush border membrane of target insect midguts, *Proc. Natl. Acad. Sci. U.S.A.,* 85, 7844, 1988.

59. **Van Rie, J., Jansens, S., Höfte, H., Degheele, D., and Van Mellaert, H.,** Specificity of *Bacillus thuringiensis* delta-endotoxins: importance of specific receptors on the brush border membrane of the midgut of target insects, *Eur. J. Biochem.,* in press.

60. **Sacchi, V. F., Parenti, P., Hanozet, G. M., Giordana, B., Lüthy, P., and Wolfersberger, M. G.,** *Bacillus thuringiensis* toxin inhibits K$^+$-gradient-dependent amino acid transport across the brush border membrane of *Pieris brassicae* midgut cells, *FEBS Lett.,* 204, 213, 1986.

61. **Wolfersberger, M., Lüthy, P., Maurer, A., Parenti, P., Sacchi, F. V., Giordana, B., and Hanozet, G. M.,** Preparation and partial characterization of amino acid transporting brush border membrane vesicles from the larval midgut of the cabbage butterfly *(Pieris brassicae), Comp. Biochem. Physiol.,* 86A, 301, 1987.

62. **Ellar, D. J.,** Investigation of the molecular basis of *Bacillus thuringiensis* delta-endotoxin specificity and toxicity, presented at the Soc. Invertebr. Pathol., College Park, MD, August 20 to 24, 1989.

63. **Metcalf, R. L.,** Insect resistance to insecticides, *Pestic. Sci.,* 26, 333, 1989.

64. **Wilkinson, C. F.,** Role of mixed-function oxidases in insecticide resistance, in *Pest Resistance to Pesticides,* Georghiou, G. P. and Saito, T., Eds., Plenum Press, New York, 1983, 175.

65. **Knipple, D. C., Bloomquist, J. R., and Soderlund, D. M.,** Molecular genetic approach to the study of target-site resistance to pyrethroids and DDT in insects, in *Biotechnology for Crop Protection,* Hedin, P. A., Menn, J. J., and Hollingsworth, R. M., Eds., American Chemical Society, Washington, D.C., 1988, 199.

66. **Saxena, R. C.,** Biochemical bases of insect resistance in rice varieties, in *Natural Resistance of Plants to Pests; Roles of Allelochemicals,* Green, M. B. and Hedin, P. A., Eds., American Chemical Society, Washington, D.C., 1986, 142.

67. **Sparks, T. C., Lockwood, J. A., Byford, R. L., Graves, J. B., and Leonard, B. R.,** The role of behavior in insecticide resistance, *Pestic. Sci.,* 26, 383, 1989.

68. **Van Rie, J., McGaughey, W. H., Johnson, D. E., Barnett, B. D., and Van Mellaert, H.,** Mechanism of insect resistance to the microbial insecticide *Bacillus thuringiensis, Science,* 247, 72, 1990.

69. **Smith, C. M.,** Use of plant resistance in insect pest management systems, in *Plant Resistance to Insects,* Smith, C. M., Ed., Wiley Interscience, New York, 1989, chap. 10.

70. **Gould, F.,** Simulation models for predicting durability of insect resistant germplasm: a deterministic diploid, two-locus model, *Environ. Entomol.,* 15, 1, 1986.

71. **Gould, F.,** Simulation models for predicting durability of insect resistant germplasm: Hessian fly (Diptera:Cecidomyiidae)-resistant winter wheat, *Environ. Entomol.,* 15, 11, 1986.

72. **Smith, C. M.,** Insect biotypes that overcome plant resistance, in *Plant Resistance to Insects,* Smith, C. M., Ed., Wiley Interscience, New York, 1989, chap. 9.

73. **Kennedy, G. G., Gould, F., Deponti, O. M. B., and Stinner, R. E.,** Ecological, agricultural, genetic, and commerical considerations in the deployment of insect-resistant germplasm, *Environ. Entomol.,* 16, 327, 1987.

74. **Tabashnik, B. E.,** Managing resistance with multiple pesticide tactics: theory, evidence, and recommendations, *J. Econ. Entomol.,* 82, 1263, 1989.

75. **Hilder, V. A., Gatehouse, A. M. R., Sheerman, S. E., Barker, R. F., and Boulter, D.,** A novel mechanism of insect resistance engineered into tobacco, *Nature,* 330, 160, 1987.

76. **Shukle, R. H. and Murdock, L. L.,** Lipoxygenase, trypsin inhibitor, and lectin from soybeans: Effects on larval growth of *Manduca sexta* (Lepidoptera: Sphingidae), *Environ. Entomol.,* 12, 787, 1983.

77. **MacIntosh, S. C., Kishore, G. M., Perlak, F. J., Marrone, P. G., Stone, T. B., Sims, S. R., and Fuchs, R. L.,** Potentiation of *Bacillus thuringiensis* insecticidal activity by serine protease inhibitors, *J. Agric. Food Chem.,* 38, 1145, 1990.

# Genetic Engineering of Viruses and Nematodes

Chapter 5

# DEVELOPMENT OF GENETICALLY ENHANCED BACULOVIRUS PESTICIDES

**H. Alan Wood**

## TABLE OF CONTENTS

# I. INTRODUCTION

## A. THE PROBLEM

During the 1980s, biotechnology has had a major impact in agricultural research and development.[1] Genetic engineering has been used in plant breeding programs for increased resistance to insect damage, disease, drought, salinity, etc., as well as the improvement of crop quality and yield. Both transgenic plants and genetically engineered microbes have been used to effect pest resistance. It is clear that biotechnology will continue to offer new approaches and opportunities in the area of plant pest management.

The potential impact of biotechnology in the area of insect pest management is particularly attractive from a health and ecological standpoint.[2] Despite a growing awareness of the problems associated with exposure to pesticides, there was a tenfold increase in synthetic chemical insecticide usage in the U.S. from 1945 to 1986.[3] In 1986, annual U.S. insecticide applications exceeded 100 million kg on approximately 38 million ha.[4] Despite this increased usage of pesticides, crop losses from insect pests continue to increase. A major agricultural challenge in the 1990s will be the development of pest control strategies which are environmentally sound and limit crop losses. Biotechnology offers many new techniques and opportunities for augmenting current pest control strategies and for developing effective and safe alternatives to chemical pesticides.

## B. BACULOVIRUS PESTICIDES

Among these alternative pesticide strategies is the use of baculoviruses to control agricultural and forest insect pests.[5,6] These viruses have been isolated only from arthropods.[7] Their host ranges are characteristically very limited and do not include beneficial insects, such as honey bees and parasitic wasps. More than 300 baculoviruses have been described in Lepidoptera, of which many are major agricultural pests.[8]

In nature baculoviruses can play an important role in the regulation of insect populations. For instance, when populations of the gypsy moth become high, a naturally occurring baculovirus epizootic occurs which characteristically kills more than 95% of the larvae.[9] These epizootics appear to be the major factor controlling gypsy moth population densities. In the absence of virus epizootics, tree death might be so extensive that the absence of a suitable food source might become the major factor which controls populations levels. Viral epizootics characteristically occur in high density populations after unacceptable crop damage has occurred. A goal of biological control has been to create viral epizootics prior to extensive damage.

Extensive safety and environmental testing with baculoviruses has been conducted over the past 25 years.[10] Based on the total lack of environmental and health concerns with these natural insecticides, the U.S. Environmental Protection Agency has registered four baculoviruses as insecticides to control *Heliothis zea* (cotton bollworm), *Orgyia pseudotsugata* (Douglas-fir tussock moth), *Neodiprion sertifer* (pine sawfly) and *Lymantria dispar* (gypsy moth).

Based on the impact of natural epizootics and the environmental safety associated with viral pesticides, baculoviruses are attractive alternatives to chemical pesticides. Yet in 1988, only three baculoviruses were commercially available.[11] There are presently several experimental products but no additional commercial sources of viral pesticides. One reason for this is the high cost of producing these pesticides *in vivo*. However, the most significant problem associated with the commercial use of baculovirus pesticides is their slow speed of action. It may take from 5 to 15 d to kill a host larva following infection; the time period is influenced by factors such as temperature, dosage, and larval age. Early instar larvae are generally more susceptible to infection and die sooner than older instar larvae. If late instar larvae are infected with a low virus dose, they can proceed with pupation, adult emergence, mating, and egg laying. Therefore, feeding damage after virus application is often inevitable, and, in many instances, virus application soon after egg hatch is essential for adequate protection.

Because of the high cost of production and slow speed of action, baculovirus pesticides have not been competitive. If environmental and health costs associated with the production and use of insecticides are ignored, chemical insecticides provide a lower cost/benefit ratio than microbial insecticides. This difference is becoming smaller as the development of insect resistance requires higher rates of application and the cost of discovery and registration of new chemical insecticides escalates. New state and federal regulations aimed at reducing ground water pollution and food residues continue to restrict the use of many pesticides. These factors will probably continue to bring the costs associated with the use of chemical pesticides closer to those for microbial pesticides.

Another factor which has limited the commercial use of viral pesticides is their host specificity. Many baculoviruses are specific to a single or a few host species within a genus. The *Autographa californica* nuclear polyhedrosis virus (AcMNPV) is unusual, having a host range which includes more than 30 species of Lepidoptera in 10 families.[10] Although a narrow host range is a positive environmental attribute, it decreases the market value of each product.

Biotechnology may provide the tools needed to overcome many of the shortcomings associated with the use of naturally occurring viral pesticides. Through genetic engineering, the insecticidal properties and host range of these viruses can be improved. This may be accomplished by exchange of genetic material between baculoviruses or insertion of foreign genes into baculovirus genomes.

## II. STRATEGIES FOR RECOMBINANT PESTICIDES

### A. ENVIRONMENTAL ISSUES

Decisions on the best strategy for genetic improvement(s) must take into account economic as well as ecological and environmental factors, such as those outlined by Tiedje et al.[12] In the U.S., small-scale field releases of these viruses need to satisfy any state requirements as well as the U.S. Environmental Protection Agency requirements under the Federal Insecticide, Fungicide and Rodenticide Act. Additional federal regulations are currently being considered for wide scale and commercial releases of genetically engineered organisms.

In order to satisfy health and environmental concerns, a genetically altered baculovirus should have a host range which remains restricted to pest insects. While host range expansion is an attractive commercial attribute, alterations that expand the host range to include beneficial insects or vertebrates would certainly be unacceptable for field releases. The procedures used to determine the host range of genetically altered baculoviruses should be carefully evaluated. The host ranges reported for most baculoviruses include only those hosts which exhibit a pathological response. Sublethal or chronic infections have rarely been considered. More stringent testing will be required with recombinant viruses. Ideally, the genetically improved virus would act as a highly specific delivery system to a wide number of important insect pests. The pests would be controlled by the action of the foreign gene product, not by the normal processes of virus replication.

Another sensitive issue is the potential of a recombinant virus to displace wild-type virus populations in nature. The new virus should not have a selective advantage over natural virus isolates, either with respect to persistence in the environment, efficiency of infection, or amount of progeny virus. The displacement of natural baculoviruses could result in unanticipated ecological disturbances over large areas.

Despite the thorough risk assessments prior to the release of any genetically engineered organism, the possibility exists that a release organism possesses unforeseen properties which would dictate mitigation. Decontamination would be difficult because baculoviruses can survive in soils for years.[13] Therefore, engineering strategies which limit recombinant virus survival in nature would be appropriate.

Another area of concern is genetic transfer or reassortment. When new genetic material is introduced into a viral genome, the engineered product should have a low potential to alter its properties through genetic transfer or reassortment. In addition, the potential of gene transfer to related and unrelated viruses along with potential consequences of such transfers should also be evaluated.

## B. ENGINEERING STRATEGIES

Several engineering strategies have been proposed for occluded baculoviruses. The initial strategies have involved the polyhedrin gene of the AcMNPV. Early in the replication cycle of the occluded baculoviruses, nonoccluded virions bud through the plasma membrane into the hemocoel of larvae, thereby effecting systemic infections. Late in the replication cycle, the virions become membrane bound within the nucleus and are then occluded within a paracrystalline structure referred to as a polyhedron. The polyhedron is primarily composed of a single viral protein, polyhedrin. The occlusion of virus particles within polyhedra is required for protection and survival of the virus in nature. Following ingestion of the polyhedra, the alkaline pH in the midgut of host larvae results in dissolution of the polyhedron crystal, thereby releasing the virions for infection of midgut cells. The polyhedrin gene is nonessential (not required for nonoccluded virus replication) and is under the control of a strong promoter. Smith et al.[14] first illustrated the utility of replacing the polyhedrin coding sequences with the coding sequence of a foreign gene under the control of the polyhedrin gene promoter. Subsequently, hundreds of foreign genes[15] have been expressed in the baculovirus expression vector system.

The properties of the polyhedrin gene and the virion occlusion process are ideal for the development nonpersistent, genetically improved viral pesticides. In 1986 and 1987, researchers at the Natural Environment Research Council's (NERC) Institute of Virology in Oxford conducted a series of field tests which documented the safety and nonpersistence of certain types of genetically altered baculoviruses.[16,17] In 1986, they infected *Spodoptera exigua* larvae with a marked strain of AcMNPV which contain a unique 80-base-pair insert downstream from the polyhedrin coding region. The infected larvae were introduced into a field facility consisting of a netted compound to restrict dispersal and the opportunity for viral amplification. Their field and laboratory data illustrated that the marked virus was identical to the parental virus with respect to host range, genetic stability, and persistence. In the summer of 1987, they repeated this experiment using a similarly marked virus from which they had removed the polyhedrin promoter and coding sequences (poly-minus virus). As expected, in the absence of the polyhedrin gene and virus particle occlusion, virus infectivity was not detectable at 1 week post death of the test larvae. These studies illustrated that the insertion of foreign DNA sequences downstream from the polyhedrin gene resulted in no detectable alteration in biological properties. In addition, they documented that poly-minus baculovirus mutants which can only produce nonoccluded virions are rapidly inactivated in the soil and on plant tissues.

The nonpersistence of a poly-minus baculovirus mutant is a highly desirable ecological trait. If the polyhedrin gene were replaced with DNA coding for a pesticidal protein, the new viruses could quickly control targeted pests through the actions of the new pesticidal gene(s) and would then rapidly lose their infectivity. Accordingly, the improved virus would pose minimal ecological risks. The problem with this strategy is that the nonoccluded virions are so unstable that they could not be delivered to the field in an active state. To deal with this problem, three strategies have been proposed and are discussed below.

In order to occlude poly-minus baculoviruses in polyhedra and thereby stabilize their infectivity, host cells can be co-infected with both wild-type and poly-minus virus isolates.[18-20] Under these conditions, the wild-type virus provides the polyhedrin protein which occludes both the wild-type and poly-minus virus particles. This co-occlusion process provides a means of delivering the poly-minus baculovirus to the field in an infectious form.

Using the co-infection/co-occlusion strategy, the persistence of the engineered virus in a virus population is determined by the probability of co-infection of individual larvae and cells with both virus types as the virus is passed from insect to insect. Wood et al.[20] studied the dynamics of this process under laboratory conditions by determining the amount of occluded poly-minus AcMPNV progeny produced following larval infections with different virus inoculum dosages and with inocula containing varying ratios of co-occluded poly-minus and wild-type virions. The results indicated that a poly-minus virus would not persist in a virus population under natural conditions.

Based on these data, researchers at the Boyce Thompson Institute were granted permission by the U.S. Environmental Protection Agency to conduct the first field application of a genetically engineered virus in 1989. A cabbage field infested with *Trichoplusia ni* larvae was sprayed with polyhedra containing equal amounts of co-occluded wild-type and poly-minus AcMNPV particles. The engineered virus genome had a deleted polyhedrin gene and no foreign DNA insert. The persistence of the engineered virus is currently being evaluated by measuring the percent occluded, poly-minus virus produced in successive generations.

A slightly different engineering strategy has been proposed by Vlak et al.[21] They found that the p10 gene of AcMNPV is also not essential for virus replication. When the p10 was deleted, the progeny virus $LD_{50}$ was significantly reduced and the membrane surrounding the polyhedra was absent. They hypothesized that the absence of p10 protein reduced the stability of polyhedra resulting in an increased efficiency of virion release. It was suggested that this might reduce the stability of polyhedra and, therefore, the persistence of p10-minus recombinants in the environment. This hypothesis needs to be tested under field conditions. The release of poly-plus baculoviruses which have foreign gene inserts presents a range of additional ecological considerations, particularly if the recombinant virus has persistence properties equivalent to those of the wild-type virus.

The polyhedrin and p10 genes are both highly expressed late in the replication cycle. Accordingly, any foreign genes under the control of either promoter will be expressed approximately 10 h post infection. In order to shorten this time, Miller[22] suggested that it might be preferable to utilize an early viral gene promoter to control the expression of the foreign gene and to retain polyhedrin expression. Although the polyhedrin gene promoter appears to be much stronger than any of the early gene promoters, the early expression of the pesticidal gene might compensate for the reduction in expression levels. If early promoters are used, the new gene may be expressed during semipermissive or nonpermissive infections, and the host range specificity might be lost.[23] An alternative to the use of early viral gene promoters might be to use a constitutive host promoter. This approach could be used to achieve high expression levels soon after infection.

Another containment strategy might be to replicate poly-minus recombinants in transgenic insects which contain integrated polyhedrin promoter and coding sequences. Virus progeny replicated in the transgenic host would be occluded. However, replication in field insects would produce only nonoccluded virions which would be rapidly inactivated following death of the target pest.

## C. PESTICIDAL GENES

The type of genetic engineering strategy employed will probably be dictated more by safety considerations than efficacy. The opposite bias will be true for the type of pesticidal genes which will be inserted into baculovirus genomes. The foreign pesticidal genes that have been proposed are varied in nature. They included insect-specific toxins, hormones, hormone receptors, metabolic enzymes, growth regulators, etc. It is generally agreed that their properties should include enhancing the speed with which infected larvae are killed or stop feeding.

The scorpion *Buthus eupeus* insect toxin-1 (BeIt) is an insect-specific paralytic neuro-toxin. Based on an amino acid sequence analysis,[24] Carbonell et al.[25] synthesized the BeIt gene. They inserted the BeIt gene into the AcMNPV genome under the control of the polyhedrin gene promoter. Although transcription and translation occurred, no biological activity was detected. The reason for the lack of activity remains unknown.

Maeda[26] replaced the polyhedrin gene of *Bombyx mori* nuclear polyhedrosis virus (BmMNPV) with the diuretic hormone gene from *Manduca sexta*. Diuretic and antidiuretic hormones are considered to play important roles in the closely regulated water balance exhibited by insects. Replication of the recombinant BmMNPV in silkworms resulted in an alteration in larval fluid metabolism and a 20% reduction in the time required to kill larvae as compared to wild-type BmMNPV infections. This was the first report of a foreign gene product which enhanced the pesticidal properties of a baculovirus.

Hammock et al.[27] isolated the juvenile hormone esterase (JHE) gene from the *Heliothis virescens* genome. The JHE of lepidopterous larvae is thought to play a major role in the reduction of juvenile hormone titers which leads to the cessation of feeding and initiation of pupal development. The JHE gene was inserted into a poly-minus mutant of AcMNPV under the control of the polyhedrin gene promoter. Infection of first instar *Trichoplusia ni* larvae with the JHE-virus resulted in a reduction in feeding and growth as compared to the uninfected or wild-type virus infected larvae. These differences, however, were not observed following infection of other than first instar larvae.

Hammock et al.[27] speculated that the AcMPNV-induced production of the ecdysteroid UDP-glucosyl transferase (EGT) may have reduced the effect of JHE at later instar stages. O'Reilly and Miller[28] reported that the *Autographa californica* MNPV genome contained a gene coding for an EGT which catalyzes the transfer of glucose from UDP-glucose to ecdysteroids. The EGT produced during wild-type AcMNPV replication inhibited molting of host larvae. *Spodoptera frugiperda* larvae infected with EGT-minus mutants were capable of molting. The EGT gene is probably present in other baculoviruses. For instance, Dougherty et al.[29] found reduced ecdysteroid titers and prolonged larval stages in *Trichoplusia ni* larvae infected with granulosis virus. The molting of larvae infected with EGT-minus baculovirus may result in a shorter larval period and, therefore, reduced feeding damage. Accordingly, an EGT-minus baculovirus, by itself or with foreign gene inserts such as the JHE gene, may be useful in the development of improved viral pesticides.

There are a large number of gene products which may be used to enhance the pesticidal properties of baculoviruses. From an environmental standpoint, the use of host gene products that are expressed at inappropriate developmental stages, in new cell types and/or in excessive amounts, are presently attractive. Metabolic enzymes, hormones and their receptors are likely candidates. Keeley and Hayes[30] have proposed the use of "antipeptides" which would bind to neurohormones, thereby blocking their physiologic functions. The introduction of naturally occurring viral pesticides which have been engineered to express host proteins should pose few if any toxicological problems.

Insect toxins may also be effective in enhancing the pesticidal properties of baculoviruses. These toxins will have to be extremely specific. When expressed in larvae under the control of a polyhedrin promoter, the resultant toxin concentrations would surely exceed natural exposure levels by several orders of magnitude. These levels of toxin may produce unan-ticipated effects. Insertion of gene sequences of the active N-terminal fraction of the *Bacillus thuringiensis* (BT) toxin should lead to the production of a safe and potentially effective toxin. The expression of the BT toxin by recombinant baculoviruses would provide the target specificity not available with foliar applications of BT toxin.

# III. CONCLUSIONS

The assessment of potential environmental impacts of genetically improved viral pesticides will include an evaluation of the properties of the foreign gene product(s) as well as the biological properties of altered virus itself. It is anticipated that in the near future several types of foreign gene inserts will be available to enhance the pesticidal properties of baculoviruses. The current field release studies are collecting much of the information which will be needed to assess the environmental safety of these new pesticides. There are, of course, many additional issues which need resolution for successful commercialization of these pesticides. Of primary concern will be the cost-to-benefit ratios as determined by production costs, stability, application technology, and field efficacy. In the majority of instances, the use of viral pesticides will be based solely on cost-to-benefit ratios as compared to synthetic pesticides.

Despite the improvements afforded through biotechnology, it is clear that viral and other microbial pesticides will only reduce, not eliminate, the agricultural requirements for synthetic pesticides. Biological pesticides are among the best solutions to reducing crop losses in the absence of ecological disturbances and potential health hazards.

# ACKNOWLEDGMENT

The preparation of this manuscript was supported in part through grant CR-815831-01-0 from the U.S. Environmental Protection Agency.

# REFERENCES

1. **National Research Council,** *Technology and Agricultural Policy, Proceedings of a Symposium,* National Academy Press, Washington, D.C., 1990.
2. **Roberts, D. W. and Granados, R. R., Eds.,** *Biotechnology, biological pesticides and novel plant-pest resistance for insect pest management,* Insect Pathology Resource Center, Ithaca, NY, 1988.
3. **Pimentel, D. and Pimentel, M.,** The future of U.S. Agriculture, *Science,* 231, 1491, 1986.
4. **Pimentel, D. and Levitan, L.,** Pesticides: amounts applied and amounts reaching pests, *BioScience,* 36(2), 86, 1986.
5. **Burges, H. D., Ed.,** In *Microbial Control of Pests and Plant Diseases 1970—1980,* Academic Press, New York, 1981.
6. **Falcon, L. A.,** Development and use of microbial insecticides, in *Biological Control in Agricultural IPM Systems,* M. A. Hoy and D. C. Hezog, Eds., Academic Press, New York, 1985, 229.
7. **Martignoni, M. E. and Iwai, P. J.,** *A Catalog of Viral Diseases of Insects, Mites and Ticks,* 4th ed., Tech. Rep. PNW-195, U.S. Department of Agriculture, Forest Service, Pacific Northwest Research Station, Portland, OR, 1986, 51.
8. **Bilimoria, S. L.,** Taxonomy and identification of baculoviruses, in *The Biology of Baculoviruses, Vol. 1,* R. R. Granados and B. A. Federici, Eds., CRC Press, Boca Raton, FL, 1986, 37.
9. **Lewis, F. B. and Yendol, W. G.,** Gypsy moth nucleopolyhedrosis virus: Efficacy, in *The Gypsy Moth: Research Toward Integrated Pest Management,* Doane C. C. and McManus, M. L., Eds., U.S. Forest Service Tech. Bull. 1584, U.S. Department of Agriculture, Washington, D.C., 1981, 503.
10. **Groner, A.,** Specificity and safety of baculoviruses, in *The Biology of Baculoviruses, Vol. 1,* Granados, R. R. and Federici, B. A., Eds., CRC Press, Boca Raton, FL 1986, 177.
11. **Cunningham, J. C.,** Baculoviruses: Their status compared to *Bacillus thuringiensis* as microbial insecticides, *Outlook Agric.,* 17(1), 10, 1988.
12. **Tiedje, J. M., Colwell, R. K., Grossman, Y. L., Hodson, R. E., Lenski, R. E., Mack, R. N., and Regal, P. J.,** The planned introduction of genetically engineered organisms: ecological considerations and recommendations, *Ecology,* 70, 298, 1989.
13. **Thompson, C. G., Scott, D. W., and Wickham, B. E.,** Long term persistence of the nuclear polyhedrosis virus of the Douglas-fir tussock moth, *Orgyia pseudotsugata,* (Lepidoptera:Lymantriidae) in forest soil, *Environ. Entomol.,* 10, 254, 1981.

14. **Smith, G. E., Summers, M. D., and Fraser, M. J.,** Production of human beta interferon in insect cells infected with a baculovirus expression vector, *Mol. Cell. Biol.,* 3, 2156, 1983.

15. **Luckow, V. V. and Summers, M. D.,** Trends in the development of baculovirus expression vectors, *Bio/Technology,* 6, 47, 1988.

16. **Bishop, D. H. L.,** UK release of genetically marked virus, *Nature,* 323, 496, 1986.

17. **Bishop, D. H. L., Entwistle, P. F., Cameron, I. R., Allen, C. J., and Possee, R. D.,** Field trials of genetically-engineered baculovirus insecticides, in *The Release of Genetically-Engineered Micro-organisms,* Sussman, M., Collins, C. H., Skinner, F. A., and Stewart-Tull, D. E., Eds., Academic Press, New York, 1988, 143.

18. **Miller, D. W.,** Genetically engineered viral insecticides: Practical considerations, in *Biotechnology for Crop protection,* Hedin, P. A., Hollingworth, R. M. and Mann, J. J., Eds., American Chemical Society, Washington, D.C., 1988, 405.

19. **Shelton, A. M. and Wood, H. A.,** Microbial pesticides, *The World & I,* 4(10), 358, 1989.

20. **Wood, H. A., Hughes, P. R., van Beek, N., and Hamblin, M.,** An ecologically acceptable strategy for the use of genetically engineered baculovirus pesticides, in *Insect Neurochemistry and Neurophysiology 1989,* Borkovec, A. B. and Masler, E. P., Eds., The Humana Press, Clifton, NJ, 1990, 285.

21. **Vlak, J. M., Klinkenberg, F. A., Zaal, K. J. M., Usmany, M., Klinge-Roode, E. C., Geervliet, J. B. F., Roosien, J., and van Lent, J. W. M.,** Functional studies on the p10 gene of *Autographa californica* nuclear polyhedrosis virus using a recombinant expressing a p10-beta-galactosidase fusion gene, *J. Gen. Virol.,* 69, 765, 1988.

22. **Miller, M. K.,** Baculoviruses as gene expression vectors, *Annu. Rev. Microbiol.,* 42, 177, 1988.

23. **McClintock, J. T., Dougherty, E. M., and Weiner, R. M.,** Semipermissive replication of a nuclear polyhedrosis virus of *Autographa californica* in a gypsy moth cell line, *J. Virol.,* 57, 197, 1986.

24. **Grishin, E. V.,** Structure and function of *Buthus eupeus* scorpion neurotoxins, *Int. J. Quantum Chem.,* 19, 291, 1981.

25. **Carbonell, L. F., Hodge, M. R., Tomalski, M. D., and Miller, M. K.,** Synthesis of a gene coding for an insect-specific scorpion neurotoxin and attempts to express it using baculovirus vectors, *Gene,* 73, 409, 1988.

26. **Maeda, S.,** Increased insecticidal effect by a recombinant baculovirus carrying a synthetic diuretic hormone gene, *Biochem. Biophys. Res. Commun.,* 165(3), 1177, 1989.

27. **Hammock, B. D., Bonning, B. C., Possee, R. D., Hanzlik, T. N., and Maeda, S.,** Expression and effects of the juvenile hormone esterase in a baculovirus vector, *Nature,* 344, 458, 1990.

28. **O'Reilly, D. R. and Miller, L. K.,** A baculovirus blocks insect molting and producing ecdysteroid UDP-glucosyl transferase, *Science,* 245, 1110, 1989.

29. **Dougherty, E. M., Kelly, T. J., Rochford, R., Forney, J. A., and Adams, J. R.,** Effects of infection with a granulosis virus on larval growth, development and ecdysteroid production in the cabbage looper, *Trichoplusia ni, Physiol. Entomol.,* 12, 23, 1987.

30. **Keeley, L. L. and Hayes, T. K.,** Speculations on biotechnology applications for insect neuroendocrine research, *Insect. Biochem.,* 17, 639, 1987.

Chapter 6

# GENETIC ENGINEERING OF NEMATODES FOR PEST CONTROL

**George O. Poinar, Jr.**

## TABLE OF CONTENTS

# I. INTRODUCTION

Nematodes are multicellular, appendageless, nonsegmented, wormlike invertebrates containing a complete digestive tract. This alimentary canal is surrounded by a body cavity which is limited externally by a body wall covered with cuticle. The some 15,000 described species of nematodes are distributed over the globe in all imaginable habitats (e.g., soil, sea, fresh water, plants, invertebrates, and vertebrates).[1]

Many of the free-living, microbotrophic nematodes which feed on bacteria can be maintained indefinitely on agar plants as long as populations are continuously transferred to new plates with fresh bacteria every 3 to 4 weeks. Because of their rapid developmental cycle (a week or less), their ease in cultivation and observation, and form of reproduction (many employ autotoky involving hermaphroditism), some of these species have been selected as model organisms for the study of cell development and lineages.

One of such nematodes is *Caenorhabditis elegans,* which is now a standard name, along with *Drosophila* and white mice in laboratories studying the genetics and developmental biology of metazoans. A number of mutants of *C. elegans* have been induced and characterized and many are available from the *Caenorhabditis* Genetics Center at the University of Missouri (Division of Biological Sciences, Columbia, MO 65211).

The employment of genetic engineering, which is understood here as the experimental alteration of the genetic constitution of an individual[2] (involving the more traditional methods of selective breeding,[3] as well as the molecular genetics of recombinant DNA technology), has been previously attempted with metazoan animals along the lines of genetic selection (also called genetic manipulation or genetic improvement). In the field of nematology, the arena has been centered on a few species of *Caenorhabditis,* especially *C. elegans.* However, there are certainly species of nematodes involved in pest control, both entomogenous and plant-parasites, that are prime candidates, and attempts have already been made with representatives of the former group.

# II. ENTOMOGENOUS NEMATODES

## A. HETERORHABDITIDAE AND STEINERNEMATIDAE

Among the insect parasites, of which there are many representatives from several families in nature, the desire and ability to conduct genetic engineering is, at present, limited to members of two families, the Heterorhabditidae and Steinernematidae. These nematodes are capable of being mass-produced in artificial media, have a wide host range, normally kill insect hosts within 48 h after contact, possess a durable infective stage capable of storage, distribution and application, are not known to be susceptible to insect immune reactions in nature, are environmentally safe and are exempt from EPA registration.[4]

## 1. Life Cycle

The life cycles of *Heterorhabditis* (the sole genus in the Heterorhabditidae) and *Steinernema* (the sole genus in the Steinernematidae) are remarkably similar and can be compared with that of the free-living microbotroph *C. elegans.* While the dauer stages (third-stage juveniles) of *C. elegans* defensively persist in the environment, those of *Heterorhabditis* and *Steinernema* offensively begin searching for potential insect hosts. Thus, in the latter genera, the dauer stage is also an infective stage. The dauer of *C. elegans* molts to the fourth stage after reaching a suitable environment (a moist environment with an ample supply of fresh bacteria). The infective stages of *Steinernema* and *Heterorhabditis* molt only after they have reached the body cavity of an insect (their "suitable environment"). The developing stages of *C. elegans* feed on bacteria in the environment and eventually reach the adult stage and mate (or if males are absent as they normally are in hermaphroditic populations), begin oviposition. The maturing stages of *Steinernema* and *Heterorhabditis* in the insect's hemolymph have a slight lag period since development must wait until the bacteria (*Xenorhabdus* spp.) introduced by the nematodes have multiplied in the hemolymph and reached high enough numbers to serve as a food source (in combination with components in the insect's hemolymph). When this occurs, the nematodes mature to the adult stage and oviposit.

While *C. elegans* depends on bacteria in the natural habitat as a food source, both *Heterorhabditis* and *Steinernema* introduce specific bacteria which are nutritionally suitable and possess other characteristics (production of antibiotics) which ensure the nematode's successful development. *Steinernema* spp. are all amphimictic and mating must occur before young can be produced. *Heterorhabditis,* however, is unique in producing a population of only hermaphrodites from the infective stages but in having a second generation (progeny from the hermaphrodites) which is purely amphimictic and must mate before producing offspring.

With *C. elegans,* development is continuous as long as there is a food source and the habitat is hospitable. When either of these is less than sufficient, nonfeeding dauer individuals are formed, and it is these individuals that insure survival of the population over periods of adversity. With *Heterorhabditis* and *Steinernema,* development proceeds one or two generations inside the now-dead insect, and as the food reserves become depleted, infective stages are formed. These resistant, nonfeeding stages leave the insect cadaver in search of new hosts and the cycle begins anew.

Aside from the infective nature of the dauer stages, the biology of *Heterorhabditis* and *Steinernema* differ from that of *C. elegans* in another important character. The dauer of the entomogenous forms carry around specific bacteria (species of *Xenorhabdus*) which are essential for their survival. The three major roles of the bacteria as far as the nematode is concerned, are:

1.  To kill the insect before immune reactions are evoked (normally an encapsulation reaction, but humeral immunity can also be involved).
2.  To multiply and provide a source of food for the nematode.
3.  To produce a range of antibiotics which keep out or lower the number of possible contaminants which could destroy the system. The use of genetic engineering to improve these nematodes could also include the genetic engineering of their associated bacteria.

## 2. Characters to be Modified

It would appear that the millions of years of natural selection which resulted in the successful development of this intricate nematode-bacterial-insect relationship would leave little to be modified. Also, would any modification induced by artificial means be established in the genome without its own loss after extended generations or its interference with other vital characters of the organism?

Biologists working with these entomogenous rhabditoids have mentioned a number of traits which they feel could be improved through genetic engineering.[5-11]

Such traits previously mentioned that could be improved include:

1.  *Xenorhabdus*-carrying ability[5]
2.  Development of anhydrobiosis[6]
3.  Persistence in the environment[6]
4.  Storage[6]
5.  Shippage[6]
6.  Sensibility to host-released attractants[9]
7.  Increase in mass production efficiency[9]
8.  Enhanced juvenile tolerance to UV radiation[10]

9.  Host finding[10]
10. Increased infectivity[7]
11. Hardiness improvement[7]
12. Broadening of host range[7]
13. Engineering of a consistent phase one *Xenorhabdus* species[8]
14. Thermo-adaptation[11]

Other traits that could be selected for include:

1.  Tolerance to particular hostile environments such as those of animal refuse
2.  Preference to particular host genera or families
3.  Improved vertical and horizontal movement in soil
4.  Resistance to parasites and pathogens in the environment

## 3. Genetic Engineering through Selective Breeding
### a. Increased Pathogenicity

Using the Mexican strain of *Steinernema carpocapsae* as the original inoculum, Lindegren et al. selected a new strain (Kapow selection) for increased production and pathogenicity.[12] The strain was produced by repeated passage of the original inoculum through larvae of *Galleria mellonella*. Only the first progeny leaving the insect cadavers were maintained and used for subsequent inoculations. After a series of generations, the resulting nematodes were considered to be much more "active" than those of the original strain. The Kapow selection produced 34% more infective juveniles 6 d earlier than the original strain. They were visibly more active and at 50 infectives for each *G. mellonella*, the insects exhibited a 3 h earlier $LT_{50}$.[13] The Kapow selection was used in field experiments in California for control of the navel orangeworm, *Amyelois transitella* (Walker) (Tortricidae) in almond orchards.[12,14]

The Kapow selection of the Mexican strain of *S. carpocapsae* produced *in vivo* at the Fresno laboratory was significantly more efficacious than the commercially produced *in vitro* A11 strain of *S. carpocapsae* when tested against larvae of the Mediterranean fruit fly (*Ceratitis capitata* [Wiedemann] [Tephritidae]).[13]

A still earlier incidence of selection of a *S. carpocapsae* strain with increased pathogenicity to a target insect was reported in studies using gypsy moth larvae (*Lymantria dispar* [L.] [Lymantriidae]) as a host.[15] After two passages through gypsy moth larvae, the virulence increased about 2.5-fold, or from $LC_{50} = 38 \pm 7$ nematodes per ml to $LC_{50} = 15 \pm 4$ nematodes per ml.

### b. Increased Host Finding

Attempts have been made to increase the host-finding ability of *S. carpocapsae*. Trials were first undertaken to measure the amount of variation in different geographic strains of this species in response to locating captive larvae of *Galleria mellonella*.[10] Assays were conducted on 150 × 50 mm Petri dishes containing a 3-mm bottom layer of 2% agar. Holes made on opposite sides of the dish tops contained two pipette tips which, when inserted, were 5 cm apart. The pipette tips were pointing down 2 mm above the agar surface. In the experimental pipette were placed two instar *Gilleria* larvae 24 h before testing. The control pipette was left empty. The ends of both pipettes were packed with steel wool which prevented exit of the larvae in the experimental pipette. The tips of the pipettes were sealed with plastic film. Approximately 1000 infective stage juveniles were placed onto a 1 cm-diameter area in the center of the agar plate. The dishes were held at 25°C for the 1-h test period. Nematode response was determined by recording the number of infective stage nematodes taken directly below the experimental and control pipette tips. The experiment was replicated four times for each of ten strains (progeny of a single geographical isolate) which originated from North America, Europe and New Zealand. Conclusions of the tests were that none of the strains evaluated showed a useful level of host-finding ability. However there was a fourfold difference between a strain from New Zealand (approximately 2% host-finding ability) and a strain from Florida (approximately 8% host-finding ability).

It was thought that if an artificial strain could be developed which incorporated all the variation of the 10 natural strains, it would demonstrate a greater host-finding ability. Therefore a "foundation" strain was formed by randomly setting up 10 mating pairs from the 10 natural strains, then taking the progeny, mixing them together and allowing them to infect *Galleria* larvae. The resulting progeny formed the foundation strain which theoretically contained maximum genetic variability. However, the foundation strain demonstrated only an approximate 4% host finding ability, and was only average when compared with the 10 natural strains.

A separate study investigated the responses of the foundation strain to selection for host-finding ability.[16] Similar experimental methods to those described previously were used. The *Galleria* larvae were positioned 3 cm above the agar surface in a downward-pointing pipette whose tip was 2 mm above the agar surface. The nematodes were kept on the agar surface for 1 h before being counted.

Readings were taken on the number of infectives that were within the inoculation site (immobile), were near the pipette tip containing the insect (moved positively), were away from the pipette tip containing the insect (moved negatively) and were in a 1 cm-diameter area directly under the pipette tip containing the insect (found the host). A high host-finding population was selected by collecting the infective stages directly below the insect-containing pipette. A low host-finding population was selected by collecting infectives that moved negatively. There were 13 rounds of selection using only nematodes that found the host. Nematodes from each selection group were reared in *Galleria* larvae to obtain inoculum for the following round.

The selection was discontinued after 13 rounds, and host-finding was measured after cycles 2, 4, 8, and 13. Whereas only 3 to 7% of the control were recovered under the treatment pipette, nearly 80% of the 13th round population aggregated below the pipette. An increase from 30 to 85% was noted in the infectives that migrated to the positive half of the plate after 13 rounds. After 6 rounds of selection, the proportion of infectives that remained within the central inoculation zone (immobile) was reduced from 33% (original foundation strain) to 8%. When selection was discontinued after the 13th round, the nematodes from the 13th round were randomly mated inside a series of *Galleria* larvae resulting in approximately 13 to 26 nematode generations. When this final population was tested, the host selection ability had dropped by half.

Nevertheless, these studies show that host-finding of entomogenous nematodes can be experimentally enhanced through genetic selection.[10,16] It is not known whether selection in the above experiments was for an increased sensitivity to host-released chemoattractants, an increase in general nematode activity, an increase in nematode activity at a particular temperature (25°C) or to other characters. Although host-finding ability did tend to revert back to that of the original foundation strain when pressure was relaxed, it could still be an important character to possess in a population that was undergoing inundative release for immediate control of an insect pest. One practical problem would be the buildup of large numbers of inoculum for release under an absence of selective pressure for host finding. If numerous generations had to be employed to achieve high numbers, the acquired trait might be greatly reduced in the final inocula.

### c. Ultraviolet Tolerance

An attempt was made to determine the degree of genetic variability to ultraviolet tolerance in diverse populations of *S. carpocapsae*.[10] Nematode strains originating from Europe, North

America, and New Zealand were tested. Approximately 500 nematodes from each strain were tested separately by placing them on moistened filter paper placed on a series of moistened 47-mm cellulose fiber pads held in a 760-mm Petri dish bottom.

The nematodes were then exposed to a UV source which emitted medium UV radiation peaking at 302 nm (wave lengths in natural sunlight believed to be harmful to nematodes). The UV lamp was mounted 30 cm above the nematodes and produced an intensity of 60 μW/cm² with the digital radiometer. After exposure, the nematodes were exposed to last-instar larvae of *Galleria* for 4 d at 25°C in the dark. The number of dead insect larvae was recorded.

The $LD_{50}$ ranged from 5.71 min in the least tolerant strain to 7.26 min in the most tolerant strain. Although the data suggested a slight genetic difference (27%) in regards to UV tolerance, the authors considered that on the basis of large environmental variation and little genetic variation, the likelihood of improving these values through further testing of other strains was unlikely. The low genetic variability also was judged to be inadequate for forming a foundation strain for further selection.[10]

### d. Temperature Selection

Adaptation of *Steinernema carpocapsae* to different temperatures on the basis of infectivity of insects and nematode development at those specific temperatures was reported by Burman and Pye.[17] However, other researchers were unable to demonstrate an adaptation to different temperatures with *S. carpocapsae* strains and *Heterorhabditis heliothidis*.[18]

### 4. Gene Manipulation

This aspect of genetic engineering (also referred to as recombinant DNA technology) involves the insertion of genes with desirable traits into a organism which lacks these traits. The insertion of nucleic acids into *C. elegans* has been accomplished by two methods, microinjection and electric discharge.

### a. Microinjection

In 1982, researchers developed a procedure for microinjecting into the germ line of adult nematodes. They extracted tRNA from an amber suppressor strain and injected it into a nonsuppressor strain, resulting in transient suppression.[19] Subsequently, Stinchcomb et al.[20] injected a series of plasmid DNAs into *C. elegans* and by assaying with Southern or dot blots, found that a small number (<0.2%) of the progeny of the injected nematodes carried the injected DNA. It had been concatemerized *in vivo* in long tandem arrays which were extrachromosomal but still heritable. The arrays were transmitted both mitotically and meiotically with variable frequencies and behaved like pieces of an endogenous chromosome which had broken off and segregated as independent genetic elements (free duplications).

The first successful case of functional exogenous DNA introduced into a nematode was Fire's work with *C. elegans*.[21] His success was dependent, in part, on the development of a successful technique for implanting the DNA into the nucleus of the developing oocyte of the nematode. This technique deserves a rather detailed description here.

The nematodes were immobilized on a dried pad of agarose under an oil layer. They lost water to the agarose pad and became partially desiccated at the same time. It was important to choose pads with the proper thickness which depended on the size of the nematodes to be treated. The agarose pads were made on 32-mm diameter glass coverslips. Best results were achieved when the temperature was maintained at 15 to 17°C during the immobilization procedure. Nematodes were then transferred from a partially desiccated agar plate without bacteria to the agarose pad. Clarke GC120F15 needles, which had been drawn to a tip size of 1 μm (3 μm also worked) by a two-stroke electrode puller, were used for the injections. The needles were filled with 0.1 to 1.0 μl of solution with a capillary tube that had been drawn by hand to a long, narrow bore. The needles were held with a Clarke electrode holder and connected with silicon tubing to the pressure system (up to 200 psi). The pressure could be adjusted so that liquid would only flow out when the needle was inside the nematode.

The injections were done under a Zeiss 1M microscope equipped with simultaneous differential interference contrast and fluorescent Lucifer yellow with fluorescein or formaldehyde-induced fluorescence filters (with a 40 power objective).

A Zeiss/Jena micromanipulator was used for injections. The needle was pressed firmly against the cuticle, then a quick thrust of the micromanipulator or microscope drove the needle into the nematode. With this method, 10 to 20 injections could be made into each nematode without significantly decreasing viability. Injection volumes varied between 1 and 10 pl/injection. After injection, the nematodes were immersed in an isotonic recovery buffer (0.1% salmon sperm DNA, 4% glucose, 2.4 m$M$ KCL, 6 m$M$ NaCl, 3 m$M$ MgCl$_2$, 3 m$M$ CaCl$_2$, and 3 m$M$ HEPES pH 7.2) which hydrated and detached them from the agarose substrate, then they were placed in a humidified chamber. After several hours, M9 buffer was added to reduce osmotic strength and the nematodes were transferred to individual Petri plates with nutrient agar and bacteria and maintained at 20°C.

Using the above method, Fire[21] injected a cloned *C. elegans* amber suppressor tRNA gene, *sup*-7 into *C. elegans* with an amber termination mutation in gene *tra*-3 whose function is required for fertility. Transient expression of *sup*-7 was demonstrated by the presence of fertile nematodes in the generation after injection. In a few cases, the fertile nematodes gave rise to stable suppressor lines carrying injected DNA sequences. The suppressor activities were mapped to chromosomal loci, indicating that the exogenous DNA had integrated into the genome.

A later study by Fire and his colleagues presented some modifications on the original injection method.[22] Nematodes grown in a slightly hypotonic environment were more suitable for growth. They were somewhat swollen and distorted less when injected. For obtaining the highest level of integrative transformation frequency, young adults which have 1 to 3 fertilized oocytes should be used. Immersion of the nematodes in heavy paraffin oil (BDH prod 29437) prior to and during injection gave better results than the original Voltalef 39 oil. The highest rates of integrative transformation were obtained when oocyte nuclei were injected; attempts were made to inject all of the oocyte nuclei in the proximal arm, including the turn region (Figures 1 and 2). No integrative transformation occurred following cytoplasmic injection. Maintaining the injected nematodes at 25°C, rather than 19°C, or 20°C, resulted in slightly increased viability.

### b. Electric Discharge of DNA-Coated Microprojectiles

A novel method for introducing exogenous genes into a target organism involves the use of DNA coated microprojectiles which are forcibly electrically discharged. Small gold (or other metal) particles, approximately 1 to 3 μm in diameter, are coated with the desired gene to be implanted. These particles are then propelled into a target organism by an electric discharge.[23,24] A small fraction of the recipient organisms will normally carry the introduced gene.

The first use of this method for causing transformations in nematodes involved the introduction of plasmid clones of the wild-type *unc*-54 gene into a strain of *C. elegans* carrying a large *unc*-54 deletion.[25] Both transient and heritable transformations were recovered. The method involved drying plasmid DNA onto the surfaces of 1.5 to 3.0 μm gold particles which were then suspended in ethanol. Approximately 70,000 plasmid DNA molecules were placed on each particle. The high voltage discharge propelled about 1 to 2 million particles toward several thousand target animals, which constituted a synchronized population of L4 hermaphrodites collected by filtration on Millipore filters. They were briefly

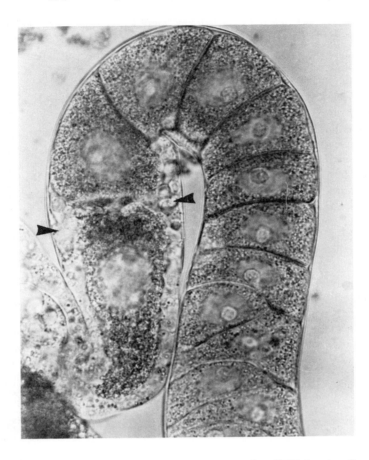

FIGURE 1.    Genetic implantation involves the insertion of DNA into the cell
nucleus. In *Heterorhabditis bacteriophora* shown here, as well as other nematodes, a choice site would be the unfertilized oocytes located in and prior
to the bend region of the oviduct (arrows show mature spermatozoa in the
fertilization zone of the oviduct). (Magnification × 500.)

air dried and mounted onto chilled plates containing isotonic recovery buffer solidified with
agar.

Approximately 80 to 90% of the nematodes died as a result of the treatment and the
survivors each contained from 4 to 8 gold particles. Both transient and heritable transformants, as identified by motility, were found in approximately 0.02% of the F1 progeny. From
10 to 30% of the transformations were passed through many generations; the remainder
expressed the transforming DNA only transiently. Since most of the heritable transformants
segregated uncoordinated offspring, the transforming DNA was likely to be extrachromosomal. This was supported by the frequency of variable phenotypes and transmission ratios.[25]

The preferred target of the particles to insure a transgenic nematode is not known. Since
the particles are only slightly smaller than the oocyte nucleus, it would appear that a direct
hit on the nucleus would cause considerable damage. Therefore, implantation into the adjacent cytoplasm may be the preferred target in these experiments[26] (Figure 3).

The above study shows that a gene can be introduced but may not implant onto the
chromosome. It may then be unstable and passed from one generation to the next in an
irregular manner. Even if the gene inserts onto the chromosome, by being introduced in
random locations, transgenic animals could function irregularly. Since the final objective of
this procedure is to obtain gene expression, it is important that various regulators (promoters,

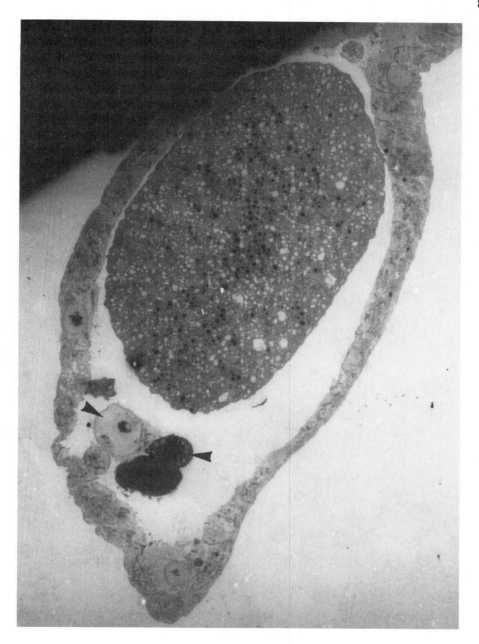

FIGURE 2. The unfertilized oocyte of *Steinernema carpocapsae* shown here would be ideal for genetic implantation experiments. Note adjacent smaller spermatozoa (arrows). (Magnification × 1000.) (Photograph courtesy of R. Hess.)

etc.) are also carried along with the introduced gene. However, because the above method is rapid, easy, and applicable to large numbers of nematodes, it will probably be the preferred method of gene transfer in the future, especially in commercial companies with a range of target organisms.

### c. *Transposons*

Transposons are short sections of DNA capable of moving to another DNA molecule

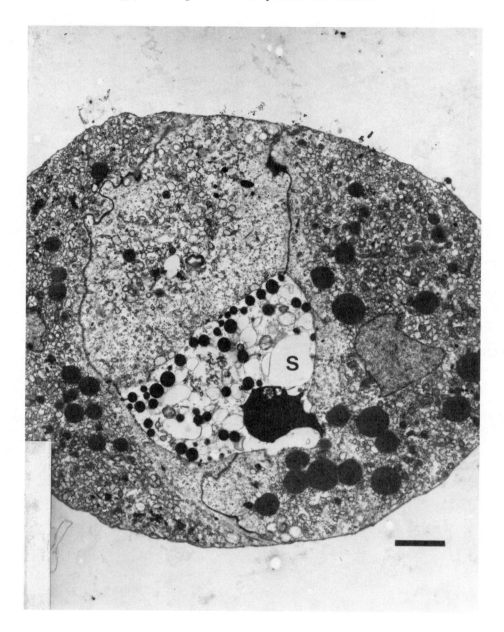

FIGURE 3.    Electric discharge involves the ''shooting'' of microprojectiles coated with DNA into cell nuclei of experimental animals. This developing oocyte of *Heterorhabditis bacteriophora* (center of photograph) would be an ideal target (darker cell in center of photo is a spermatozoan [below S]; bar = 2 μm). (Photograph courtesy of R. Hess.)

or another region of the same molecule. These nucleotide sequences occur naturally in living cells. If a transposon inserts into the coding region of a gene, the information of that gene is interrupted and it may not continue to produce a functional protein. In other cases, the transposon may activate a gene simply by its insertion near it. A number of transposable elements have been recorded in the *C. elegans* genome[27] and it is very likely that entomogenous nematodes also possess them. Transposons can be used as cloning vehicles, especially those in plasmids and it is likely that the transposon tagging approach may be suitable for use for locating and cloning genes of entomogenous nematodes.

### d. Mutagenesis

Mutagens are agents which increase the rate of mutations by causing changes in the nucleotide sequences of DNA. Mutations can be considered as errors in DNA, that result in different or incorrect amino acids being inserted into proteins.

Causing mutagenesis in *C. elegans* hermaphrodites with EMS or X-rays have been standard procedure, resulting in the formation of hundreds of mutants. The standard procedure is to remove the nematodes from the agar plates, place them in a small volume (4 ml) of buffer containing the chemical mutagen (usually 20 µl EMS) for 3 h, wash them in sterile buffer, and allow them to deposit their unmutagenized eggs over the following few hours. After this period, about 25 eggs are collected from each hermaphrodite and the $F_2$ generation is examined for mutants.

Fodor[11] was successful in producing mutants of *S. carpocapsae* that were resistant to the anthelmintics levamizole ($5 \times 10^{-4} M$) and avermectin (1 µm/ml). He achieved mutagenesis by adding the mutagen to the insect host when the nematodes were in the young adult stage. The mutagen (5 µl EMS/ml M9 buffer) was added in vapor form for 8 h to a closed desiccator jar containing the insect larvae. Female nematodes were removed from the treated *Galleria* larvae 3 d later and transferred to a 1% solution of $5 \times 10^{-4} M$ of the anthelmintic.

Other mutations which were first recognized in *C. elegans* but could have an important application with insect parasitic nematodes are those affecting the onset and termination of the dauer condition. Dauer defective (def) individuals are incapable of forming the dauer stage, either spontaneously or in the presence of a pheromone (dauer recovery inhibiting factor — DRIF). On the other hand, dauer constitutive (const) individuals produce dauer above 24°C with or without the dauer-stimulating pheromone.[28]

It would be advantageous in producing entomogenous nematodes if batches could be enticed to produce infective juveniles simultaneously. If the nematodes were infective constitutive individuals, a slight rise in temperature (to 24°C) would result in the synchronous formation of infective stages, which could then be harvested.

Fodor et al.[11] extracted a DRIF-like compound from both ''*in vivo*'' and ''*in vitro*'' cultures of *S. carpocapsae*. This compound inhibited nearly 100% infective recovery of *S. carpocapsae* but inhibited only 60% of the infectives of *S. feltiae* ($=$ *bibionis*). However, a pheromone extract from *S. feltiae* completely inhibited the recovery of *S. feltiae* infectives but inhibited only 44% of *S. carpocapsae* infectives. These pheromones appear to be species specific and could well be used in the production of these nematodes.

### 5. Genetic Engineering of *Xenorhabdus* Spp.

The pathogenesis of heterorhabditid and steinernematid nematodes is due in large part to their symbiont bacteria of the genus *Xenorhabdus* which are carried in the alimentary tract of the infective juveniles.[29] The association between nematode and bacteria is one of mutualism. *Xenorhabdus* alone has no infective abilities so is not pathogenic to insects. The infective juvenile nematodes introduce and release the bacteria in the insect hemocoel. Once in this habitat, the bacteria multiply in the host's hemocoel and establish a lethal septicemia within 24 h. Without *Xenorhabdus* in the insect's hemocoel, steinernematid and heterorhabditid nematodes do not have the available nutrients to grow and reproduce normally.[1] It is now known that *Xenorhabdus* exists in two main phase variants, both of which are carried by the infective juveniles and are pathogenic to insects. These two phase variants (earlier known as primary and secondary form) differ in many characters, including colony morphology, colony color, antimicrobial activity, and presence of inclusion bodies (in phase one only). One major difference is that phase variant one provides better conditions for nematode multiplication than phase variant two.[30,31] Frequently, under culture conditions, for no apparent reason, phase variant one will revert to phase variant two, and nematode production will drastically fall.

It has been shown that both phase one and phase two variants are genetically identical but have different regulatory systems.[32] The genes which are differentially expressed include those governing antibiotics, pigments, bioluminescence, extracellular lipase and extracellular protease. A genomic library of *X. luminescens* DNA in the vector pUC18 has been constructed and the bioluminescence genes have been cloned.[33] Attempts are now being made to screen the same library to isolate plasmids carrying genes for the other differentially expressed functions of *X. luminescens*. Thus it would be possible to obtain colonies carrying plasmids for genes controlling antibiotic production, extracellular protease and lipase. These plasmids could be introduced into phase variant two population in order to provide optimum conditions of nematode growth and reproduction. There may be additional advantages of modifying a phase variant two population of *X. luminescens* for nematode production since phase variant two populations are not susceptible to attack by the XLP bacteriophage.[34]

During the past few years, various commercial companies have been interested in utilizing *Xenorhabdus* spp. alone for insect control. This desire has been motivated by the wide host range exhibited by the *Xenorhabdus* carrying nematodes.

There are several obstacles, however, The first is that *Xenorhabdus* bacteria have no invasive powers of their own. If spread in the environment and ingested by insects, they would pass through the alimentary tract and leave the potential host unharmed. The second problem is that these bacteria are highly fastidious outside of their nematode carrier-insect hemocoel habitat and could not survive long (a few hours at most) in the environment. It is possible that with recombinant DNA techniques, one could introduce the characters of invasiveness and environmental protection into the genome of *Xenorhabdus*, but it may be easier to transfer the important characters of *Xenorhabdus* into bacteria which possess the former two characters. When injected into the hemocoel of *Galleria* larvae in small numbers, *X. nematophilus* cells were able to multiply and kill the insects without the presence of nematodes.[35] In studies with diapausing pupae of the cecropia moth (*Hyalophora cecropia*), Götz et al.[36] noted that the presence of nematodes greatly reduced the number of *X. nematophilus* cells necessary to destroy the insects. They concluded that the antibacterial activity in immune hemolymph of the cecropia pupae was destroyed by the nematode.

Such a situation has never been reported in other hosts and in general, much smaller dosages of *Xenorhabdus* than those reported for cecropia pupae, are needed to kill insects, especially larvae. Whether this high lethal dose is specific to *Cecropia* or *Cecropia* pupae, or is more widespread in insects in general, is a challenging question for future investigators.

## B. MERMITHIDA

Aside from the genera *Steinernema* and *Heterorhabditis*, there are other groups of nematodes which have potential as insect biological control agents.[37] These include members of the orders Mermithida and Tylenchida. Mermithids in general are rather host specific, often being limited to a particular host family (sometimes a genus) or occasionally several families within an order. Only rarely can they successfully complete their development in hosts belonging to two or more orders. However, in those forms which have a host range crossing ordinal boundaries (*Agamermis decaudata, Filipjevimermis leipsandra, Mermis nigrescens*), genetic engineering could be used to select strains more infective toward particular hosts.

Some of the negative features of the mosquito mermithid, *Romanomermis culicivorax* which was produced commercially for several years in the 1970s, was its ineffectiveness at low temperatures (15°C or below), its intolerance of mild salinity, its intolerance of lowered oxygen tensions in polluted waters and its inability to be mass produced *in vitro*.[4] This latter feature is characteristic of mermithids in general. Genetic engineering would probably be useful in producing strains of *R. culicivorax* that could tolerate low temperatures, low oxygen tensions and even mild salinity, but genetically modified populations that could be mass produced "*in vitro*" would be a greater challenge. Attempts at using *R. culicivorax* in genetic engineering experiment would depend on its demand (now low) for actual and potential use against mosquito populations.

## C. TYLENCHIDA

A successful nematode biological control agent which is being utilized today in Australia is the tylenchid, *Delandenus siricidicola*, a parasite of the sirex wood wasp, *Sirex noctilio*. This nematode is unusual because it has two separate life cycles, one involving insect parasitism, the other parasitism of the fungal symbiont of the insect. Although not especially insect-host specific, *D. siricidicola* is highly fungal specific and will only reproduce on *Amylostereum arealatum* which is symbiotically associated with the wasp, *S. noctilio*.[33] It is possible that genetic engineering could produce strains of *D. siricidicola* which could feed on related species of *Amylostereum* associated with other *Sirex*, thus allowing this nematode to be used against a wider range of *Sirex* species.

Other characteristics (e.g., increased pathogenicity to *Sirex* species, lack of infectivity to important insect parasitoids in the host habitat, increased storage ability, increased survival ability in the field, etc.) could also be targets in a genetic engineering program of *Deladenus* species.

## D. HETEROXENOUS PARASITES OF VERTEBRATES

A number of important nematode parasites of domestic animals and humans are heteroxenous (the life cycle involves an intermediate host). In many cases the intermediate hosts are insects and since the nematodes undergo some development (usually from egg or first stage juvenile to infective third stage juvenile) in the insect, they can be considered as insect-parasitic nematodes. There are many modifications that could be brought about to lessen the effect of these nematodes on vertebrates, but one possibility involving development of the parasite in the intermediate host was exemplified by the discovery of the first case of a heteroxenous nematode completing its development in a terrestrial invertebrate.[39] The seuratid nematode, *Rabbium paradoxus*, was able to complete its development in an ant host (*Camponotus castaneus*). All other adult members of this group of nematodes occur in the intestinal tract of reptiles.

This behavioral modification involving the completion of the life cycle of a normally heteroxenous nematode in its intermediate host is rare and has been reported previously in a few cases involving aquatic invertebrates.[1] However, its occurrence shows that for some reason (scarcity of the definitive host?) genes can be activated which allow the infective stage to continue its development in the intermediate host. If strains of important human filarial parasites could be genetically modified to allow a proportion or all of the progeny to continue development in the insect host, the intermediate hosts (normally mosquitoes, blackflies and other small flies) would undoubtedly be killed or so damaged (as the parasites increased in size) that they could not serve as effective vectors. Thus, the normally heteroxenous nematodes would become biological controlling agents of their normal vectors.

# III. PLANT PARASITIC NEMATODES

The use of nematodes for pest control is not limited to entomogenous forms. Out of the vast number of plant parasitic nematodes, a few have been tested (and many have potential) as biological control agents of weeds.

Russian knapweed (*Acroptilon repens* [L.]) (*Centaurea repens* [L.]) is a composite native to Mongolia, Western Turkestan, Iran, Armenia, and Asia Minor, and was accidentally introduced into Canada around 1900 as a contaminant of alfalfa seed. It has now spread into five provinces in Canada.[40] Its intense competitive ability suppresses and eliminates other plant growth in pastures. In addition, it is poisonous to horses.

In portions of its native area, a nematode (*Paranguina picridis* Kirj. and Ivan.) produces galls on the stems, leaves, and root collar of infected plants, and normally kills about 20% of the plants. This nematode is now used as a biological control agent in the USSR and has been tested against Russian knapweed in Canada. The nematode's usefulness in Canada was limited by slow movement and dependency on moist spring conditions. In addition, *P. picridis* will also attack a range of closely related plants including desirable knapweeds and globe artichoke.

A second nematode, *Ditylenchus phyllobius* (Thorne) (*Anguillulina phyllobia* Thorne; *Nothanguina phyllobia* [Thorne]; *Orrina phyllobia* [Thorne]) is a native of western North America, where it attacks the silver-leaf nightshade, *Solanum elaeagnifolium* Cav. (Solanaceae). The plant infests some 1.2 million ha in Texas, where it competes with crop plants and also is poisonous to animals. The nematode causes extensive hypertrophy, thickening and convoluting of infested leaves, frequently killing the plant. Studies showed that high infections could be obtained by artificially inoculating *S. elaeagnifolium* populations with infected plant tissue.[41,42] The program was somewhat set back when it was found that *D. phyllobius* could also infect eggplant (*Solanum melongena esculentum*).

In both of the above cases, genetic engineering could be used to restrict the host range of these nematodes to a few closely related species, excluding those beneficial to man. Genetic engineering could also be used to increase host-finding activity and growth on modified host tissue (callus or hydroponics) in order to expedite mass production for biological control purposes.

## IV. POSSIBLE PITFALLS

The ultimate aim of genetic engineering is to improve the phenotype of an organism. There is always the risk that by genetically modifying an organism, some desirous characters already present may be modified or lost. Even with genetic manipulation, Gaugler's G13 *Galleria*-selected and S-20 scarab-selected strains of *S. carpocapsae* did not have the storage persistence at 4 to 7°C characteristic of the foundation strain.[43] With gene implantation techniques, which can be rather traumatic and result in high nematode mortality, damage to genes governing desirous traits may be high. Thus, the resulting nematode population should be tested for infectivity, storage, bacterial retention, etc.

Other concerns center around the persistence of the modified character. In genetic manipulation studies involving the production of host seeking strains, this character was greatly reduced when selection pressure was removed.[16] With gene implantation experiments, the introduced genes may not be implanted on the chromosome, so that the desired characters are lost after a number of generations. Or they may be introduced on the chromosomes at random locations resulting in the irregular function of transgenic animals. Or the introduced genes may not be accompanied by their regulators (promoters, etc.) and therefore not function normally.

## V. ENVIRONMENTAL CONCERNS

What will be the result of placing genetically engineered nematodes in the environment? The chances that a genetically modified steinernematid or heterorhabditid nematode will attack crop plants is low because they lack the basic mouthparts (no stylet or spear) necessary to penetrate plant cells and they certainly could not multiply if they did enter wounded plant tissue. Could a genetically modified insect parasite attack vertebrates? The infective stage juveniles of *Heterorhabditis* and *Steinernema* will attempt to penetrate into any invertebrate they encounter and even into some poikilothermic vertebrates. The infective juveniles of *S. glaseri* and *S. carpocapsae* have been shown to survive for 5 and 2 d, respectively, when introduced intraperitoneally into rats.[44] In the above cases, none of the nematodes developed to the following stage or caused any injury to the host. Could a "mistake" in genetic engineering allow this to happen?

In rare cases where heterorhabditids and steinernematids developed to the adult stage in the coelom of amphibian tadpoles, development was halted because the tadpole cadaver was filled with foreign bacteria. Could genetically modified strains be able to continue their development on foreign bacteria?

It is unlikely that genetic engineering experiments aimed at developing strains of entomogenous nematodes better at host seeking for specific insects or any of the other characters cited above would be any more infective to vertebrates than present investigations already have demonstrated. However it certainly would be prudent to conduct tests with nontarget organisms, including vertebrates, on any genetically modified nematodes that were intended for field release.

The same precaution should be taken with genetically modified strains of the symbiotic bacteria, *Xenorhabdus*. A recent study reported on the presence of a new group of five strains of *X. luminescens* DNA hybridization group 5 recovered from wounds and blood of humans.[45] Differences between these 5 groups of bacteria and the *Xenorhabdus* carried by heterorhabditid and steinernematid nematodes are significant enough to place the former groups in a separate species or possibly genus. However, the clinical populations do show some similarities with *X. luminescens* and probably represent soil contaminants that entered the wounds directly or the blood stream via wounds. Therefore any genetically modified strains of *Xenorhabdus* should always be tested for vertebrate infectivity.

## VI. CONCLUSIONS

In discussing the topic of genetic engineering of nematodes for pest control, we can safely say that the field is still in its infancy. However, as our techniques and facilities become more sophisticated for gene implantation into metazoans, improvement should come slowly and steadily. One major problem in dealing with insect and plant-parasitic nematodes is that so little about their basic genetics is known. Even with *C. elegans,* which has been studied for some 25 years by a battery of highly competent researchers, only some 75% of the genome has been physically mapped and cloned.

We have seen that genetic manipulation has already been conducted with entomogenous nematodes in the genus *Steinernema*. Although recombinant DNA technology is still a preliminary technique with nematodes, it certainly will be attempted with *Steinernema* and *Heterorhabditis* in the near future.

An outline of the various genetic engineering pathways for nematodes is presented in Figure 4. As discussed by Otsuka,[7] genetic selection alone can be expected to produce a desired effect in an organism only if (1) a small degree of genetic change is required, and (2) if only one or a few genes are responsible for controlling the phenetic function. Genetic manipulation can be hindered by a long generation time, the inability to produce progeny throughout the year, a small brood size, an inadequate population size, a large number of chromosomes, a large genome, the inability to mate specific individuals, an insensitivity to mutagens or inadequate laboratory culturing methods. Recombinant DNA technology can be attempted when subjects are encountered which contain the above-mentioned characteristics.

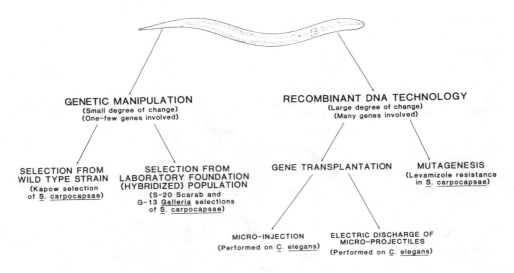

FIGURE 4.    Genetic engineering pathways for nematode candidates for pest control.

# REFERENCES

1. **Poinar, G. O., Jr.,** *The Natural History of Nematodes,* Prentice-Hall, Englewood Cliffs, N.J., 1983, 323.
2. **Lincoln, R. J., Boxshall, G. A., and Clark, P. F.,** *A Dictionary of Ecology, Evolution and Systematics,* Cambridge University Press, Cambridge, 1982, 298.
3. **Drlica, K.,** *Understanding DNA and Gene Cloning,* John Wiley & Sons, New York, 1984, 205.
4. **Poinar, G. O., Jr.,** Entomophagous nematodes, *Fortsch. Zool.,* 32, 95, 1986.
5. **Akhurst, R. J.,** *Neoaplectana* species: Specificity of association with bacteria of the genus *Xenorhabdus, Exp. Parasitol.,* 55, 258, 1983.
6. **Kaya, H. K.,** Entomogenous nematodes for insect control in IPM systems, in *Biological Control in Agricultural IPM Systems,* Hoy, A. M. and Herzog, D. C., Eds., Academic Press, New York, 1985, 283.
7. **Otsuka, A. J.,** Possible directions for genetic analysis and modification of nematode species, in *Fundamental and Applied Aspects of Invertebrate Pathology,* Samson, R. A., Vlak, J. M., and Peters, D., Eds., Foundation 4th Int. Colloq. Invert. Pathol., Wageningen, 1986, 296.
8. **Gaugler, R.,** Feasibility of genetically improving steinernematid and heterorhabditid nematodes, in *Fundamental and Applied Aspects of Invertebrate Pathology,* Samson, R. A., Vlak, J. M., and Peters, D., Eds., Foundation 4th Int. Colloq. Invert. Pathol., Wageningen, 1986, 300.
9. **Gaugler, R.,** Entomogenous nematodes and their prospects for genetic improvement, in *Biotechnology in Invertebrate Pathology and Cell Culture,* Maramorosch, K., Ed., Academic Press, New York, 1987, 457.
10. **Gaugler, R., McGuire, T., and Campbell, J.,** Genetic variability among strains of the entomopathogenic nematode *Steinernema feltiae, J. Nematol.,* 21, 247, 1983.
11. **Fodor, A.,** *Caenorhabditis elegans* as a model for the study of entomopathogenic nematodes, unpublished data, 1989.
12. **Lindegren, J. E., Agudelo-Silva, F., Valero, K. A., and Curtis, C. E.,** Comparative small-scale field application of *Steinernema feltiae* for navel orangeworm control, *J. Nematol.,* 19, 503, 1987.
13. **Lindegren, J. E.,** personal communication, 1989.
14. **Agudelo-Silva, F., Lindegren, J. E., and Valero, K. A.,** Persistence of *Neoaplectana carpocapsae* (Kapow selection) infectives in almonds under field conditions, *Fla. Entomol.,* 70, 288, 1987.
15. **Shapiro, M., Poinar, G. O., Jr., and Lindegren, J. E.,** Suitability of *Lymantria dispar* (Lepidoptera: Lymantriidae) as a host for the entomogenous nematode, *Steinernema feltiae* (Rhabditida: Steinernematidae), *J. Econ. Entomol.,* 78, 342, 1985.
16. **Gaugler, R., Campbell, J. F., and McGuire, T. R.,** Selection for host-finding in *Steinernema feltiae, J. Invert. Pathol.,* 54, 363, 1989.
17. **Burman, M. and Pye, A. E.,** *Neoaplectana carpocapsae:* Movements of nematode populations on a thermal gradient, *Exp. Parasitol.,* 48, 258, 1980.

18. **Dunphy, G. B. and Webster, J. M.,** Temperature effects on the growth and virulence of *Steinernema feltiae* strains and *Heterorhabditis heliothidis*, *J. Nematol.,* 18, 270, 1986.
19. **Kimble, J., Hodgkin, J., Smith, T., and Smith, J.,** Suppression of an amber mutation by microinjection of suppressor tRNA in *C. elegans, Nature*, 299, 456, 1982.
20. **Stinchcomb, D. T., Shaw, J. E., Carr, S. H., and Hirsh, D.,** Extrachrosomal DNA transformation of *Caenorhabditis elegans, Mol. Cell. Biol.,* 5, 3484, 1985.
21. **Fire, A.,** Integrative transformation of *Caenorhabditis elegans, EMBO J.,* 5, 2673, 1986.
22. **Fire, A., Moerman, D., Harrison, S. W., Albertson, D., and Waterston, R.,** Nonsense and antisense: tools for nematode DNA transformation, unpublished data, 1989.
23. **Winston Brill,** personal communication, 1988.
24. **McCabe, D. and Christore, P.,** Stable transformation of soybean by particle transformation acceleration, *Biotechnology*, 6, 923, 1988.
25. **Rushforth, A., McCabe, D., Collins, J., and Anderson, P.,** Transformation of *C. elegans* by DNA coated microprojectiles, in Abst. *C. elegans* Meeting, Cold Spring Harbor, New York, May 10 to 14, 1989, 1989, 223.
26. **Philip Anderson,** personal communication, 1989.
27. **Emmons, S. W.,** The genome, in *The Nematode Caenorhabditis elegans*, Wood, E. B., Ed., Cold Spring Harbor Laboratory, Cold Spring Harbor, New York, 1989, 17.
28. **Riddle, D. L.,** The dauer larva, in *The Nematode Caenorhabditis elegans*, Wood, E. B., Ed., Cold Spring Harbor Laboratory, Cold Spring Harbor, New York, 1989, 373.
29. **Thomas, G. M. and Poinar, G. O., Jr.,** *Xenorhabdus* gen. nov., a genus of entomopathogenic nematophilic bacteria of the family Enterobacteriaceae, *Int. J. System. Bact.,* 29, 352, 1979.
30. **Akhurst, R. J.,** Morphological and functional dimorphism in *Xenorhabdus* spp., bacteria symbiotically associated with the insect pathogenic nematodes *Neoaplectana* and *Heterorhabditis, J. Gen. Microbiol.,* 121, 303, 1980.
31. **Akhurst, R. J.,** Biology and taxonomy of *Xenorhabdus,* presented at Int. Symp. Entomopathogenic Nematodes in Biological Control, Pacific Grove, CA, August 20 to 22, 1989.
32. **Nealson, K.,** Physiology and biochemistry of *Xenorhabdus,* presented at Int. Symp. Entomopathogenic Nematodes in Biological Control, Pacific Grove, CA, August 20 to 22, 1989.
33. **Frackman, S.,** Molecular genetics of *Xenorhabdus,* presented at Int. Symp. Entomopathogenic Nematodes in Biological Control, Pacific Grove, CA, August 20 to 22, 1989.
34. **Poinar, G. O., Jr., Hess, R. T., Lanier, W., Kinney, S., and White, J. H.,** Preliminary observations of a bacteriophage infecting *Xenorhabdus luminescens* (Enterobacteriaceae), *Experientia,* 45, 191, 1989.
35. **Poinar, G. O., Jr. and Thomas, G. M.,** The nature of *Achromobacter nematophilus* as an insect pathogen, *J. Invert. Pathol.,* 91, 510, 1967.
36. **Götz, P., Boman, A., and Boman, H. G.,** Interactions between insect immunity and an insect-pathogenic nematode with symbiotic bacteria, *Proc. R. Soc. London,* B212, 333, 1981.
37. **Poinar, G. O., Jr.,** *Nematodes for Biological Control of Insects,* CRC Press, Boca Raton, FL, 1979, 277.
38. **Bedding, R. A.,** Nematode parasites of Hymenoptera, in *Plant and insect Nematodes,* Nickle, W. R., Ed., Marcel Dekker, New York, 1984, 925.
39. **Poinar, G. O., Jr., Chabaud, A. G., and Bain, O.,** *Rabbium paradoxus* sp.n. (Seuratidae: Skrjabinelaziinae) maturing in *Camponotus castaneus* (Hymenoptera: Formicidae), *Proc. Helminthol. Soc. Wash.,* 56, 120, 1989.
40. **Watson, A. K. and Harris, P.,** *Acroptilon repens* (L.) DC., Russian knapweed (Compositae), in *Biological Control Programmes against Insects and Weeds in Canada 1969—1980,* Kelleher, J. S. and Hulme, M. A., Eds., Commonwealth Agricultural Bureau, Slough, 1984, 410.
41. **Robinson, A. F., Orr, C. C., and Abernathy, J. R.,** Distribution of *Nothanguina phyllobia* and its potential as a biological control agent for silver-leaf nightshade, *J. Nematol.,* 10, 362, 1978.
42. **Northam, F. E. and Orr, C. C.,** Effects of a nematode on biomass and density of silverleaf nightshade, *J. Range Manag.,* 35, 536, 1982.
43. **Gaugler, R.,** Genetic approaches to enhancing biological control effectiveness, presented at Int. Symp. Entomopathogenic Nematodes in Biological Control, Pacific Grove, CA, August 20 to 22, 1989.
44. **Poinar, G. O., Jr.,** Non-insect hosts for the entomogenous rhabditoid nematodes *Neoaplectana* (Steinernematidae) and *Heterorhabditis* (Heterorhabditidae), *Rev. Nématol.,* 12, 923, 1989.
45. **Farmer, J. J., III et al.,** *Xenorhabdus luminescens* (DNA Hydridization Group 5) from human clinical specimens, *J. Clin. Microbiol.,* 27, 1594, 1989.

*Bioengineering of Plants*

Chapter 7

# ENGINEERING OF INSECT RESISTANCE IN PLANTS WITH *BACILLUS THURINGIENSIS* GENES

**Marnix Peferoen and Herman Van Mellaert**

## TABLE OF CONTENTS

# I. *BACILLUS THURINGIENSIS*, THE BACTERIUM AND THE INSECTICIDE

## A. THE DISCOVERY

In 1908, Dr. Berliner isolated a sporulating bacterium, containing an inclusion body next to the spore, from diseased Mediterranean flour moth larvae.[1] Since it was isolated in Thuringen, he named the bacterium *Bacillus thuringiensis*. However, this was probably not the first recording of *Bacillus thuringiensis*. In 1901, Dr. Ishiwata reported to have isolated a bacterium from diseased silkworm larvae during an outbreak of flacherie. Unfortunately, the isolate was lost in the following years, and later became known as *Bacillus sotto*.

For a long time, it was unclear why the bacterium was toxic to insects, but in 1953, Hannay suggested that the toxicity resides in the crystals.[2] This was confirmed by Angus in bioassays with silkworm larvae.[3] In 1955, Hannay and Fitz-James[4] demonstrated that the crystals contained proteins, later called δ-endotoxins.[5] A major breakthrough was made when a gene encoding a crystal protein was cloned into *E. coli*.[6] The recombinant protein proved to be highly active against tobacco hornworm larvae. This firmly established that the insecticidal activity of *B. thuringiensis* was primarily determined by its crystal proteins.

## B. MORE STRAINS

Since its discovery, researchers have continued to isolate *B. thuringiensis* strains. Obviously, there was a need to find methods to determine whether these were reisolations of the same strain, or whether the isolates were really different. One of the most successful methods was the serotyping, introduced by de Barjac and Bonnefoi.[7] *B. thuringiensis* isolates are differentiated by comparing agglutination reactions of their motile cells with antisera, generated against flagellae on vegetative *B. thuringiensis* cells. Today more than 30 different serotypes have been described. Although the biological significance for the serotype is unclear, the method has proved to be useful for general classification of *B. thuringiensis* strains.

Until the late 1970s, it was generally accepted that *B. thuringiensis* crystal proteins were only active against lepidopteran larvae. However, in 1977 Goldberg and Margalit isolated a strain which was highly active against mosquito and blackfly larvae;[8] and in 1983, Krieg and co-workers found a strain highly toxic to some coleopteran larvae such as the Colorado potato beetle.[9] These two discoveries stimulated researchers, especially from industry, to screen for new strains with activity against a wider variety of insects. As a consequence, in addition to existing collections of Dr. Dulmage and Dr. Aizawa, thousands of strains have been isolated and are kept in academic and industrial collections.

*Bacillus thuringiensis* has been used as an insecticide for more than three decades. These bioinsecticides were mostly based on the same strain (HD-1). For a few years now, new strains have been used for the development of bioinsecticides with different insecticidal spectra.

# II. *BACILLUS THURINGIENSIS* CRYSTAL PROTEINS

## A. DIFFERENT STRAINS, DIFFERENT TOXINS

The key to understand differences in insecticidal spectrum of different *B. thuringiensis* strains, is to understand the crystal protein composition. The work of de Barjac led to the classification of *B. thuringiensis* strains on the basis of their serotypes.[10] But it also became clear that the serotype of a strain has very little to do with the insecticidal activity and spectrum of the strain. The most striking example can be found in the *morrisoni* serotype, containing strains active against lepidopteran, dipteran, and coleopteran larvae (HD-12, PG-14, and *B. tenebrionis*, respectively). Höfte et al.[11] analyzed the crystal protein composition of several strains and found that many strains produced several crystal proteins (Cry) and

## TABLE 1
### Insecticidal Crystal Proteins of
### *Bacillus thuringiensis*

| Cry type | Predicted mol mass | *Bt* strain | Ref. |
|---|---|---|---|
| CryIA(a) | 133.2 | HD-1 | 16 |
| CryIA(b) | 131.0 | berl. 1715 | 17 |
| CryIA(c) | 133.3 | HD-73 | 18 |
| CryIB | 138.0 | HD-2 | 19 |
| CryIC | 134.8 | HD-110 | 20 |
| CryID | 132.5 | HD-68 | 63 |
| CryIIA | 70.9 | HD-263 | 21 |
| CryIIB | 70.8 | HD-1 | 21 |
| CryIIIA | 73.1 | *tenebrionis* | 22 |
| CryIVA | 134.4 | *israelensis* | 23 |
| CryIVB | 127.8 | *israelensis* | 24 |
| CryIVC | 77.8 | *israelensis* | 25 |
| CryIVD | 72.4 | *israelensis* | 26 |
| CytA | 27.4 | *israelensis* | 27 |

that the same crystal proteins can be found in strains of different serotypes. This mobility of crystal protein genes (*cry*) among *B. thuringiensis* strains was to be expected, since most *cry* genes are situated on large conjugative plasmids (Lereclus et al.[12]), and since several of these genes have been shown to be bordered by transposon-like elements.[13-15] Moreover, Höfte and co-workers found a manifest correlation between the presence of a certain crystal protein and the insecticidal spectrum of the isolate.[11] So, the insecticidal activity and spectrum of a strain is primarily determined by the crystal proteins.

Thousands of strains have been isolated, and today 42 nucleotide sequences of *cry* genes have been published, which represent 14 clearly distinct crystal protein genes (Table 1, Höfte and Whiteley[28]). The Cry proteins are specifically toxic to different insect larvae. Based on their toxicity spectrum, the crystal proteins are grouped in four classes, Lepidoptera-specific, Lepidoptera- and Diptera-specific, Coleoptera-specific, and Diptera-specific. The *cry* genes clearly constitute a family of genes. All these genes encode proteins either of some 130 kDa or some 70 kDa, which are proteolytically converted to a toxic core fragment of some 60 kDa, except for CryIVD.[29] In the 130-kDa proteins, the toxic fragment is localized in its N-terminal half, so that apparently the highly conserved C-terminal half is not essential for toxicity. It is speculated that the C-terminal half, containing several cysteine residues, may be involved in the crystal formation. The smallest toxic fragment of CryIA(b) was determined in deletion analysis and its 12 C-terminal amino acids proved to be highly conserved among Cry toxins, except in CryII proteins.[30] Other studies suggest that this sequence does indeed define the C-terminus of toxic fragments of crystal proteins.[31,32] In addition to that region, there are four more domains of highly conserved sequences in the toxic core fragments of all but three crystal proteins (CryIIA, CryIIB, and CryIVD). The 120-amino acid region at the N-terminus is highly hydrophobic in all crystal proteins except CryII proteins and CryIVD. It is, however, the hydrophobic character which is conserved, not the amino acid sequence itself. A secondary structure model, based on the sequences of seven different toxins, shows the N-terminus of the toxins to be rich in α-helix structure and the C-terminus to contain alternating β-strands and coil structures, which are characteristic for a β-sheet conformation.[33]

In the lepidopteran pathotype (CryI proteins), 6 genes have been described, coding for protoxins of 130 to 140 kDa which accumulate in bipyramidal crystals. Three crystal proteins (CryIA[a], CryIA[b], and CryIA[c]) seem to belong to a subfamily because they are very closely related (82 to 90% amino acid identity). Each CryI protein is not only characterized by its amino acid sequence, but also by its particular insecticidal activity and spectrum.[28]

The second group of crystal proteins (CryII), formerly named P2 proteins, are toxic to both lepidopteran and dipteran larvae (Cry IIA) and to lepidopteran larvae (CryIIB).[21] The 65-kDa CryII proteins crystallize in cuboidal inclusions, and have been identified in different strains. The homology of these two genes to the other *cry* genes is rather limited.

The CryIIIA protein is the only crystal protein which has been described to be toxic to coleopteran larvae such as the Colorado potato beetle.[22] The protein is synthesized as a 72-kDa protein with a second translation initiation site at amino acid 48, and both products are stored in rhomboid crystals.[34]

Four crystal proteins are toxic to dipteran larvae (CryIV), two of which are produced as protoxins of some 130 kDa (CryIVA and CryIVB) while the other two proteins (CryIVC and CryIVD) of 78 kDa and 72 kDa, respectively, seem to be truncated on their C-terminal end (see Table 1 for references). Again, the protoxins are proteolytically cleaved to a toxic core fragment, although there is some confusion as to the right molecular weight of this fragment.[24,29]

CytA is a cytotoxin rather than an insecticide and has no homology with the insecticidal crystal proteins.[27]

A number of crystal proteins have been reported to be nontoxic.[35] This, however, may be the result of bioassays mostly done with economically important insects, that can be easily tested in feeding assays. From an evolutionary point of view, it is argued that *B. thuringiensis* produces crystal proteins, in order to establish a niche for the bacteria in insects. If some of the crystal proteins are not toxic to some insects, they must give the bacteria another competitive edge. They may be toxic to other organisms, or they may have a totally different, but advantageous, function.

## B. THE QUEST FOR NEW TOXIN GENES

The search for new *B. thuringiensis* strains and genes has been boosted by the interest in using microbial agents in the control of insect pests, and by the discovery of strains with activities against insects other than the Lepidoptera. Numerous *B. thuringiensis* isolation and screening programs have been initiated, especially in industries. These screening programs are of course biased towards the economically important pest insects. Table 1 summarizes all genes published before 1990, but we expect new *cry* genes to be published in the coming years.

Another approach in the quest for new toxin genes was to engineer new genes. Mutagenesis experiments have been performed by different research groups, but very few experiments have been published,[36] and there are no reports of synthetic genes coding for crystal proteins with increased insecticidal activity or with a different insecticidal spectrum. This does not necessarily reflect the failure of these experiments, since most of this work has been, and still is, mainly done in industrial research labs, which may not always publish their results. However, it is clear that so long as we do not understand the mechanism of toxicity, the knowledge base for an intelligent engineering program is still too small.

# III. MECHANISM OF ACTION

## A. THE INSECT MIDGUT

Most studies on the mode of action of *B. thuringiensis* toxins have been done with lepidopteran larvae. However, the mechanism of action in Lepidoptera may be representative for all insects susceptible to *B. thuringiensis* toxins. There are several steps involved in the mechanism of action.[37] Upon ingestion, the crystal proteins are dissolved by the alkaline gut juices in the midgut lumen of these larvae. In the next step, the protoxins are converted by gut proteases into toxic core fragments. Shortly thereafter, the midgut epithelial cells swell and eventually burst. Because this is a very rapid event, it appears that the toxin does not have to be internalized to provoke lysis.

Sacchi et al.[38] studied the effect of toxins on the potassium-amino acid cotransport in brush border membrane vesicles prepared from midguts of *Pieris brassicae*. Upon toxin treatment, they observed the dissipation of the potassium gradient by the formation of channels in the brush border membranes. They concluded that the channels were potassium specific. In contrast, Knowles and Ellar proposed the mechanism of colloid-osmotic lysis, which involves the formation of small (0.1 to 1 nm) aspecific pores in the cell membrane of the epithelial cells, resulting in an influx of ions accompanied by an influx of water.[39] The cells swell and eventually burst. Experiments done by Hendrick et al.[40] seem to confirm that there is indeed a formation of nonspecific pores, rather than potassium selective channels. It is unclear whether *B. thuringiensis* toxins induce pore formation indirectly through interaction with membrane proteins, or directly through insertion into the membrane.

## B. TOXIN RECEPTORS AND RESISTANCE DEVELOPMENT

Two main factors seem to be responsible for the extreme specificity of *Bacillus thuringiensis* toxins for different insects: (1) solubilization and proteolytic activation conditions may vary from insect to insect, (2) the presence of specific toxin-binding sites in the gut epithelium.

Most studies indicate that the specificity of the crystal proteins is independent of proteolytic activation.[37,41] Yet, it was shown by Haider et al.[42] that depending on the proteolytic enzymes, a protoxin can be activated into either a dipteran or a lepidopteran toxin. Work by Arvidson et al.[43] suggests that specificity is lost upon reduction of the cysteine residues in the protoxin, but can be restored by reoxidation of the cysteines and/or proteolytic removal of the C-terminus containing most cysteine residues.

Although differential processing may be involved, recent experiments show that the interaction of toxins with high affinity binding sites in the insect midgut predominantly determine the specificity. Binding studies with brush border membrane vesicles from *Manduca sexta* and *Pieris brassicae* midgut epithelial cells and labeled toxins demonstrated that there is a distinct correlation between toxicity and specific binding.[44] Experiments with *Heliothis virescens*, confirming this correlation, suggest that this could be a general principle in the mechanism of action.[45]

There are a few reports on the development of resistance against *B. thuringiensis* toxins, by continuous exposure to certain toxins under laboratory conditions.[46,47] Although the *Plodia interpunctella* larvae developed resistance to certain *B. thuringiensis* strains, they were still susceptible to some other strains.[48] Recently, it was demonstrated that the *P. interpunctella* strain resistant to the CryIA(b) toxin had become more susceptible to the CryIC toxin.[49] Resistance to the CryIA(b) correlates with a 50-fold reduction in affinity of the membrane receptor for this protein. Increased susceptibility to the CryIC toxin is reflected in an increase of the CryIC binding sites in the midgut epithelial cells. All this suggests that a combination or alteration of different toxins could help to prevent the development of resistance in insects towards *B. thuringiensis* toxins.

## IV. PLANT ENGINEERING WITH *B. THURINGIENSIS* GENES

### A. PLANT TRANSFORMATION

Since 1983, more than 20 different plant species have been successfully transformed using different systems.[50] Today, the *Agrobacterium tumefaciens* transformation system is still most commonly used, because it is a very efficient vector to stably introduce genes in plants. *Agrobacterium tumefaciens*, being a plant pathogen causing tumorous crown galls on infected dicotyledonous plants, proved unsuccessful in transformation of most monocotyledonous plants. Free DNA delivery systems such as microinjection, electroporation, and particle gun technology are being developed for those plants resistant to *Agrobacterium*-mediated transformation.

In the past 10 years, considerable progress has been made in the identification of genes conferring desirable agronomic traits. Coincidently, plant molecular research has lead to the better understanding of gene regulation. Foreign genes can only be expressed in plants when the gene is surrounded by appropriate signals. Several promoters have been described, some of which conferring expression in a wide variety of organs during most stages of development, others being very specific for a particular tissue, a certain stimulus, and/or a developmental stage. Promoters functional in plant cells do not have to be from plant origin to exert a signaling function in plant cells.

## B. *B. THURINGIENSIS* TOXIN EXPRESSION

Two promoters have been used in the engineering of insect resistance in plants: the TR promoter from the Ti-plasmid of *Agrobacterium tumefaciens*[51] and the cauliflower mosaic virus (CaMV) 5S promoters.[52] The TR promoter, a dual promoter directing expression in both directions, is a wound-stimulated promoter. It confers a low level of expression in intact organs, while upon wounding the TR promoter is highly stimulated. The 35S promoter is active during most stages of development and in most plant tissues. Yet, variation of expression levels among individual transgenic plants, and variation in tissue-specific expression patterns between transgenic hosts have been reported.[53]

Compared to other foreign genes, *B. thuringiensis* toxin genes preceded by the TR or 35S promoter are poorly expressed in transgenic plants. Especially constructions containing the full length sequence encoding the protoxin are expressed at levels hardly detectable. In fact, no plants resistant to insect attacks have been engineered by using full length genes. Expression levels are significantly enhanced by transforming with the truncated sequence encoding the toxic fragment only.

The *neo* gene encoding neomycin phosphotransferase II (NPTII) conferring resistance to neomycin/kanamycin, is commonly used in plant engineering as a selectable marker. The *neo* gene is not only used to sort out transformed from nontransformed plants, but can also be used to select engineered plants with high levels of expression. Using the dual TR promoter, the *neo* gene and the toxin gene can be cloned on both sides of the TR promoter. Transcription is simultaneously initiated in both directions, so that selection for high kanamycin resistance does also single out plants with high levels of *B. thuringiensis* toxin. Moreover, it proved possible to make a translational fusion between the gene encoding the toxic fragment and the *neo* gene, resulting in a fusion protein toxic to insects with NPTII activity.[54] High kanamycin resistance would then evidently select for high level of *B. thuringiensis* toxin.

# V. INSECTICIDAL ACTIVITIES

## A. GREENHOUSE EVALUATION

In 1987, the first report on the engineering of insect resistance in tobacco plants with *Bacillus thuringiensis* toxin genes was published by Vaeck et al.[55] Using plant expression vectors with the full-length and truncated *cryIA(b)* gene and the TR promoter, it was demonstrated that truncated genes are expressed at significantly higher levels. With truncated genes, up to 42 ng CryIA(b) per mg total protein were measured, while with full-length genes, some 1 to 2 ng could be detected. The insecticidal activity of the plants was tested with tobacco hornworm larvae. There was a clear correlation between the level of expression of the CryIA(b) protein and the insecticidal activity. Plants expressing some 0.004% of their total protein as the insecticide caused 100% mortality within 6 d. Shortly thereafter similar work was reported in tomatoes by Fischoff et al.,[56] and in tobacco by Barton et al.[57] Both studies concluded that, with the 35S promoter, expression levels of truncated genes are significantly higher than of full-length toxin genes. It is not clear why full-length genes are expressed at a lower level but Barton et al.[57] suggested that the full-length endotoxin is lethal to plant cells. In tomatoes, insecticidal effects were also noted on tomato fruitworms. Recently, it was reported that resistance against Colorado potato beetle larvae was engineered in potatoes.[58] Potato leaves and tubers have also been made resistant against potato tuber moth larvae.[59]

## B. FIELD TRIALS

The first field trial was done in 1986 in North Carolina with tobacco plants engineered by Plant Genetic Systems. For 3 consecutive years now, tobacco plants and their offspring have been tested in the field for their resistance against major insect pests such as the tobacco hornworm and the tobacco budworm. Plants that only had an intermediate insecticidal activity in the greenhouse seemed to be highly effective against insect attacks in the field. Although we cannot exclude that plants in the field produce higher levels of *B. thuringiensis* toxins, we suspect a synergistic effect between the toxin expressed by the plant and the presence of lepidopteran predators and parasites. It was indeed observed that many of the lepidopteran larvae were heavily infested with parasitic wasps.

Recently, the first report was published on the field performance of plants engineered with the *B. thuringiensis* toxin.[60] For 2 years, transformed tomato plants were tested to evaluate the potential use of these plants in the control of important lepidopteran pests on tomatoes in the U.S. Over the 2 years, tomato plants proved highly resistant to feeding damage by tobacco hornworm larvae. However, tobaco hornworm larvae are only a minor pest on tomatoes and are more susceptible to the *B. thuringiensis* toxin than tomato fruitworm and tomato pinworm larvae. Both the fruitworm and the pinworm larvae initially feed on leaves, but eventually they bore through the tomato skin and enter the fruit. The damage tolerance to fruits is very low, and a single hole can make the tomato unmarketable. Therefore, the focus of the field trials was on the level of fruit damage. As well for the fruitworm as for the pinworm, it was concluded that there was significant reduction in fruit damage, even under heavy infestation conditions. However, the damage to the fruits was still commercially unacceptable. Again, improved insect control was observed in the field compared to the greenhouse.

Other crops, such as cotton, engineered for insect resistance are being tested in the field, but no data have been published yet.

# VI. PERSPECTIVES

The practical feasibility of engineering insect resistance in plants with *Bacillus thuringiensis* toxin genes has been demonstrated. It is expected that the first generation of commercial plants will be marketed in the mid 1990s. These products will have to clear the way for registration and to gain acceptance by the breeders, the farmers, and the consumers.

For a second generation of insect-resistant plants, we expect progress in different areas. In several instances, expression levels of *B. thuringiensis* toxins in plants are still too low for control of the target pest at a commercially acceptable level. Therefore, expression levels have to be boosted. On the other hand, insecticidal proteins may not be needed or wanted in all tissues or during all stages of development. Regulatory agencies may require levels to be low in edible parts of the plant. In addition, Gould suggested that, in order to delay development of resistance, selection pressure should be kept low.[61] The insecticide could be targeted only to certain tissues, allowing controlled levels of insect pressure on economically less important organs. Plant promoters specific for stem, leaf, root, etc. have been identified.[62]

There are still several important insect pests that are not susceptible to *B. thuringiensis* toxins. Screening programs are established to identify new toxins which are active against other insect pests. With additional toxins, it would then become possible to control the entire complex of insect pests on a certain crop. Moreover, when different target sites are involved in the action of different toxins, it may be useful to combine *B. thuringiensis* toxins to prevent development of resistance in insects.

Several of the most important crops are still recalcitrant to current transformation techniques. With substantial efforts concentrated on crop transformation, we expect that these problems will eventually be overcome, so that engineering *Bacillus thuringiensis* insecticidal proteins will be a major tool in the control of insect pests on agronomically important crops.

# REFERENCES

1. **Berliner, E.**, Über die Schlaffsucht der Mehlmottenraupe (*Ephestia kuehniella* Zell), und ihren Erreger *Bacillus thuringiensis* n. sp., *Z. Angew. Entomol.*, 2, 29, 1915.
2. **Hannay, C. L.**, Crystalline inclusions in aerobic spore-forming bacteria, *Nature*, 172, 1004, 1953.
3. **Angus, T. A.**, A bacterial toxin paralyzing silkworm larvae, *Nature*, 173, 545, 1954.
4. **Hannay, C. L. and Fitz-James, P.**, The protein crystals of *Bacillus thuringiensis* Berliner, *Can. J. Microbiol.*, 1, 674, 1955.
5. **Heimpel, A. M.**, A taxonomic key proposed for the species of the "crystalliferous bacteria", *J. Invertebr. Pathol.*, 9, 364, 1967.
6. **Schnepf, H. E., and Whiteley, H. R.**, Cloning and expression of the *Bacillus thuringiensis* crystal protein gene in *Escherichia coli*, *Proc. Natl. Acad. Sci. U.S.A.*, 78, 2893, 1981.
7. **de Barjac, H. and Bonnefoi, A.**, Essai de classification biochimique et sérologique de 24 souches de *Bacillus* du type *B. thuringiensis*, *Entomophaga*, 1, 5, 1962.
8. **Goldberg, L. J. and Margalit, J.**, A bacterial spore demonstrating rapid larvicidal activity against *Anopheles serengetii*, *Uranotaenia unguiculata*, *Culex univittatus*, *Aedes aegypti* and *Culex pipiens*, *Mosq. News*, 37, 355, 1977.
9. **Krieg, A., Huger, A. M., Langenbruch, G. A., and Schnetter, W.**, *Bacillus thuringiensis* var. *tenebrionis*: ein neuer gegenüber Larven von Coleopteren wirksamer Pathytyp, *Z. Angew. Entomol.*, 96, 500, 1983.
10. **de Barjac, H.**, Identification of H-serotypes of *Bacillus thuringiensis*, in *Microbial Control of Pests and Plant Diseases 1970-1980*, Burges, H. D., Ed., Academic Press, London, 1981, 35.
11. **Höfte, H., Van Rie, J., Jansens, S., Van Houtven, A., Vanderbruggen, H., and Vaeck, M.**, Monoclonal antibody analysis and insecticidal spectrum of three types of lepidopteran-specific insecticidal crystal proteins on *Bacillus thuringiensis*, *Appl. Environ. Microbiol.*, 54, 2010, 1988a.
12. **Lereclus, D., Bourgouin, C., Lecadet, M. M., Klier, A., and Rapoport, G.**, Role, structure, and molecular organization of the genes coding for the parasporal delta-endotoxins of *Bacillus thuringiensis*, in *Regulation of Procaryotic Development*, Smith, I., Slepecky, R. A., and Setlow, P., Eds., American Society for Microbiology, Washington D.C., 1989, 255.
13. **Kronstad, J. W. and Whiteley, H. R.**, Inverted repeat sequences flank the *Bacillus thuringiensis* crystal protein gene, *J. Bacteriol.*, 154, 419, 1984.
14. **Mahillon, J., Seurinck, J., Van Rompuy, V., Delcour, J., and Zabeau, M.**, Nucleotide sequence and structural organization of an insertion sequence element (IS231) from *Bacillus thuringiensis* strain Berliner 1715, *EMBO J.*, 4, 3895, 1985.
15. **Bourgouin, C., Delecluse, A., Ribier, J., Klier, A., and Rapoport, G.**, A *Bacillus thuringiensis* subsp. *israelensis* gene encoding a 125-kilodalton larvicidal polypeptide is associated with inverted repeat sequences, *J. Bacteriol.*, 170, 3575, 1988.
16. **Schnepf, H. E., Wong, H. C., and Whiteley, H. R.**, The amino acid sequence of a crystal protein from *Bacillus thuringiensis* deduced from the DNA base sequence, *J. Biol. Chem.*, 260, 6264, 1985.
17. **Wabiko, H., Raymond, K. C., and Bulla, L. A., Jr.**, *Bacillus thuringiensis* entomocidal protoxin gene sequence and gene product analysis, *DNA*, 5, 305, 1986.
18. **Adang, L. F., Staver, M. J., Rocheleau, T. A., Leighton, J., Barker, R. F., and Thompson, D. V.**, Characterized full-length and truncated plasmid clones of the crystal protein of *Bacillus thuringiensis* subsp. *kurstaki* HD-73 and their toxicity to *Manduca sexta*, *Gene*, 36, 289, 1985.

19. **Brizzard, B. L. and Whiteley, H. R.,** Nucleotide sequence of an additional crystal protein gene cloned from *Bacillus thuringiensis* subsp. *thuringiensis, Nucleic Acid Res.,* 16, 4168, 1988.

20. **Honnée, G. T., Van Der Salm, T., and Visser, B.,** Nucleotide sequence of crystal protein gene isolated from *B. thuringiensis* subspecies entomocidus 60.5 coding for a toxin highly active against *Spodoptera* species, *Nucleic Acid Res.,* 16, 6240, 1988.

21. **Widner, W. R. and Whiteley, H. R.,** Two highly related insecticidal crystal proteins of *Bacillus thuringiensis* subsp. *kurstaki* possess different host range specificities, *J. Bacteriol.,* 171, 965, 1989.

22. **Herrnstadt, C., Gilroy, T. E., Sobieski, D. A., Bennet, B. D., and Gaertner, F. H.,** Nucleotide sequence and deduced amino acid sequence of a coleopteran-active delta-endotoxin gene from *Bacillus thuringiensis* subsp. *san diego, Gene,* 57, 37, 1987.

23. **Ward, E. S. and Ellar, D.,** Nucleotide sequence of a *Bacillus thuringiensis* var. *israelensis* gene encoding a 130 kDa delta-endotoxin, *Nucleic Acid Res.,* 15, 7195, 1987.

24. **Chungjatupornichai, W., Höfte, H., Seurinck, J., Angsuthanasombat, C., and Vaeck, M.,** Common features of *Bacillus thuringiensis* toxins specific for Diptera and Lepidoptera, *Eur. J. Biochem.,* 173, 9, 1988.

25. **Throne, L., Garduno, F., Thompson, T., Decker, D., Zounes, M., Wild, M., Walfield, A. M., and Pollock, T.,** Structural similarity between the Lepidoptera- and Diptera-specific insecticidal endotoxin genes of *Bacillus thuringiensis* subsp. *kurstaki* and *israelensis, J. Bacteriol.,* 166, 801, 1986.

26. **Donovan, W. P., Dankocsik, C. C., and Gilbert, M. P.,** Molecular characterization of a gene encoding a 72-kilodalton mosquito-toxic crystal protein from *Bacillus thuringiensis* subsp. *israelensis, J. Bacteriol.,* 170, 4732, 1988.

27. **Waalwijck, C., Dullemans, A. M., Van Workum, M. E. S., and Visser, B.,** Molecular cloning and the nucleotide sequence of the $M_r$ 28,000 crystal protein gene of *Bacillus thuringiensis* subsp. *israelensis, Nucleic Acids Res.,* 13, 8206, 1985.

28. **Höfte, H. and Whiteley, H. R.,** Insecticidal crystal proteins of *Bacillus thuringiensis, Microbiol. Rev.,* 53, 242, 1989.

29. **Chilcott, C. N. and Ellar, D. J.,** Comparative study of *Bacillus thuringiensis* var. *israelensis* crystal proteins *in vivo* and *in vitro, J. Gen. Microbiol.,* 134, 2551, 1988.

30. **Höfte, H., De Greven, H., Seurinck, J., Jansens, S., Mahillon, J., Ampe, C., Vandekerckhove, J., Vanderbruggen, H., Van Montagu, M., Zabeau, M., and Vaeck, M.,** Structural and functional analysis of a cloned delta endotoxin of *Bacillus thuringiensis* Berliner 1715, *Eur. J. Biochem.,* 161, 273, 1986.

31. **Schnepf, H. E. and Whiteley, H. R.,** Delineation of a toxin-encoding segment of a *Bacillus thuringiensis* crystal protein gene, *J. Biol. Chem.,* 260, 6273, 1985.

32. **Pao-intara, M., Angsuthanasombat, C., and Panyim, S.,** The mosquito larvicidal activity of 130 kDa delta-endotoxin of *Bacillus thuringiensis* var. *israelensis* resides in the 72 kDa amino-terminal fragment, *Biochem. Biophys. Res. Commun.,* 153, 294, 1988.

33. **Convents, D., Houssier, C., Lasters, I., and Lauwereys, M.,** The *Bacillus thuringiensis* delta-endotoxin—Evidence for a two domain structure of the minimal toxic fragment, *J. Biol. Chem.,* 265, 1369, 1990.

34. **McPherson, S. A., Perlak, F. J., Fuchs, R. L., Marrone, P. G., Lavrik, P. B., and Fischoff, D. A.,** Characterization of the coleopteran-specific protein gene of *Bacillus thuringiensis* var. *tenebrionis, Biol Technology,* 6, 61, 1988.

35. **Martin, P. A. W. and Travers, R. S.,** Worldwide abundance and distribution of *Bacillus thuringiensis* isolates, *Appl. Environ. Microbiol.,* 55, 2437, 1989.

36. **Ge, A. Z., Shivarova, N. I., and Dean, D. H.,** Location of the *Bombyx mori* specificity domain on a *Bacillus thuringiensis* δ-endotoxin protein, *Proc. Natl. Acad. Sci. U.S.A.,* 86, 4037, 1989.

37. **Lüthy, P. and Ebersold, H. R.,** *Bacillus thuringiensis* delta-endotoxin: histopathology and molecular mode of action, in *Pathogenesis of Invertebrate Microbial Diseases,* Davidson, E. W., Ed., Allenheld Osmun, Totawa, N.J., 1981, 235.

38. **Sacchi, V. F., Parenti, P., Hanozet, G. M., Giordana, B., Lüthy, P., and Wolfersberger, M. G.,** *Bacillus thuringiensis* toxin inhibits K$^+$- gradient-dependent amino acid transport across the brush-border membrane of *Pieris brassicae* midgut cells, *FEBS Lett.,* 204, 213, 1986.

39. **Knowles, B. H. and Ellar, D. J.,** Colloid-osmotic lysis is a general feature of the mechanism of action of *Bacillus thuringiensis* δ-endotoxins with different insect specificities, *Biochim. Biophys. Acta,* 924, 509, 1986.

40. **Hendrickx, K., De Loof, A., and Van Mellaert, H.,** Effects of *Bacillus thuringiensis* delta-endotoxin on the permeability of brush border membrane vesicles from tobacco hornworm, *(Manduca sexta)* midgut, *Comp. Biochem. Physiol.,* 95C, 241, 1990.

41. **Jaquet, F., Hütter, R., and Lüthy, P.,** Specificity of *Bacillus thuringiensis* delta-endotoxin, *Appl. Environ. Microbiol.,* 53, 500, 1987.

42. **Haider, M. Z., Knowles, B., and Ellar, D. J.,** Specificity of *Bacillus thuringiensis* var. *colmeri* insecticidal delta-endotoxin by differential processing of the protoxin by larval gut proteases, *Eur. J. Biochem.,* 156, 531, 1986.

43. **Arvidson, H., Dunn, P. E., Strnad, S., and Aronson, A. I.**, Specificity of *Bacillus thuringiensis* for lepidopteran larvae: factors involved *in vivo* and in the structure of a purified protoxin, *Mol. Microbiol.*, 3, 1533, 1989.

44. **Hofmann, C., Vanderbruggen, H., Höfte, H., Van Rie, J., Jensens, S., and Van Mallaert, H.**, Specificity of *Bacillus thuringiensis* δ-endotoxins is correlated with the presence of high-affinity binding sites in the brush border membrane of target insect midguts, *Proc. Natl. Acad. Sci. U.S.A.*, 85, 7844, 1988.

45. **Van Rie, J., Jensens, S., Höfte, H., Degheele, D., and Van Mellaert, H.**, Specificity of *Bacillus thuringiensis* δ-endotoxins, Importance of specific receptors on the brush border membranes of the midgut of target insects, *Eur. J. Biochem.*, 186, 239, 1990.

46. **McGaughey, W. H.**, Insect resistance to the biological insecticide *Bacillus thuringiensis*, *Science*, 229, 193, 1985.

47. **Stone, T. B., Sims, S. R., and Marrone, P. G.**, Selection of tobacco budworm for resistance to a genetically engineered *Pseudomonas fluorescens* containing the delta-endotoxin of *Bacillus thuringiensis* subsp. *kurstaki*, *J. Invertebr. Pathol.*, 53, 228, 1989.

48. **McGaughey, W. H. and Johnson, D. E.**, Toxicity of different serotypes and toxins of *Bacillus thuringiensis* to resistant and susceptible Indeanmeal moths (Lepidoptera: Pyralidae), *J. Econ. Entomol.*, 80, 1122, 1987.

49. **Van Rie, J., McGaughey, W. H., Johnson, D. E., Barnett, B. D., and Van Mellaert, H.**, Mechanism of insect resistance to the microbial insecticide, *Bacillus thuringiensis*, *Science*, 247, 72, 1990.

50. **Gasser, C. S. and Fraley, R.**, Genetically engineering plants for crop improvement, *Science*, 244, 1293, 1989.

51. **Velten, J., Velten, L., Hain, R., and Schell, J.**, Isolation of a dual plant promoter fragment from the Ti plasmid of *Agrobacterium tumefaciens*, *EMBO J.*, 12, 2723, 1984.

52. **Odell, J. R., Nagy, F., and Chua, N. H.**, Identification of DNA sequences required for activity of the cauliflower mosaic virus 35S promoter, *Nature*, 313, 810, 1985.

53. **Benfey, P. N. and Chua, N.-H.**, Regulated genes in transgenic plants, *Science*, 244, 174, 1989.

54. **Höfte, H., Buyssens, S., Vaeck, M., and Leemans, J.**, Fusion proteins with both insecticidal and neomycin phosphotransferase II activity, *FEBS Lett.*, 226, 364, 1988b.

55. **Vaeck, M., Reynaerts, A., Höfte, H., Jansens, S., De Beuckeleer, M., Dean, C., Zabeau, M., Van Montagu, M., and Leemans, J.**, Transgenic plants protected from insect attack, *Nature*, 328, 33, 1987.

56. **Fischoff, D. A., Bowdish, K. S., Perlak, F. J., Marrone, P. G., McCormick, S. M., Niedermeyer, J. G., Dean, D. A., Kusano-Kretzmer, K., Mayer, E. J., Rochester, D. E., Rogers, S. G., and Fraley, R. T.**, Insect tolerant transgenic tomato plants, *Bio/Technology*, 5, 807, 1987.

57. **Barton, K., Whiteley, H., and Yang, N.-S.**, *Bacillus thuringiensis* δ-endotoxin in transgenic *Nicotiana tabacum* provides resistance to lepidopteran insects, *Plant Physiol.*, 85, 1103, 1987.

58. **McPherson, S., Perlak, F., Fuchs, R., MacIntosh, S., Dean, D., and Fischoff, D.**, Expression and analysis of the insect control protein from *Bacillus thuringiensis* var. *tenebrionis*, in *Abstracts 1st Int. Symp. Molecular Biology of the Potato*, Bar Harbor, Maine, 1989, 51.

59. **Peferoen, M., Jansens, S., Reynaerts, A., Leemans, J.**, Potato plants with engineered resistance against insect attack, in *Molecular and Cellular Biology of the Potato*, Vayda, M. E. and Park, W. C., Eds., CAB International, Wallingford, 1990, 193.

60. **Delannay, X., LaVallée, B. J., Proksch, R. K., Fuchs, R. L., Sims, S. R., Greenplate, J. T., Marrone, P. G., Dodson, R. B., Augustine, J. J., Layton, J. G., and Fischoff, D. A.**, Field performance of transgenic tomato plants expressing the *Bacillus thuringiensis* var. *kurstaki* insect control protein, *Bio/Technology*, 7, 1265, 1989.

61. **Gould, F.**, Genetic engineering, integrated pest management and the evolution of pests, *Trends in Biotech.*, 6, 15, 1988.

62. **Kuhlemeier, C., Green, P. J., and Chua, N.-H.**, Regulation of gene expression in higher plants, *Annu. Rev. Plant Physiol.*, 38, 221, 1987.

63. **Höfte, H.**, unpublished.

Chapter 8

# PLANT TRANSFORMATION BY PARTICLE BOMBARDMENT

## Theodore M. Klein and Pal Maliga

## TABLE OF CONTENTS

# I. INTRODUCTION

Advances in the genetic engineering of economically important crop species are dependent on the development of efficient and routine technologies for gene transfer. In a growing number of species, including tobacco, tomato, potato, and oilseed rape, *Agrobacterium*-mediated gene transfer represents a standard route toward the production of transgenic plants.[1] Recent results indicate that *Agrobacterium* can effectively transfer genes to some varieties of soybean.[2] In addition, there is evidence that *Agrobacterium* can deliver genes to corn[3] and rice[4] cells, although transgenic plants of these species have not been produced by these techniques. It is clear that *Agrobacterium* is currently not universally applicable to all plant species and its ineffectiveness is particularly apparent for cereal crops.[5] Even in species that are amenable to transformation by *Agrobacterium,* genotype-specific responses continue to limit its application. Because of the difficulties encountered in applying *Agrobacterium* for the transformation of many plant species, approaches for direct DNA transfer into plant cells have been explored.[6-8]

Either protoplasts or intact cells and tissues can be the recipients for DNA by direct transfer. Protoplasts are obtained by the enzymatic removal of the cell wall. DNA can then be transferred across the cell membrane by a variety of techniques including electroporation or chemically induced uptake by polyethylene glycol (PEG). A portion of the protoplasts integrate the foreign DNA, regenerate cell walls, and divide to form transgenic calli. Transformed plants can then be regenerated from the calli. The genetic transformation of both indica[9] and japonica[10,11] rice cultivars has been achieved by DNA transfer to protoplasts. In addition, nonfertile transgenic maize plants have been recovered from protoplasts.[12] The major limitation of this approach is the necessity for cell cultures from which viable protoplasts can be obtained and the difficulties associated with the regeneration of fertile plants from these protoplasts. It appears that the appropriate cell cultures are difficult to produce and currently can only be derived from a limited set of genotypes.

Delivering foreign DNA indirectly into intact plant cells eliminates the need for protoplasts. Advances have been reported in using microinjection,[13] electroporation,[14] lasers,[15] and silica carbide fibers[16] for DNA delivery into intact plant cells. The most successful direct delivery approach to date involves the use of DNA-coated particles that are accelerated to velocities permitting their penetration of cell walls and membranes.[17,18] Both corn[19,20] and soybean[21] have been transformed by delivering DNA directly into intact tissues by particle bombardment.

The ability to deliver DNA into intact tissues and then monitor its expression shortly after delivery has been applied as a tool in the study of gene regulation in plants.[22] Transient assays based on DNA delivery by particle bombardment are accelerating investigations of the factors that control plant gene expression by reducing the need to produce stably transformed plants. Particle bombardment is also proving essential in the transfer of foreign genes to plastids[23] and mitochondria.[24] In this review, we will summarize the recent advances in plant biology that have been made possible by particle bombardment.

# II. PARTICLE BOMBARDMENT TECHNOLOGY

## A. ACCELERATION DEVICES

The concept of DNA transfer by particle bombardment centers upon the acceleration of DNA-bearing particles (also called microprojectiles or microcarriers) to velocities that permit their penetration and entry into intact cells. The process requires a device (particle gun) for microprojectile acceleration. Several conceptual designs for the particle gun have been described.[18] The basic system that has received the most attention employs a macroprojectile (or macrocarrier), a mechanism for accelerating the macroprojectile, and a means of stopping

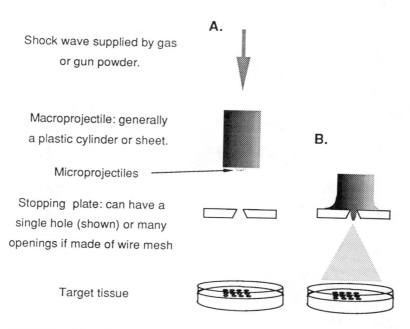

**A.**

Shock wave supplied by gas
or gun powder.

Macroprojectile: generally
a plastic cylinder or sheet.

**B.**

Microprojectiles

Stopping plate: can have a
single hole (shown) or many
openings if made of wire mesh

Target tissue

FIGURE 1. Schematic representation of the particle bombardment process showing the components of a system employing a macroprojectile: (A) before acceleration of the macroprojectile; (B) following acceleration of the macroprojectile and its impact with the stopping plate.

the macroprojectile (Figure 1). The DNA-bearing particles are placed on the leading surface of the macroprojectile and are released upon impact with a stopping plate or screen. The stopping plate or screen is designed to halt the forward motion of the macroprojectile while permitting the passage of the microprojectiles. The particles then travel through a partial vacuum until they reach the target tissue. The partial vacuum is used to reduce the aero-dynamic drag upon the microprojectiles and to decrease the force of the shock wave created when the macrocarrier impacts the stopping plate. Acceleration of the macrocarrier is achieved by the shock wave generated from a gun powder charge,[17,25,26] the capacitance discharge through a drop of water,[27] or the sudden release of compressed air,[28] nitrogen,[29] or helium.[30]

There are many physical and biological parameters that influence the efficiency of the process. The primary variables include the velocity of the particles and the number of particles that impact the target tissue per unit area. The velocity of the particles in flight has been determined for only one system. Using laser interferometry, the particle acceleration device that utilizes gunpowder[25] accelerates microprojectiles to a velocity of about 3000 feet per second.[30a] However, the actual velocity of the microprojectile upon impact and penetration of a cell or the velocity necessary for the penetration of several cell layers are not known. Control of velocity is potentially important for DNA delivery to meristematic cells that reside below the epidermis.

## B. MICROPROJECTILES FOR DNA DELIVERY

The material utilized as the source of microprojectiles represents a critical component that can influence the efficiency of the process. Powders of high-density metals such as gold[21,27] or tungsten[17,25] are available in a range of average diameters and have been used for gene transfer. Unfortunately, the size uniformity of the powders that are currently available is poor. For example, the gold powder used for the transformation of soybean[21] has an average diameter of 1.5 to 3 μm, with many particles being either smaller or larger

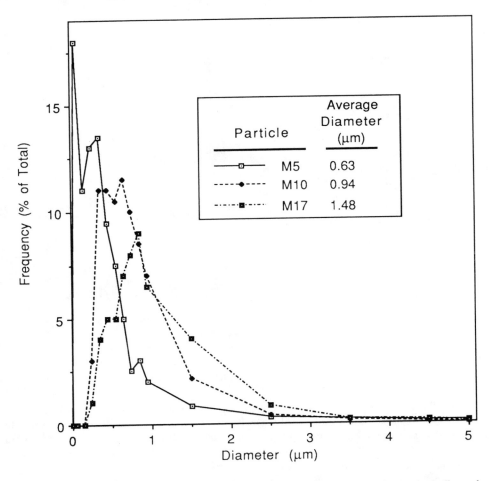

FIGURE 2.    Particle size distribution of some tungsten powders (Sylvania, Inc.) that are typically used for microprojectiles.

than this average. Therefore, many of the particles in this powder may be ineffective because they are too small and do not penetrate the target tissue or because they are too large and cause extensive damage. Figure 2 illustrates the size distribution of some tungsten powders that are frequently used for DNA delivery. It is probable that improvements in the uniformity of these materials would increase efficiencies of gene tansfer. Efforts are currently under way to produce uniform metal powders of defined sizes for use as microprojectiles.

More research is needed to define other materials that can be used as microprojectiles and to determine the optimal size and shape of these particles for the penetration of various cell and tissue types. Another critical factor that requires more attention is the means used to adsorb DNA to the particles. The ideal microprojectile would retain sufficient intact DNA upon acceleration, impact, and penetration so that each cell surviving penetration also receives DNA. It is not clear if the calcium-spermidine precipitation procedure[17,25] that most researchers currently use to coat the particles is optimal. Other methods which utilize PEG or polylysine have been reported, but a comprehensive study comparing these various procedures has not been done. Methods to encapsulate the DNA within the microprojectile or to provide a protective coating to the adsorbed DNA may protect the DNA during the bombardment process. In addition, the method used to precipitate or encapsulate DNA may influence the rate at which it is released from the microprojectile. Controlling the rate of release of DNA from the microprojectile may influence the likelihood of its integration into the host genome by increasing the probability that DNA would be present when the cells are competent for transformation.

# III. TRANSFER OF GENES TO THE NUCLEUS

## A. TRANSIENT GENE EXPRESSION

The expression of foreign genes can be monitored in a short period of time (several hours to days) following their delivery to cells. In such experiments, the status of the introduced DNA (i.e., integrated into the host genome or existing as free plasmid in the nucleus) is not of concern. Transient assays are useful for the optimization of parameters associated with direct delivery systems. Expression of the β-glucuronidase (GUS) gene has been used to study some of the variables affecting DNA transfer to corn cells by particle bombardment.[25] However, efficiencies of gene transfer as estimated by transient expression should be treated with caution, since the precise relationship between the levels of transient expression and stable integration of the foreign DNA is not clear. Changes in certain parameters (i.e., velocity or quantity of the microprojectiles) may enhance the number of cells that transiently express the introduced DNA, but may decrease the number of stable transformants due to increased damage to the cells, preventing their division. Various reporter genes (i.e., GUS, chloramphenicol acetyltransferase [CAT], luciferase [LUX]) have been used to determine the range of cell and tissue types into which DNA can be delivered by the process.[22] It appears that particle bombardment can be used to transfer genes into most plant tissues. These include pollen,[31] leaf,[32,33] and various seed[25,33-35] and seedling[33,36] tissues.

Transient expression systems have been utilized to study gene function and to define critical *cis*- and *trans*-acting factors essential for regulated expression. There are several advantages to using transient expression systems based on particle bombardment. The process can be used to target cell types that otherwise would not be accessible to direct DNA transfer, such as pollen.[31] Another advantage of particle bombardment is that the need to produce protoplasts from a particular tissue is eliminated. As is true of other transient expression systems, particle bombardment eliminates the need to generate large numbers of transgenic plants, thereby saving both space and time.

The ability to monitor the expression of genes shortly after their direct delivery to intact tissues is proving valuable for the study of gene regulation by genetic,[34] tissue-specific,[31,34] and environmental[36] factors. For example, Bruce and co-workers[36] have developed a system to study the regulation of phytochrome genes. The phytochrome gene in many monocots is subject to negative feedback by phytochrome. A chimeric gene consisting of the 5′ region from the oat phytochrome gene and a CAT coding region was introduced into rice seedlings by particle bombardment. The gene was expressed following exposure of the seedlings to far red light while expression was repressed after exposure to red light. It was demonstrated that the chimeric and endogenous phytochrome genes exhibit a similar pattern of response to light. This system is being exploited to determine the critical elements of the phytochrome gene that are necessary for response to light related stimuli.

Another area that has benefited from particle bombardment is the study of genes that regulate anthocyanin production in maize. Important *cis*-acting regions of the *Bz1* gene (a structural gene of the anthocyanin pathway) have been defined by introducing chimeric reporter genes containing various 5′ deletions and mutations into maize tissues.[37] Two potential binding sites for the products of the *C1* and *B-1* genes, known by genetic evidence to control pigment production, have been identified using this system.[38] Several genes that regulate anthocyanin production have also been introduced into intact maize tissues.[33,38] This has permitted an assessment of the functional relationship between some of the members of the *R* gene family.[38,39] Some of the other studies using particle bombardment for the analysis of the factors that control gene expression have been reviewed.[25]

## TABLE 1
## Species That Have Been Stably Transformed by
## Particle Bombardment

| Plant species | Tissue bombarded | Plants recovered | Refs. |
|---|---|---|---|
| Tobacco | Leaves | Yes | 32,40 |
| | Suspension culture | Non-regenerable[a] | 32,41,42 |
| Soybean | Embryos | Non-regenerable[a] | 27 |
| | Apical meristem | Yes | 21 |
| | Suspension culture | Yes | 43 |
| Corn | Suspension culture | Non-regenerable[a] | 44,45 |
| | Suspension culture | Yes | 19,20 |
| | Apical meristem | Yes[b] | 46 |
| Cotton | Suspension culture | Yes[c] | 47 |
| Papaya | Zygotic embryos | Yes[d] | 48 |
| | Somatic embryos | | |

[a]  Plants cannot be regenerated from these cells even prior to bombardment.
[b]  Transgenic anthocyanin marker expressed in somatic tissue; foreign gene not present in the progeny.
[c]  Fertility of plants was not determined.
[d]  Fertility of the plants was not determined; Southern analysis of these plants was not performed.

## B. STABLE TRANSFORMATION OF THE NUCLEAR GENOME

Two broad approaches for recovering transformants by particle bombardment have been employed. These approaches differ in the tissues that are targeted for DNA transfer and in the genes that are used as markers to detect the transformants. One approach involves the bombardment of tissues that give rise directly to plants and the subsequent screening of the resulting plants for the presence of the introduced gene. The other approach, which has been more broadly applied, involves bombarding tissues or cell cultures with a gene that confers resistance to a particular drug (i.e., kanamycin or hygromycin) or herbicide (i.e., phosphinothricin or chlorsulfuron). Transformed cells are then selected by their ability to grow on medium containing the inhibitor, and plants are regenerated from these clones. Table 1 lists the species that have been genetically transformed by particle bombardment.

### 1. Screening for the Identification of Transformed Plants

The bombardment of soybean provides an example of the use of screening procedures for the recovery of transformed plants. Transgenic soybean plants have been recovered by bombardment of shoot apical meristems with microprojectiles carrying genes encoding the GUS or neomycin phosphotransferase enzymes.[21] The bombarded tissue was cultured on medium containing high levels of cytokinin to induce the formation of about 8 to 10 shoots from an individual meristem. Histochemical localization of GUS activity showed that most of these plants had a portion of their stem, branches and leaves transformed. Many of the chimeric plants possessed GUS sectors that were restricted to the epidermis. In some cases the GUS sector extended into subepidermal layers. Delivery of the foreign gene to these lower layers is required for transmittance to germinal cells. About 0.05% of the shoots that were recovered from bombarded meristems gave rise to plants that carried the foreign gene in the pollen or ovary.

Recent results indicate that transgenic corn may also be recovered using a screening approach. Corn meristems were bombarded with a gene that activates the anthocyanin pathway.[39] One plant containing a purple sector that ran the entire length of the plant was recovered. However, progeny of this plant were not transgenic, indicating that neither the pollen or ovules of the plant grown from the bombarded meristem carried the foreign gene.[39a,46] Apparently the gene was integrated into the genome of a meristematic cell that gave rise to only somatic cells and not reproductive tissue. More research is necessary to determine if this is a practical approach for the transformation of corn.

The importance of the above results is that the requirement for *in vitro*-grown tissues that provide the appropriate morphogenic response is eliminated. Therefore, virtually any genotype can be transformed by this approach. In addition, since extensive tissue culture manipulations are avoided, the potential for the generation of somaclonal variation is reduced. The primary limitation of this approach is the large number of apical meristems that must be isolated and bombarded and the need to analyze large numbers of plants for the presence of the introduced gene.

## 2. Identification of Transgenic Clones by Selective Markers

Selection procedures rely on the differential growth of cells expressing an introduced gene that confers resistance to a particular compound. Stable transformation by particle bombardment was first demonstrated in *Nicotiana tabacum* by bombardment of leaf or suspension cultures with a gene that confers resistance to kanamycin.[32] The utility of selection schemes for the transformation of more recalcitrant crops has been demonstrated in corn[19,20] and soybean.[43,43a]

The transformation of corn has been achieved by bombardment of cell cultures with genes that confer resistance to chlorsulfuron[19] or phosphinothricin.[20] In these studies, embryogenic suspension culture cells were bombarded with tungsten microprojectiles coated with plasmid DNA. After a period of growth on nonselective medium, the bombarded cells were transferred to solid medium containing the appropriate herbicide. In general, the growth of herbicide-resistant calli could be distinguished from the limited growth of the nontransformed cells 6 to 8 weeks after bombardment. The introduced plasmids also carried a gene that codes for an enzyme that is easily assayed (i.e., GUS or LUX) so that the transformed nature of the colonies that developed on the selective media could be readily verified by enzyme assays. The expression of these genes could also be assessed in the plants regenerated from the transgenic calli. Transgenic soybean[43] and cotton[47] have also been produced, using selection approaches involving the bombardment of embryogenic suspensions with a gene that confers resistance to hygromycin.

The recovery of transformants from bombarded tissue culture cells requires the utilization of efficient selection systems. Attempts to recover transgenic clones of embryogenic corn cultures from suspension cells bombarded with the widely used kanamycin-resistant marker has thus far proven unsuccessful. Levels of kanamycin that are normally used for recovering transformed lines of tobacco do not adequately inhibit the growth of these maize cells. Higher levels of the drug appear too toxic and lead to necrosis of the cells and potentially the release of compounds that could inhibit the growth of the relatively few transformed cells in the bombarded sample. However, moderate levels of chlorsulfuron that sufficiently inhibit the growth of nontransformed cells do not cause browning and tissue necrosis. Testing a range of selectable markers when initiating a project for the transformation of a new species is advisable.

The major limitation to obtaining transgenic plants by selection following particle bombardment is the necessity for appropriate morphogenic cultures. In cereals, the required embryogenic cultures can only be obtained from a limited number of genotypes. However, these embryogenic cultures are relatively simple to produce when compared to the effort necessary to produce cultures from which regenerable protoplasts can be obtained. Research is needed to optimize the transformation of cells within primary explants that can give rise directly to somatic embryos. For example, epidermal cells of the scutellum of a number of corn and rice genotypes can give rise to somatic embryos, and plants can be regenerated from these embryos. This approach therefore would broaden the number of genotypes which could be transformed by particle bombardment.

# IV. TRANSFER OF GENES TO PLASTIDS AND MITOCHONDRIA

An important application of particle bombardment is the delivery of genes to cytoplasmic organelles. A major impediment to the delivery of DNA to plastids and mitochondria appears to be the double membrane surrounding these organelles. Microprojectiles can breach this barrier either by directly penetrating or "nicking" these membranes or by delivering DNA to the cytoplasm in a form that is suitable for uptake by the organelles. Regardless of the exact mechanism by which the DNA is delivered, particle bombardment will permit the detailed molecular analysis of the structure and function of plastid and mitochondrial genes. This section will discuss research directed toward the transformation of chloroplasts of alga and higher plants and the mitochondria of *Saccharomyces cerevisiae*. Some of the recent progress in this area has been reviewed.[23,24]

## A. TRANSFORMATION OF CHLOROPLASTS
### 1. Transient Gene Expression Assays
Daniel and colleagues[49] reported the transient expression of chimeric plastid genes following their delivery to suspension culture cells of *Nicotiana tabacum* by particle bombardment. Chimeric reporter genes were obtained by fusing the 5′ end of the *psb*A (pea) or *rbc*L (maize) plastid genes with a CAT coding region. CAT activity was detected after bombardment of the tobacco cells with these constructs. The argument for expression of these genes in plastids was based on the finding that the activity of the reporter gene was maintained over a longer period when a replication origin from the genome of the pea plastid was included in the plasmid vector than when this replication origin was omitted. However, the plastid *psb*A promoter has been shown to direct transcription of several heterologous coding regions following their transfer to the nucleus.[50] Direct evidence for CAT activity in chloroplasts was not presented. Therefore the possibility of nuclear expression of the introduced gene could not be excluded. In a follow-up work,[51] *N. tabacum* suspension culture cells were bombarded with a reporter gene consisting of a GUS coding region under the control of the *psb*A promoter. Expression of the gene in chloroplasts was claimed on the basis of microscopic observation of the cells following their treatment with a substrate that permits histochemical detection of GUS activity. High osmolarity in the medium was considered important for obtaining efficient gene delivery.

### 2. Stable Transformation of the Plastid Genome
The stable transformation of plastids in higher plants was first claimed by DeBlock and co-workers[52] who delivered a CAT gene to tobacco cells by *Agrobacterium*. Although the CAT gene was under the control of a promoter that should only be expressed in the nucleus, the plants regenerated from the chloramphenicol-resistant cell line possessed the foreign gene in plastids as indicated by Southern hybridization analysis and the presence of CAT activity in the chloroplasts. The authors explain their observations by the fortuitous integration of the gene into the plastid genome resulting in its expression from a plastid promoter. However, the line was reportedly unstable and the research has yet to be reproduced.

Reproducible stable transformation of chloroplasts has been obtained in the unicellular alga, *Chlamydomonas reinhardtii*.[53,54] Mutants unable to grow photosynthetically as a result of a deletion in the *atp*B gene were bombarded with plasmid DNA carrying a functional *atp*B gene. Transformed lines were recovered by restoration of photosynthetic competency. These initial studies have been extended to transformation with the *psb*A gene encoding a DCMU-resistance mutation, and with streptomycin and spectinomycin resistance markers carried by the 16S ribosomal RNA genes.[55,56] Study of transgenic plastid genomes confirmed that transformation of one of the repeated regions is quickly followed by correction of the second copy. The mechanism for DNA integration into the plastid genome was by homologous recombination. Foreign DNA could be readily inserted into the genome as long as it was flanked by homologous chloroplast DNA seqeunces.[54] Homologous recombination can be used to replace wild-type *Chlamydomonas* chloroplast genes with mutant genes. This has been demonstrated by cotransformation of *in vitro*-mutated forms of the *atp*B gene with a spectinomycin-resistance allele (K. L. Kindle and D. B. Stern, cited in Reference 24).

Stable transformation of plastids in a higher plant, *Nicotiana tabacum*, was obtained after selection for the plastid 16S rRNA resistance marker following bombardment of leaves with DNA-coated tungsten particles.[57] In the 16S rRNA gene, a spectinomycin resistance mutation is flanked on the 5′ side by a streptomycin resistance mutation, and on the 3′ side by a *Pst* I site generated by ligating an oligonucleotide in the intergenic region. Transgenic cell lines were selected by spectinomycin resistance and plants were regenerated from these cells lines. Stable transformants were distinguished from spontaneous mutants by the flanking, cotransformed streptomycin resistance and *Pst* I markers. In plastid DNA from *Nicotiana*, the 16S rRNA gene is located in the repeated region. Transformation of one of the 16S rRNA genes was followed by correction of the second 16S rRNA gene copy, as had been reported for *Chlamydomonas* plastids. Transformation of *Nicotiana* plastid DNA is about 100-fold less efficient than transformation of the nucleus. This contrasts with the *Chlamydomonas* system in which plastid transformation[55] appears to be at least as efficieint as gene transfer to the nucleus.[58] In *N. tabacum,* the 3,000 to 12,000 copies[59,60] are localized in up to a hundred plastids. This contrasts with the 80 ptDNA copies, carried by a single plastid in *Chlamydomonas*.[61] Spectinomycin can be used in *Nicotiana* as a nonlethal color marker, similar to resistance to streptomycin and lincomycin.[62] Selection for the nonlethal spectinomycin resistance marker was considered important since it allows sufficient time for the resistant plastid genome copies to increase in number and allow phenotypic expression.

## B. TRANSFORMATION OF MITOCHONDRIA

In *Saccharomyces cerevisiae,* standard transformation procedures directed toward nuclear transformation have failed to yield mitochondrial transformants.[24] Microprojectiles, however, can apparently pierce the mitochondrial membrane and deliver DNA directly into the organelle thereby overcoming the major barrier to transformation. Mitochondrial transformants were selected by complementing a respiratory-deficient mutant with the wild-type *oxi*3 gene.[63] This was accomplished by co-bombardment of the *oxi*3 gene with a gene that complimented an auxotrophic mutation causing a requirement for uracil. Nuclear transformants were recovered by their ability to grow on media lacking uracil. Of these nuclear transformants, a small proportion were found to contain the introduced mitochondrial gene. In another set of experiments, yeast cells were bombarded with a mixture of plasmids carrying either a selectable nuclear gene (URA3) or a mitochondrial gene (*oxi*1).[64] The recipient cells were *rho*⁰ and therefore do not possess mitochondrial DNA. When the *oxi*1 gene was introduced, the *rho*⁰ was converted to a *rho*⁻ strain by the concatamerization of the *oxi*1 DNA within the mitochondria. Yeast clones expressing the transgenic URA3 marker were selected after bombardment: among these the synthetic *rho*⁻ strains were identified after marker rescue by recombination with *rho*⁺ mtDNA carrying an appropriate *oxi*1 point mutation, or in *trans* during the growth of diploids heteroplasmic for both the plasmid-derived *oxi*1 sequences and *rho*⁺ mtDNA with *oxi*1 deleted. Plasmid-borne sequences recombine homologously with mtDNA like naturally occurring *rho*⁻ mtDNAs. Mitochondrial transformants were identified by screening nuclear transformants at the frequency of $10^{-3}$. Repopulation of mitochondria with DNA is therefore relatively inefficient and is similar in frequency to complementing the respiratory deficient mutant.

# V. POTENTIAL FOR FUTURE APPLICATIONS

Many of the advances that have been made possible by particle bombardment were thought to be simple pipe dreams when research in this area was initiated. Therefore, it is difficult to predict the full potential of this new technology. However, recent progress should provide a glimpse of some of the new targets at which the particle gun will be aimed. It appears likely that the most immediate impact of particle bombardment will be in the development of crop plants that carry introduced genes of commercial interest.[65] For example, it will be possible to transfer genes that confer resistance to viruses, insects, and herbicides into crops such as corn that have previously been difficult or impossible to transform.

The ability to manipulate the plastid genome by molecular techniques will have an impact on both basic and applied research. The gene encoding the large subunit of Rubisco, a key enzyme of photosynthesis, is a particularly interesting target for these studies. Plants with improved photosynthesis may be obtained by modifying the catalytic properties of this enzyme. The chloroplast genome may also be a target for insertion of genes of proteins that are to be produced in large quantities since the introduced genes, present in each of the plastid genomes, would be greatly amplified. In addition, it will be interesting to see plastid and mitochondrial transformation applied to the improvement of major crop species.

Particle bombardment will be helpful in several areas relating to plant pathology. Viral genomes can be introduced into plant cells by bombardment.[66] Therefore, it may be possible to directly introduce viruses that are normally transmitted to host plants by insect vectors. This capability would greatly streamline studies of many plant viruses by eliminating the need to maintain the insect vector. The recent finding that microprojectiles can be used for the transformation of some bacteria[66a] and plant pathogenic fungi[66b] indicates that the process may be of value for microbes that are difficult to transform by more established procedures. Several important plant pathogens fall into this category, including *Phytophthora* spp. and the oomycete responsible for downy mildew.[67]

Transient assays of DNA introduced into intact tissues by particle bombardment have provided a powerful system for the functional analysis of genes that code for transcription factors and the promoter regions with which these factors potentially interact. The recent findings that genes can be delivered directly into embryos of *Drosophila*[68] and the skin and liver of mice[69] indicate that the *in situ* analysis of gene expression made possible by particle bombardment should be useful in animal systems as well. In addition, there are many types of animal cells maintained in culture that are proving recalcitrant to DNA transfer by existing techniques. Particle bombardment may be useful for gene delivery to such cells as primary liver and neuronal cultures and to cells involved with the immune system. There are already several reports indicating that genes can be delivered to animal cell cultures.[70,71] The use of the particle gun should also expand as a research tool as attempts are made to deliver proteins, RNA, and other macromolecules to plant, animal, and microbial cells to study their biological activity.

# REFERENCES

1. **Klee, H., Horsch, R., and Rogers, S.,** *Agrobacterium*-mediated plant transformation and its further applications to plant biology, *Annu. Rev. Plant Physiol.,* 38, 467, 1987.
2. **Hinchee, M. A. W., Cannor-Ward, D. V., Newell, C. A., McDonnell, R. E., Sato, S. J., Gasser, C. S., Fischoff, D. A., Re, D. B., Fraley, R. T., and Horsch, R. B.,** Production of transgenic soybean plants using *Agrobacterium*-mediated DNA transfer, *Bio/Technology,* 6, 915, 1988.
3. **Grimsley, N. H., Hohn, T., Davies, J. W., and Hohn, B.,** *Agrobacterium*-mediated delivery of infectious maize streak virus into maize plants, *Nature,* 325, 177, 1987.

4. **Raineri, D. M., Bottino, P., Gordon, M. P., and Nester, E.,** *Agrobacterium*-mediated transformation of rice (*Oryza sativa* L.), *Bio/Technology,* 8, 33, 1990.
5. **Potrykus, I.,** Gene transfer to cereals, *Bio/Technology,* 8, 535, 1990.
6. **Klein, T. M., Roth, B. A., and Fromm, M. E.,** Advances in direct gene transfer into cereals, in *Genetic Engineering,* Vol. 11, Setlow, J., Ed., Plenum Press, New York, 1989, 13.
7. **Paszkowaski, J., Saul, M. W., and Potrykus, I.,** Plant gene vectors and genetic transformation: DNA-mediated direct gene transfer, in *Cell Culture and Somatic Cell Genetics,* Vol. 6, Schell, J., and Vasil, I. K., Eds., Academic Press, San Diego, 1989, 52.
8. **Davey, M. R., Rech, E. L., and Mulligan, B. I.,** Direct DNA transfer to plant cells, *Plant Mol. Biol.,* 13, 273, 1989.
9. **Datta, S. K., Peterhans, A., Datta, K., and Potrykus, I.,** Genetically engineered fertile indica-rice recovered from protoplasts, *Bio/Technology,* 8, 736, 1990.
10. **Shimamoto, K., Terada, R., Izawa, T., and Fujimoto, H.,** Fertile rice plants regenerated from trans-formed protoplasts, *Nature,* 338, 274, 1989.
11. **Toriyama, K., Arimoto, Y., Uchimiya, H., and Hinata, K.,** 1988 Transgenic rice plants after direct gene transfer to protoplasts, *Bio/Technology,* 6, 1072, 1988.
12. **Rhodes, C. A., Pierce, D. A., Mettler, I. J., Mascarenhas, D., and Detmer, J. J.,** Genetically transformed maize plants from protoplasts, *Science,* 240, 204, 1988.
13. **Neuhaus, G., Spangenberg, G., Mittelsten,-Scheid, O., and Schweiger, H. G.,** Transgenic rapeseed plants obtained by microinjection of DNA into microspore derived proembryos, *Theor. Appl. Genet.,* 75, 30, 1987.
14. **Dekeyser, R. A., Claes, B., De Rycke, R. M. U., Habets, M., Van Montagu, M. C., and Caplan, A. B.,** Transient gene expression in intact and organized rice tissue, *Plant Cell,* 2, 591, 1990.
15. **Weber, G., Monajembashi, S., Greulich, K. O., and Wolfrum, J.,** Microperforation of plant tissue with a UV laser beam and injection of DNA into cells, *Naturwissenschaften,* 75, 35, 1988.
16. **Kaeppler, H. F., Gu, W., Somers, D. A., Rines, H. W., and Cockburn, A. F.,** Silica carbide fiber-mediated DNA delivery into plant cells, *Plant Cell Rep.,* 9, 415, 1990.
17. **Klein, T. M., Wolf, E. D., Wu, R., and Sanford, J. C.,** High-velocity microprojectiles for delivery of nucleic acids into living cells, *Nature,* 327, 70, 1987.
18. **Sanford, J. C., Klein, T. M., Wolf, E. D., and Allen, N.,** Delivery of substances into cells and tissues using a particle bombardment process, *J. Part. Sci. Technol.,* 5, 27, 1987.
19. **Fromm, M. E., Morrish, F., Armstrong, C., Williams, R., Thomas, J., and Klein, T. M.,** Inheritance and expression of chimeric genes in the progeny of transgenic maize, *Bio/Technology,* 8, 833, 1990.
20. **Gordon-Kamm, W. J., Spencer, T. M., Mangano, M. L., Adams, T. R., Daines, R. J., Start, W. G., O'Brien, J. V., Chambers, S. A., Adams, W. R., Willetts, N. G., Rice, T. B., Mackey, C. J., Krueger, R. W., Kausch, A. P., and Lemaux, P. G.,** Transformation of maize cells and regeneration of fertile transgenic plants, *Plant Cell,* 2, 603, 1990.
21. **McCabe, D. E., Swain, W. F., Marinell, B. J., Christou, P.,** Stable transformation of soybean (*Glycine max*) by particle bombardment, *Bio/Technology,* 6, 923, 1988.
22. **Klein, T. M., Goff, S. A., Roth, B. A., and Fromm, M. E.,** Applications of the particle gun in plant biology, in *Progress in Plant Cellular and Molecular Biology,* Nijkamp, H. J. J., van der Plas, L. H. W., and van Aartrijk, J., Eds., Kluwer Academic Publishers, Dordrecht, 1990, 56.
23. **Boynton, J. E., Gillham, N. W., Harris, E. H., Newman, S. M., Randolph-Anderson, B. L., Johnson, A. M., and Jones, A. R.,** Manipulating the chloroplast genome of *Chlamydomonas:* molecular genetics and transformation, in *Current Research in Photosynthesis,* Vol. 3, Baltscheffsky, M., Ed., Kluwer Academic Publishers, Dordrecht, The Netherlands, 1990, 509.
24. **Butow, R. A. and Fox, T. D.,** Organelle transformation: shoot first; ask questions later, *Trends in Biochemical Sciences,* 15, 465, 1990.
25. **Klein, T. M., Gradziel, T., Fromm, M. E., and Sanford, J. C.,** Factors influencing gene delivery into Zea mays cells by high-velocity microprojectiles, *Bio/Technology,* 6, 559, 1988.
26. **Zumbrunn, G., Schneider, M., and Rochaix, J.-D.,** A simple particle gun for DNA-mediated cell transformation, *Technique,* 1, 204, 1989.
27. **Christou, P., McCabe, D. E., and Swain, W. F.,** Stable transformation of soybean callus by DNA-coated gold particles, *Plant Physiol.,* 87, 671, 1988.
28. **Oard, J. H., Paige, D. F., Simmonds, J. A., and Gradziel, T. M.,** Transient gene expression in maize rice and wheat cells using an airgun apparatus, *Plant Physiol.,* 92, 334, 1990.
29. **Morikawa, H., Iida, A., and Yamada, Y.,** Transient expression of foreign genes in plant cells and tissues obtained by a simple biolistic device (particle gun), *Appl. Microbiol. Biotechnol.,* 31, 320, 1989.
30. **Johnston, S. A.,** Biolistic transformation: microbes to mice, *Nature,* 346, 776, 1990.
30a. **Lewis, P. and Arentzen, R.,** personal communication.
31. **Twell, D., Klein, T. M., Fromm, M. E., and McCormick, S.,** Transient expression of chimeric genes delivered into pollen by microprojectile bombardment, *Plant Physiol.,* 91, 1270, 1989.

32. **Klein, T. M., Harper, E. C., Svab, Z., Sanford, J. C., Fromm, M. E., and Maliga, P.,** Stable genetic transformation of intact *Nicotiana* cells by the particle bombardment process, *Proc. Natl. Acad. Sci. U.S.A.,* 85, 8502, 1988.

33. **Ludwig, S. R., Bowen, B., Beach, L., and Wessler, S. R.,** Regulatory gene as a novel visible marker for maize transformation, *Science,* 247, 449, 1990.

34. **Klein, T. M., Roth, B. A., and Fromm, M. E.,** Regulation of anthocyanin biosynthetic genes introduced into intact maize tissues by microprojectiles, *Proc. Natl. Acad. Sci. U.S.A.,* 86, 6681, 1989.

35. **Londsdale, D., Sertaç, Ö, and Cuming, A.,** Transient expression of exogenous DNA in intact, viable wheat embryos following particle bombardment, *J. Exp. Bot.,* 41, 1161, 1990.

36. **Bruce, W. B., Christensen, A. H., Klein, T. M., Fromm, M. E., and Quail, P. H.,** Photoregulation of a phytochrome gene promoter from oat transferred into rice by particle bombardment, *Proc. Natl. Acad. Sci. U.S.A.,* 86, 9692, 1989.

37. **Roth, B. A., Goff, S. A., Klein, T. M., and Fromm, M. E.,** *C1* and *R* dependent expression of the maize *Bz1* gene requires sequences with homology to mammalian *myb* and *myc* binding sites, *The Plant Cell,* in press.

38. **Goff, S. A., Klein, T. M., Roth, B. A., Fromm, M. E., Cone, K. C., Radicella, J. P., and Chandler, V. L.,** Transactivation of anthocyanin biosynthetic genes following transfer of *B* regulatory genes into maize tissues, *EMBO J.,* 9, 2517, 1990.

39. **Ludwig, S. R. and Wessler, S. R.,** Maize *R* gene family: tissue-specific helix-loop-helix proteins, *Cell,* 62, 849, 1990.

39a. **Tomes, D., Ross, M., and Bowen, B.,** personal communication.

40. **Tomes, D. T., Weissinger, A. K., Ross, M., Higgins, R., Drummond, B. J., Schaaf, S., Malone-Schoneberg, J., Staebell, M., Flynn, P., Anderson, J., and Howard, J.,** Transgenic tobacco plants and their progeny derived by microprojectile bombardment of tobacco leaves, *Plant Mol. Biol.,* 14, 261, 1990.

41. **Timmermans, M. C. P., Maliga, P., Vieira, J., and Messing, J.,** The pFF plasmids: cassettes utilizing CaMV sequences for expression of foreign genes in plants, *J. Biotech.,* 14, 333, 1990.

42. **Iida, A., Morikawa, H., and Yamada, Y.,** Stable transformation of cultured tobacco cells by DNA-coated gold particles accelerated by gas pressure-driven particle gun, *Appl. Microbiol. Biotech.,* 33, 5, 1990.

43. **McMullen, M. D. and Finer, J.,** Stable transformation of cotton and soybean embryogenic suspension cultures via microprojectile bombardment, *J. Cell. Biol. UCLA Symposium on Molecular and Cellular Biology, Abstracts,* 14E, 285, 1990.

43a. **Finer, J.,** personal communication.

44. **Klein, T. M., Kornstein, L., Sanford, J. C., and Fromm, M. E.,** Genetic transformation of maize cells by particle bombardment, *Plant Physiol.,* 91, 440, 1989.

45. **Spencer, T. M., Gordon-Kamm, W. J., Daines, R. J., Start, W. G., and Lemaux, P. G.,** Bialaphos selection of stable transformants from maize cell culture, *Theor. Appl. Genet.,* 79, 625, 1990.

46. **Tomes, D. T.,** Transformation in corn: nonsexual gene transfer, in 26th Illinois Corn Breeders School, University of Illinois at Urbana, Dudley, J., Ed., 1990, 7.

47. **Finer, J. J. and McMullen,** Transformation of cotton (*Gossypium hirsutum* L.) via particle bombardment, *Plant Cell Rep.,* 8, 586, 1990.

48. **Fitch, M. M., Manshardt, R. M., Gonsalves, D., Slightom, J. L., and Sanford, J. C.,** Stable transformation of papaya via microprojectile bombardment, *Plant Cell Rep.,* 9, 189, 1990.

49. **Daniell, H., Vivekananda, J., Nielsen, B. L., Ye, G. N., Tewari, K. K., and Sanford, J. C.,** Transient foreign gene expression in chloroplasts of cultured tobacco cells after biolistic delivery of chloroplast vectors, *Proc. Natl. Acad. Sci. U.S.A.,* 87, 88, 1990.

50. **Cornelissen, M., Vandewiele, M.,** Nuclear transcriptional activity of the tobacco plastid *psb*A promoter, *Nucleic Acids Res.,* 17, 19, 1989.

51. **Ye, G.-N., Daniell, H., and Sanford, J. C.,** GUS as a marker for higher plant chloroplast transformation, *Plant Mol. Biol.,* 15, 809, 1990.

52. **De Block, M., Schell, J., and Van Montagu, M.,** Chloroplast transformation by *Agrobacterium tumefaciens, EMBO J.,* 4, 1367, 1985.

53. **Boynton, J. E., Gillham, N. W., Harris, E. H., Hosler, J. P., Johnson, A. M., Jones, A. R., Randolph-Anderson, B. L., Robertson, D., Klein, T. M., Shark, K. B., and Sanford, J. C.,** Chloroplast transformation in *Chlamodomonas* with high-velocity microprojectiles, *Science,* 240, 1534, 1988.

54. **Blowers, A. D., Bogorad, L., Shark, K. B., Sanford, J. C.,** Studies on *Chlamydomonas* chloroplast transformation: foreign DNA can be stably maintained in the chromosome, *Plant Cell,* 1, 123, 1989.

55. **Boynton, J. E., Gillham, N. W., Harris, E. H., Newman, S. M., Randolph-Anderson, B. L., Johnson, A. M., and Jones, A. R.,** Manipulating the chloroplast genome of *Chlamydomonas*: molecular genetics and transformation, in *Current Research in Photosynthesis,* Vol. 3, Baltscheffsky, M., Ed., Kluwer Academic Publishers, Dordrecht, The Netherlands, 1990, 509.

56. **Newman, S. M., Boynton, J. E., Gillham, N. W., Randolph-Anderson, B. L., Johnson, A. M., and Harris, E. H.,** Transformation of chloroplast ribosomal RNA genes in *Chlamydomonas:* molecular and genetic characterization of integration events, *Genetics,* 126, 875, 1990.

57. **Svab, Z., Hajdukiewitz, P., and Maliga, P.,** Stable transformation of plastids in higher plants, *Proc. Natl. Acad. Sci. U.S.A.,* 87, 8526, 1990.

58. **Kindle, K. L., Schnell, R. A., Fernandez, E., and Lefebvre, P. A.,** Stable nuclear transformation of *Chlamydomonas* using the *Chlamydomonas* gene for nitrate reductase, *J. Cell Biol.,* 109, 2589, 1989.

59. **Cannon, G., Heinhorst, S., Siedlecki, J., and Weissbach, A.,** Chloroplast DNA synthesis in light and dark grown cultured *Nicotiana tabacum* cells as determined by molecular hybridization, *Plant Cell Rep.,* 4, 41, 1985.

60. **Yasuda, T., Kuroiwa, T., Nagata, T.,** Preferential synthesis of plastid DNA and increased replication of plastids in cultured tobacco cells following medium renewal, *Planta,* 174, 235, 1988.

61. **Harris, E. H.,** *The Chlamydomonas Sourcebook,* Academic Press, San Diego, 1989, 354.

62. **Moll, B., Polsby, L., and Maliga, P.,** Streptomycin and lincomycin resistances are selective plastid markers in cultured *Nicotiana* cells, *Mol. Gen. Genet.,* 221, 245, 1990.

63. **Johnston, S. A., Anziano, P. Q., Shark, K., Sanford, J. C., and Butow, R. A.,** Mitochondrial transformation in yeast by bombardment with microprojectiles, *Science,* 240, 1538, 1988.

64. **Fox, T. D., Sanford, J. C., and McMullin, T. W.,** Plasmids can stably transform yeast mitochondria lacking endogenous mtDNA, *Proc. Natl. Acad. Sci. U.S.A.,* 85, 7288, 1988.

65. **Gaser, C. S. and Fraley, R. T.,** Genetically engineering plants for crop improvement, *Science,* 224, 1293, 1989.

66. **Creissen, G., Smith, C., Francis, R., Reynolds, H., and Mullineaux, P.,** *Agrobacterium*—and micro-projectile—mediated viral DNA delivery into barley microspore-derived cultures, *Plant Cell Rep.,* 8, 680, 1990.

66a. **Sanford, J.,** personal communication.

66b. **Chumley, F., Ho, M., and Klein, T.,** unpublished results.

67. **Wang, J. and Leong, S. A.,** DNA-mediated transformation of phytopathogenic fungi, in *Genetic Engineering,* Vol. 11, Setlow, J., Ed., Plenum Press, New York, 1989, 127.

68. **Baldarelli, R. M. and Lengyel, J. A.,** Transient expression of DNA after ballistic introduction into *Drosophila* embryos, *Nucleic Acids Res.,* 18, 5903, 1990.

69. **Williams, R. S., Johnston, S. A., Riedy, M., DeVit, M. J., McElligott, S. G., and Sanford, J. C.,** Introduction of transgenes into tissues of living mice by DNA-coated microprojectiles, *Proc. Natl. Acad. Sci. U.S.A.,* in press.

70. **Zelenin, A. V., Titomirov, A. V., and Kolesnikov, V. A.,** Genetic transformation of mouse cultured cells with the help of high-velocity mechanical DNA injection, *FEBS Lett.,* 244, 65, 1989.

71. **Fredericksen, T. L., Rowe, D. G., Armaleo, D., Hebrank, J., Thaxton, J. P., and Ricks, C. A.,** Gene delivery into chick cells via biolistic process, *Poultr. Sci.,* 69 (Suppl. 1), 53, 1990.

Chapter 9

# MONOCLONAL ANTIBODIES FOR DETECTION OF RICE VIRUSES: GRASSY STUNT, STRIPE, DWARF, GALL DWARF, AND RAGGED STUNT

## Pepito Q. Cabauatan and Hiroyuki Hibino

## TABLE OF CONTENTS

# I. INTRODUCTION

Virus diseases are among the major constraints in rice production, particularly in Asia where 11 out of 15 known rice viruses occur.[1-3] All of the 11 viruses except rice necrosis mosaic virus are insect-borne. The damage by virus diseases and often by virus and their vectors combined causes immeasurable losses in rice crop production every year. For the past several decades, major outbreaks of rice virus diseases have occurred in several Asian countries.[1-3]

Development of rapid and accurate diagnosis of the virus agents is essential to establish proper control measures of the virus diseases. At present, rice virus diseases in Asia are diagnosed mainly on the basis of symptoms caused by the diseases and on their relations to their specific vectors. However, virus identification by symptoms in rice is not always conclusive. Some viruses cause similar symptoms, some are symptomless or do not show definite symptoms, and some show different symptoms on different rice cultivars.[1-3] Besides, genetic and physiological disorders also cause abnormalities similar to the virus-induced symptoms. On the other hand, diagnosis by virus transmission test using specific vectors are laborious and difficult to handle many samples, and require elaborate greenhouse facilities and continuous supply of the vectors.

Serology provides a more reliable, sensitive, and rapid diagnosis of rice viruses. Serology can also be applied in studying virus epidemiology and evaluating rice cultivars for resistance to viruses, which generally involve a large number of samples to be tested. In the early 1960s, rabbit polyclonal antisera (PCA) were produced for rice dwarf virus (RDV)[4] and rice stripe virus (RSV),[5] and then in the early 1970s for rice black-streaked dwarf virus.[6] So far, antisera to all other rice viruses except rice bunchy stunt virus have been produced, and reliable serological techniques have been developed to detect rice viruses.[7-14] However, large-scale application of serology to rice virus diagnosis has been limited by the availability of antisera. Good quality PCA to rice viruses are generally difficult to obtain. Several rice viruses are phloem-limited and their concentrations in infected plants are low. Hence, purification of these viruses is difficult and virus yield after the purification is often insufficient for immunization. Use of antisera is especially advantageous in testing routinely a large number of plants and insect vectors for viruses. Serology has become a common practice in rice virus researches, though its intensive use so far has been only in the disease forecasting programs for RSV,[12,15] and in epidemiology and resistance studies for tungro virus disease.[16-18] Production of monoclonal antibodies (MCA) to rice viruses seems to provide a good alternative.

Serology in virus research has made remarkable progress since Kohler and Milstein[19] introduced hybridoma technology. This technology not only revolutionized the method of antibody production but also opened new areas in virus research. Halk et al.[20] enumerated several advantages of MCA over PCA, which apply very well to rice viruses. The advantages are

1.  A small quantity of antigen is required to stimulate an immune response in mice.
2.  The immunogen need not be highly purified to obtain antibodies specific to a single determinant.
3.  Hybridomas can be preserved by freezing in liquid nitrogen, thereby assuring a continuous supply of antibody.
4.  Highly specific MCAs may reveal serological relationships previously unrecognized with PCA.
5.  Qualitative and quantitative variabilities often encountered with PCA are eliminated with the use of MCA.

Rapid progress has been achieved also in plant virology ever since the hybridoma technology was applied to plant viruses.[20,21] However, application of hybridoma technology to rice viruses lags behind other plant viruses. By 1985, MCAs to about 30 plant viruses had been produced, and of this number, only one MCA was directed against rice dwarf virus-nucleic acid.[23] To date, MCAs against five rice viruses have been reported,[22-30] and much information about them has yet to be published. This chapter reviews the recent progress made on the application of hybridoma technology to these rice viruses and possible areas of application of MCA to solve virus disease problems in Asia. As additional information, this review also includes a brief description of the five rice viruses on which MCAs have been generated.

## II. RICE VIRUSES WITH MONOCLONAL ANTIBODIES GENERATED

### A. PROPERTIES

#### 1. Rice Grassy Stunt Virus (RGSV)

RGSV[31] is a member of the Tenuiviruses. RGSV-infected plants are severely stunted and tiller profusely.[32] Leaves of diseased plants are short, narrow, erect, stiff, and pale green. They often have numerous small, dark-brown spots of various shapes. Rice plants infected at an early stage of the growth produce no yield. RGSV strains which cause more severe symptoms including leaf yellowing have been reported in India, Indonesia, the Philippines, Thailand, and Taiwan.[31,33,34]

RGSV[31] is transmitted in a persistent manner by the brown planthopper, *Nilaparvata lugens,* and other *Nilaparvata* spp. RGSV multiplies in planthoppers but is not transmitted via the eggs. RGSV particles are filamentous, 6 to 8 nm in width with a modal contour length of 950 to 1850 nm.[35] RGSV is composed of four single-stranded (ss) RNAs and one major and one minor protein.[31,35,36]

#### 2. Rice Stripe Virus (RSV)

RSV[37] is another member of the Tenuiviruses. RSV develops characteristic chlorotic stripes on rice leaves.[37,38] Other diagnostic symptoms include mottling, drooping, and unfolding of leaves. Infected plants lack vigor, show general chlorosis, and produce few poor panicles. The disease may cause premature death if plants are infected at an early stage of growth.

The major vector of RSV in rice fields is the smaller brown planthopper, *Laodelphax striatellus.*[37,38] The virus persists in the vector and is transmitted from infective females to their progeny via the eggs.

RSV particles are circular filaments with a helical configuration about 8 nm in width.[35,37,39-41] RSV has four components, each of which is associated with one each ss and double-stranded (ds) RNA.[40] The fifth component (smallest one), which does not form a distinct band after density gradient centrifugation of purified virus, is associated with a ss-RNA.[41] RSV has one major protein.[42] RSV-infected plants produce a large amount of nonstructural protein which is serologically unrelated to the coat protein.[37,42]

#### 3. Rice Dwarf Virus (RDV)

RDV[43] is a Phytoreovirus. RDV-infected plants develop numerous white specks on leaves[38,43] which may coalesce to form continuous streaks along the veins. Infected plants are severely stunted and develop numerous diminutive tillers with darker green leaves.

RDV is transmitted in a persistent manner by leafhoppers, *Nephotettix cincticeps, N. nigropictus, N. virescens,* and *Recilia dorsalis.*[38,43] RDV multiplies in the leafhoppers and is transmitted via the eggs. RDV particles are icosahedral, about 70 nm in diameter.[43,44] RDV contains 12 ds-RNA and 7 major proteins.[45,46]

#### 4. Rice Gall Dwarf Virus (RGDV)

RGDV[47] is another member of the Phytoreoviruses. Gall dwarf symptoms include small whitish galls on the underside of the leaf blades and outer side of leaf sheaths, severe stunting, and dark-green foliage.[47,48] Infected plants produce poor panicles which do not bear grains. RGDV particles are limited in the phloem and gall tissues in infected plants.[47]

RGDV is transmitted in a persistent manner by leafhoppers, *N. nigropictus, N. virescens, N. cincticeps, N. malayanus,* and *R. dorsalis.* RGDV multiplies in the leafhoppers and is transmitted via the eggs. RGDV particles are polyhedral, about 65 nm in diameter.[49] RGDV consists of 12 ds-RNA and 7 major proteins.[50,51]

#### 5. Rice Ragged Stunt Virus (RRSV)

RRSV[52] is a possible Fijivirus. Diagnostic symptoms of RRSV are stunting of plants, twisting of leaf tips, ragged leaves, and vein swelling or galls on the underside of leaf blades and leaf sheaths.[53,54] Infected plants produce poor panicles which bear discolored unfilled grains. RRSV particles are limited in the phloem and gall tissues in infected plants.[52,55]

The major vector of RRSV is the brown planthopper, *N. lugens.*[52-54] RRSV persists in the planthopper but without transovarial passage. RRSV particles are isometric, 63 to 65 nm in diameter.[52,55,56] RRSV consists of 10 ds-RNA[56,57] and 5 proteins.[58]

### B. PURIFICATION AND SEROLOGY

The first step in the production of MCA against a plant virus is the preparation of a suitable antigen for immunization. As in PCA production, it is advisable to immunize with highly purified virus; then a high proportion of virus-specific antibody-secreting hybridoma can be expected, and screening of these hybridoma will be easier.[21] Occasionally, nonmouse PCAs are used in ELISA for screening of MCA-secreting hybridoma. Basically, most serological techniques developed for PCAs are applicable in detecting rice viruses using MCAs.

#### 1. RGSV

Several procedures have been employed to purify RGSV.[31,34,35,59] As RGSV particles are similar to RSV in morphology,[35] purification procedures recommended[31] for RGSV basically follow those developed for RSV.[39]

PCAs with titers of 1/200 to 1/2000 in the ring interface precipitin tests have been obtained from rabbits immunized with purified virus preparations.[34,35,59] The latex test detected RGSV in infected leaf and insect vector extracts at 1/5000 and 1 to 2/1000 dilutions, respectively.[10,35] In enzyme-linked immunosorbent assay (ELISA), RGSV was detected in leaf and vector extracts at 1/100,000 and 1/5000 dilutions.[34,35,59] RGSV is serologically distantly related to RSV.[35]

#### 2. RSV

Purification methods employed[26,37,40] basically followed the procedures described by Koganezawa et al.[39] Rabbit PCAs with titers of 1/500 to 1/2000 have been obtained.[5,7,10,12,15,60] RSV was detected in infected leaf extracts at 1/10,000 dilution in latex test, 1/6000 to 1/100,000 dilutions in ELISA, and 1/20,000 dilution in the hemagglutination test. The three tests also detect RSV in individual virus carrier insects. The dilution end point of insect extracts for the detection is 1/10,000 to 1/20,000 in the hemagglutination test.

#### 3. RDV

Since Fukushi et al.[44] isolated RDV, extensive efforts have been devoted for the development of better purification precedures.[4,61-63] The improved procedures have been used to purify RDV from infected rice plants.[64-67]

Rabbit PCA with titers of 1/4000 to 1/10,000 have been obtained.[4,64,65,67] The latex test detects RDV in infected plant and insect vector extracts at 1/5000 and 1/100 dilutions, respectively. ELISA detects RDV in plant extracts at 1/30,000 dilution. An antiserum to SDS-dissociated RDV was found to be reactive to one of the RGDV structural proteins.[68] Antiserum to RDV-RNA reacted also to synthetic ds-RNAs, poly(A) poly(U), and poly(I) poly(C).[69]

## 4. RGDV

RGDV was purified basically following the procedures developed for RDV.[49] Rabbit PCA with a titer of 1/2000 in double gel diffusion test has been obtained.[49] RGDV is detected in infected leaf extracts at 1/200 dilution in the latex test and 1/5000 dilution in ELISA.[10] The latex test also detects RGDV in individual virus carrier insects.[10] The antiserum also reacts to ds-RNAs of RGDV and RDV and poly(I) poly(C).[49]

## 5. RRSV

RRSV was also purified basically following the procedures developed for RDV and RGDV.[8,52,56,57,70,71] Rabbit PCAs with titers of 1/300 to 1/1300 have been obtained.[8,70,71] ELISA detects RRSV in infected plant and insect vector extracts at 1/300 and 1/5000 dilutions, respectively,[8] but the latex test is not applicable to detect the virus because of its low sensitivity.[10]

## C. PRODUCTION OF MONOCLONAL ANTIBODIES

Review articles that provide detailed procedures for MCA production against plant viruses have been published.[20,21,27] For rice viruses, MCA production was generally patterned after those that were applied to other plant viruses, with only minor modifications.[26,27] In many cases, little information is available on properties of antibodies obtained.

## 1. RGSV

Hsu et al.[25] obtained hybridoma cell lines secreting immunoglobulin (IgG)-G antibodies reactive to RGSV following the standard procedures.[72] BALB/c mice were separately immunized with two strains, RGSV1 and RGSV2,[34] which were purified from infected rice plants. After immunization, mice were sacrificed, spleens were harvested, and cell suspensions were prepared. Fusion of spleen cells with myeloma cells (P3/WS1/1-AG4-1) was mediated with 45% PEG (4000 mol wt) at 37°C. Fused cells were grown initially in hypoxanthine-aminopterine-thymidine (HAT) selective medium, then in HT medium. Indirect ELISA was used to screen hybridoma cell lines secreting RGSV specific antibodies. Thirty-five hybridoma cultures positive for RGSV were initially selected. After further selection and limiting dilution cloning, single stable hybridoma cell line was obtained. The MCA was not strain specific.

## 2. RSV

High consumption of antiserum for use in the disease-forecasting program for RSV prompted Omura et al.[26] to develop MCA specific to RSV. For immunization, they used RSV purified from artificially infected maize plants. BALB/c mice were immunized first by intraperitoneal injections followed by 2 i.v. injections 3 and 7 weeks after the first injection. Spleen cells were harvested 3 d after the last injection. Fusion of spleen cells with myeloma cells (P3-X63-Ag8-U1) was performed by gently adding 50% PEG 4000. Indirect ELISA using the avidine-biotin system was used for screening hybridoma cell lines secreting RSV specific antibodies. A single stable cell line with the highest antibody titer was selected after limiting dilution cloning. Hybridoma cells were injected into mice for ascites induction.

A single injected mouse accumulated ascitic fluid of 11 to 23 ml, with precipitin ring interface titers of 1/800 to 1/1600.

### 3. RDV

MCAs specific against intact virus particles and ds-RNA of RDV have been produced.[22,23,28,30] In each case, the procedures for immunization, cell fusion, screening, selection, and cloning of MCA-secreting hybridomas were basically similar to those described for RGSV and RSV.[25,26] Kitagawa et al.[22] obtained 7 hybridoma cell lines secreting MCAs to RDV. Specificity of MCAs was compared, using the immunoblotting method. One MCA reacted to four RDV-proteins, and two others reacted to two proteins each. By affinity chromatography using the MCA reactive to two proteins, the two proteins were isolated after dissociation of purified RDV particles.

Matsumoto and Kitagawa[23] obtained a single hybridoma cell line secreting MCA, after immunizing mice with RDV-RNA combined with methylated-bovine serum albumin. Specificity of MCA was analyzed by ELISA, the passive hemagglutination test, and the agar gel diffusion test. MCA reacted with RDV ds-RNA and synthetic ds-RNAs, poly(A) poly(U) and poly(I) poly(C), but not with ss-RNA, ss- and ds-DNA, and synthetic ss-RNA. In the agar gel diffusion test, a single clear reaction band was formed between MCA and RDV-RNA.

Fan et al.[28] obtained 41 hybridoma cell lines secreting MCAs to RDV, 8 of which had high titers. Specificity of the 8 MCAs to 5 major RDV-proteins was determined. Three MCAs reacted to all proteins, two each reacted to two different proteins, and one reacted to four proteins. This indicated that each protein had at least two or three epitopes, some of which were common among proteins.

Harjosudarmo et al.[30] compared four ELISA methods in screening MCA-secreting hybridoma cell lines. Of cell lines obtained, 70% gave positive ELISA to RDV in the four methods. In direct double-antibody-sandwich (DAS) ELISA, three MCAs efficiently detected RDV in leaf extracts. One MCA had an especially high titer.

### 4. RGDV

Several hybridoma cell lines secreting MCAs specific to RGDV were established by Nozu et al.[24,27] In ELISA, these MCAs had titers $10^{-6}$ to $10^{-8}$ against purified RGDV. Specificity of MCAs was tested by the immunoblotting method after gel electrophoresis of dissociated RGDV. All the MCAs recognized a single protein (45K) which was located on the surface of the virus particles, although mouse PCA to RGDV recognized all of the seven RGDV proteins. The failure in obtaining MCA specific to proteins that are located inside RGDV particles was probably due to the screening method (direct ELISA) used wherein intact RGDV particles were bound to the plates. If dissociated RGDVs were used in the screening instead, MCAs to other proteins might be also obtained.

### 5. RRSV

Harjosudarmo et al.[29] obtained three stable hybridoma clones secreting MCAs specific to RRSV. In direct DAS-ELISA, one MCA used as the first and second antibodies detected RRSV, but other two MCAs used failed to detect the virus. However, all three MCAs reacted to RRSV in indirect ELISA, using plates treated with purified RRSV at pH 9.6, and indirect DAS-ELISA using PCA as the first antibody and MCA as the second antibody.

## III. APPLICATION OF MONOCLONAL ANTIBODIES IN RICE VIRUS RESEARCH

A great interest in MCAs to date has been in their use to detect rice viruses either in plant or insect vector tissue extracts. For practical application of MCAs in rice virus researches, development of reliable and simple detection methods is essential. Virus epidemiology and resistance researches generally consume large quantities of antisera for routinely testing thousands of plant or insect samples. With the advent of hybridoma technology, a virtually unlimited supply of highly specific antibodies is assured for these purposes.

## A. DETECTION METHODS
### 1. ELISA

MCAs have been used in ELISA for the detection of RGSV, RSV, RDV, RGDV, and RRSV in suspension or infected tissue extracts.[22-30] For ELISA, IG is purified from the mouse ascitic fluid, and used for coating ELISA plates, and for its conjugation either with alkaline phosphatase or peroxidase.[73] Direct DAS-ELISA[73] is generally used for the detection of plant viruses. Optimum concentration for coating IgG and dilution of IgG-enzyme conjugate were 1 μg/ml and 1/1000, respectively, for a MCA to RGSV.[25] They were 1 μg/ml and 1/800 for a MCA to RSV.[26] The whole procedures of direct DAS-ELISA takes 2 d. By simultaneous incubation of samples and the conjugate with plates and shortening of the incubation time, one cycle of procedures can be completed in 2 h without major reduction in its detection efficiency.[12]

For indirect DAS-ELISA, the plate is generally coated by nonmouse PCA-IgG. After trapping the virus antigen in samples on the plate, it is treated with MCA-IgG, and then enzyme antimouse IgG conjugate. In indirect DAS-ELISA, the immunosorbent reaction is amplified and so its detection efficiency is generally higher than that of direct DAS-ELISA. Nonspecific reaction in indirect DAS-ELISA can be also higher. In indirect ELISA, an antimouse IgG-enzyme conjugate can be used for the detection of many viruses by using IgG specific to each virus as the first and second antibodies. This advantage is especially helpful in disease diagnosis where diseased plant samples are to be tested for several possible viruses. MCAs may fail to detect the virus antigen in direct DAS-ELISA using MCAs as the first and second antibodies.[29] Indirect DAS-ELISA using the same MCAs as the second antibody may detect the corresponding antigen.

### 2. Latex Test

The latex (agglutination or flocculation) test[10,74] is simple and economical. It is reliable and sensitive enough to detect some of the rice viruses in tissue extracts.[10] MCA has been used in the latex test for detecting RGSV in plant extracts,[25] and RSV in plant and insect extracts. Latex particles are sensitized with the mouse ascitic fluid at 1/1000 dilution for RGSV and at 1/800 for RSV. RGSV is detected in infected plant saps at 1/1000 dilution, while RSV is detected in saps at 1/10,000 dilution. The latex test detects RSV in individual insect vectors, but for RGSV, the detection is erratic in its vector *N. lugens,* because the extracts at lower dilution interfere with the agglutination of latex particles. The latex test is especially useful for the diagnosis of RGSV and RSV. The test takes 20 to 60 min. It is sensitive enough to detect the antigens in 20 to 25 μl of infected leaf extracts at 1/10 to 1/100 dilution. For the detection of RGSV, a 15-ml preparation of sensitized latex suspension can test 700 to 750 samples in just about 4 h, if the appropriate homogenizer is available. Sensitized latex suspension can be stored several months in the refrigerator.

### 3. Dot-Blot Immunoassay (DBI)

DBI also permits rapid identification of viruses in small amounts of leaf extracts. In DBI, MCA to RGSV effectively detects the virus in as little as 2.5 μl of infected leaf extracts at 1/1000 dilution.[25,75] When horseradish peroxidase was used in DBI, strong positive reactions were obtained in both infected and healthy extracts, whereas, alkaline phosphatase gave no background reactions in healthy controls. When alkaline phosphatase was used, MCAs gave distinct positive and negative reactions for infected and healthy samples, respectively, while with PCAs, both samples showed intense color reactions.

#### 4. Hemagglutination Test

The hemagglutination test[7] has been widely adopted in the RSV disease forecasting programs for the detection of RSV in insect vectors in overwintering populations.[5,15] As shown for PCAs, MCA detects RSV in plant and insect vector extracts with about equal sensitivity both in ELISA and the latex test.

#### 5. Others

MCAs against RDV-RNA have been used to detect RDV in the agar gel double diffusion test.[23] A single clear reaction band was formed between MCA and RDV-RNA.

### B. DIAGNOSIS AND DISEASE SURVEY

Disease diagnosis is essential for establishing proper control measures. As described above, diagnosis of rice virus diseases based on symptoms is not always conclusive. This is particularly true for the tungro disease, which is the most destructive virus disease for rice in Asia.[1-3] Tungro is a composite disease associated with two viruses, one of which does not develop clear symptoms and serves as a helper for the transmission of the other virus by leafhopper vectors.[3,76] Although rabbit PCAs to most rice viruses have been produced, the use of antisera is still limited, and false diagnosis of rice virus diseases is common in most Asian countries. Serology provides a reliable diagnosis of virus diseases. Serology also provides a precise analysis of virus disease development in the fields and an accurate virus disease survey. MCA production provides a large quantity of specific antibody which avails disease survey programs to cover a large area. MCA recognize a single determinant (epitope) on the antigen, while PCA is a mixture of antibodies, each of which has its own recognition site. Because of high specificities, MCA against the type strain of a virus may fail to detect other virus strains of the same virus in infected plant saps.[20,21,77] However, MCA specific to a virus strain provides a tool for precise diagnosis of the strain. For general survey, use of MCAs which react with a wide range of virus strains is desired. MCA to RGSV developed was not strain specific,[25] and detected all RGSV isolates in the Philippines, and one isolate each, in Indonesia and Thailand.

### C. VIRUS EPIDEMIOLOGY AND DISEASE FORECASTING

Most rice viruses which cause serious diseases are insect-borne.[3] Incidences of these virus diseases depend on susceptibility of rice to the viruses, vector population, and density of virus sources in the area, or percentage of infective insects in vector populations before or at the early crop season. Sensitive serological assays using PCA for detecting rice viruses in individual insect vectors have been developed.[5,8,10,12-15,35,59,60,64,65,67] In ELISA, however, *N. lugens* females carrying the eggs gave strong nonspecific reactions.[8] The nonspecific reactions were eliminated when extraction buffer at lower pH (6.5) was used to macerate insects. Use of an acidic buffer may also help in lowering the level of nonspecific reactions on other vector species.

Models to assess RSV incidence in the coming crop season have been developed and adapted in forecasting epidemics of the virus in a large area.[15] In the disease forecasting program, percentage of RSV-carrying vectors in overwintering populations is monitored by testing insects individually for RSV. First, the hemagglutination test and then the latex test, and ELISA were adopted widely in the disease forecasting programs.[5,10,12,26,59] MCA to RSV developed also detects the virus in individual insects, both in ELISA and the latex test.[26] MCA is used widely as is PCA in the disease forecasting programs and is expected to replace PCA eventually. MCA as well as PCA to RSV and their detection kits are commercially available.

*N. lugens,* a vector of RGSV and RRSV, is capable of long-distance migratory flight, even across the ocean.[78] In Central China, Japan, and Korea, *N. lugens* does not overwinter, and every year in the early summer, it migrates from endemic areas where rice is grown throughout the year. ELISA has been used in monitoring RGSV carriers in the immigrant population.[59] In the tropics, *N. lugens* also disperses the two viruses through long distance flight from one area to others, and probably from one island to others.[79] In the Philippines, ELISA has been used in monitoring the percentage of *N. lugens* carrying RGSV and RRSV, either together or separately in light-trapped populations.[80] After 1985, the MCA obtained has been used in the monitoring program.[81] During these periods, incidences of RGSV and RRSV were low in the areas, and prediction of possible virus disease epidemics was not practiced. In the tropics, prediction of virus disease epidemics is difficult and has never been successful, since rice is generally grown continually, and virus and its vectors are present throughout the year, if water is available. Monitoring of migrating *N. lugens* for its density and virus carriers is expected to provide fundamental information on *N. lugens* ecology and virus epidemiology, which would be used in modeling for the prediction of possible outbreak of the insect and viruses.

## D. SCREENING OF LINES FOR RESISTANCE

Rice cultivars resistant to virus have been widely planted in Asia.[82,83] Virus resistance is one of the major objectives in the rice breeding programs in many countries. The latex test and ELISA have been widely applied in evaluating rice cultivars for resistance to the tungro-associated viruses,[16,17] RRSV,[84] and RSV. There are cultivars which show field resistance to rice viruses and develop very mild symptoms when infected. Quantitative assay of the viruses in infected tissues is essential for evaluating cultivars of this type.[84,85] Serology has become an essential tool in screening rice lines for resistance to some viruses. Generally, the screening involves a large number of samples to be tested routinely. Hence, constant supply of antibody of similar quality is needed to pursue the breeding of rice cultivars for resistance to the respective viruses. Use of MCA is advantageous in this area of research.

# IV. DISCUSSION AND CONCLUSION

To this date, application of hybridoma technology to rice viruses is still a relatively new research area. Published information on MCA applications to rice virus research is still rather scanty. Except for MCAs directed against RGSV and RSV, use of MCAs in rice virus researches is limited. Even though such is this case, the advantages of MCAs over PCA to rice viruses have been well demonstrated. For one thing, MCAs are most advantageous to rice viruses which are difficult to purify in quantities sufficient for PCA production. But with MCA production, only a fraction of the antigen required for one injection into rabbits for PCA production is needed, yet an unlimited supply of antibodies with uniform specificity is obtained. This advantage is of great importance to rice virus research programs where large quantities of antibodies are needed.

MCA specific to a virus can be obtained by using the immunogen with impurity or even a mixture of viruses. Tungro disease is associated with two viruses[3,76] and the two viruses are difficult to separate from each other biologically by vector transmission or their host ranges, and physically during the purification.[9] With the hybridoma technology, it is now possible to produce MCA specific to either virus from partially purified virus or from a mixture of the two viruses. So far, results from detection and diagnostic application of MCAs against RGSV and RSV have been encouraging. The ease with which MCAs can be applied to detect rice viruses with the efficiency and sensitivity the same as if not better than, the PCAs, indicates that MCAs may eventually replace PCAs in large-scale diagnostic programs in the future.

Another advantage of MCA for rice virus diagnosis is the elimination of false results arising from contamination with host antigens. Background reactions in healthy controls, which is commonly associated with PCAs in ELISA, were completely eliminated with the use of MCA. This advantage was convincingly demonstrated when PCA was compared with MCA in DBI using alkaline phosphatase for RGSV detection in plant saps.[25,75] With PCAs, the healthy controls developed color reactions as intense as the positive controls, but with MCAs the nonspecific reactions were eliminated.

MCAs to other plant viruses have been used to differentiate strains of the same virus.[72,77] Although there have been only a few rice virus strains reported, this is another area where MCAs can be applied in the future. MCA against RGSV obtained was not strain specific.[25] Apparently, RGSV strains tested have a common epitope recognizable by the MCA. The MCA is being used for diagnosis, survey, and epidemiology.

Indeed, the hybridoma technology has opened a new era in rice virus research; however, there are still areas of MCA application that remain to be explored. At present, some of the MCAs generated against rice viruses or their proteins or RNA have not been well characterized. At the rate at which hybridoma technology is being applied to rice viruses, we may expect further progress in rice virus research.

# REFERENCES

1. **Ling, K. C.,** *Rice Virus Diseases,* International Rice Research Institute, Los Banos, Philippines, 1972.
2. **Ou, S. H.,** *Rice Diseases,* 2nd ed., Commonwealth Mycological Institute, Surrey, U.K., 1985.
3. **Hibino, H.,** Insect-borne viruses of rice, in *Advances in Disease Vector Research,* Vol. 6, Harris, K. F., Ed., Springer-Verlag, New York, 1990, 209.
4. **Kimura, I.,** Further studies on the rice dwarf virus II, *Ann. Phytopathol. Soc. Jpn.,* 27, 204, 1962 (in Japanese).
5. **Yasuo, S. and Yanagita, K.,** Serological studies on rice stripe and dwarf virus diseases: II. Hemagglutination test for rice stripe virus *Ann. Phytopathol. Soc. Jpn.,* 28 (Abstr.), 84, 1963 (in Japanese).
6. **Luisoni, E., Lovisolo, O., Kitagawa, Y., and Shikata, E.,** Serological relationship between maize rough dwarf virus and rice black-streaked dwarf virus, *Virology,* 52, 281, 1973.
7. **Saito, Y.,** Hemagglutination of leafhopper borne viruses, in *Viruses, Vectors and Vegetation,* Maramorosh, K., Ed., Interscience, New York, 1969, 463.
8. **Hibino, H. and Kimura, I.,** Detection of rice ragged stunt virus in insect vectors by enzyme-linked immunosorbent assay, *Phytopathology,* 72, 656, 1982.
9. **Omura, T., Saito, Y., Usugi, T., and Hibino, H.,** Purification and serology of rice tungro spherical and rice tungro bacilliform viruses, *Ann. Phytopathol. Soc. Jpn.,* 49, 73, 1983.
10. **Omura, T., Hibino, H., Usugi, T., Inoue, H., Morinaka, T., Tsurumachi, S., Ong, C. A., Putta, M., Tsuchizaki, T., and Saito, Y.,** Detection of rice viruses in plants and individual insect vectors by latex flocculation test, *Plant Dis.,* 68, 374, 1984.
11. **Bajet, N. B., Daquioag, R. D., and Hibino, H.,** Enzyme-linked immunosorbent assay to diagnose rice tungro, *J. Plant Prot. Tropics,* 2, 125, 1985.
12. **Takahashi, Y., Omura, T., Shohara, K., and Tsuchizaki, T.,** Rapid and simplified ELISA for routine field inspection of rice stripe virus, *Ann. Phytopathol. Soc. Jpn.,* 53, 254, 1987.
13. **Woo, Y. B. and Lee, K. W.,** Detection of rice black-streaked dwarf virus in rice, maize and insect vectors by enzyme-linked immunosorbent assay, *Korean J. Plant Pathol.,* 3, 108, 1987 (in Korean).
14. **Takahashi, Y., Omura, T., Hayashi, T., Shohara, K., and Tsuchizaki, T.,** Detection of rice transitory yellowing virus (RTYV) in infected rice plants and insect vectors by simplified ELISA, *Ann. Phytopathol. Soc. Jpn.,* 54, 217, 1988.
15. **Kishimoto, R. and Yamada, Y.,** A planthopper-rice virus epidemiology model : rice stripe and small brown planthopper, *Laodelphax striatellus* Fallen, in *Plant Virus Epidemics—Monitoring and Predicting Outbreaks,* Mclean, G. D., Garret, R. G., and Ruesink, W. G., Eds., Academic Press, Sydney, 1986, chap. 16.
16. **Hibino, H., Tiongco, E. R., Cabunagen, R. C., and Flores, Z. M.,** Resistance to rice tungro-associated viruses in rice under experimental and natural conditions, *Phytopathology,* 77, 871, 1986.
17. **Hibino, H., Daquioag, R. D., Cabauatan, P. Q., and Dahal, G.,** Resistance to rice tungro spherical virus in rice, *Plant Dis.,* 72, 843, 1988.
18. **Bajet, N. B., Aguiero, V. M., Daquioag, R. D., Jonson, G. B., Cabunagan, R. C., Mesina, E. M., and Hibino, H.,** Occurrence and spread of rice tungro spherical virus in the Philippines, *Plant Dis.,* 70, 971, 1986.

19. **Kohler, G. and Milstein, C.,** Continuous cultures of fused cells secreting antibody of predefined specificity, *Nature,* 256, 495, 1975.

20. **Halk, E. L. and De Boer, S. H.,** Monoclonal antibodies in plant disease research, *Annu. Rev. Phytopathol.,* 23, 321, 1985.

21. **Sander, E. and Dietzgen, R. G.,** Monoclonal antibodies against plant viruses, *Adv. Virus Res.,* 29, 131, 1984.

22. **Kitagawa, Y., Okuhara, E., and Shikata, E.,** Specificity of monoclonal antibodies to rice dwarf virus and purification of antigen proteins by affinity chromatography, *Ann. Phytopathol. Soc. Jpn.,* 50 (Abstr.), 436, 1984 (in Japanese).

23. **Matsumoto, T. and Kitagawa, Y.,** Specificity of monoclonal antibody to rice dwarf virus-RNA *Ann. Phytopathol. Soc. Jpn.,* 50 (Abstr.), 436, 1984 (in Japanese).

24. **Nozu, Y., Hagiwara, K., Omura, T., and Nishimori, T.,** Production of monoclonal antibodies against rice gall dwarf virus, *Ann. Phytopathol. Soc. Jpn.,* 51 (Abstr.), 359, 1985 (in Japanese).

25. **Hsu, H. T., Cabauatan, P. Q., and Hibino, H.,** Mouse monoclonal antibody to rice grassy stunt virus and its use for disease detection, *Phytopathology,* 76 (Abstr.), 1132, 1986.

26. **Omura, T., Takahashi, Y., Shohara, K., Minobe, Y., Tsuchizaki, T., and Nozu, Y.,** Production of monoclonal antibodies against rice stripe virus for the detection of virus antigen in infected plants and viruliferous insects, *Ann. Phytopathol. Soc. Jpn.,* 52, 270, 1986.

27. **Nozu, Y.,** Monoclonal antibodies against plant viruses—procedure and some applications, *Jpn. Agric. Res. Quart.,* 21, 96, 1987.

28. **Fan, Y. J., Namba, S., Yamashita, S., and Doi, Y.,** Antigen analysis of rice dwarf virus coat proteins using monoclonal antibodies, *Ann. Phytopathol. Soc. Jpn.,* 53 (Abstr.), 124, 1987 (in Japanese).

29. **Harjosudarmo, J., Yamada, N., Oshima, K., Uyeda, I., and Shikata, E.,** Monoclonal antibodies to rice ragged stunt virus, *Ann. Phytopathol. Soc. Jpn.,* 55 (Abstr.), 110, 1989 (in Japanese).

30. **Harjosudarmo, J., Oshima, K., Uyeda, I., and Shikata, E.,** Screening of hybridoma secreting monoclonal antibodies to rice dwarf virus, *Ann. Phytopathol. Soc. Jpn.,* 55 (Abstr.), 540, 1989 (in Japanese).

31. **Hibino, H.,** Rice grassy stunt virus, *CMI/AAB Descriptions of Plant Viruses,* No. 320, 1986.

32. **Rivera, C. T., Ou, S. H., and Iida, T. T.,** Grassy stunt disease of rice and its transmission by the planthopper, *Nilaparvata lugens* Stal, *Plant Dis. Rep.,* 50, 453, 1966.

33. **Chen, C. C. and Chiu, R. J.,** Three symptomatologic types of rice virus diseases related to grassy stunt in Taiwan, *Plant Dis.,* 66, 15, 1982.

34. **Hibino, H., Cabauatan, P. Q., Omura, T., and Tsuchizaki, T.,** Rice grassy stunt virus strain causing tungro-like symptoms in the Philippines, *Plant Dis.,* 69, 538, 1985.

35. **Hibino, H., Usugi, T., Omura, T., Tsuchizaki, T., Shohara, K., and Iwasaki, M.,** Rice grassy stunt virus: A planthopper-borne circular filament, *Phytopathology,* 75, 894, 1985.

36. **Toriyama, S.,** Ribonucleic acid polymerase activity in filamentous nucleoproteins of grassy stunt virus, *J. Gen. Virol.,* 68, 925, 1987.

37. **Toriyama, S.,** Rice stripe virus, *CMI/AAB Descriptions of Plant Viruses,* No. 269, 1983.

38. **Iida, T. T.,** Dwarf, yellow dwarf, stripe, and black-streaked dwarf diseases of rice, in *The Virus Diseases of the Rice Plant,* Johns Hopkins Press, Baltimore, 1969, chap. 1.

39. **Koganezawa, H., Doi, Y., and Yora, K.,** Purification of rice stripe virus, *Ann. Phytopathol. Soc. Jpn.,* 41, 148, 1975 (in Japanese).

40. **Ishikawa, K., Omura, T., and Tsuchizaki, T.,** Association of double- and single-stranded RNAs with each four components of rice stripe virus, *Ann. Phytopathol. Soc. Jpn.,* 55, 315, 1989.

41. **Ishikawa, K., Omura, T., and Hibino, H.,** Morphological characteristics of rice stripe virus, *J. Gen. Virol.,* 70, 3465, 1989.

42. **Koganezawa, H.,** Purification and properties of rice stripe virus, *Trop. Agric. Res. Series,* 10, 151, 1977.

43. **Iida, T. T., Shinkai, A., and Kimura, I.,** Rice dwarf virus, *CMI/AAB Descriptions of Plant Viruses,* No. 102, 1972.

44. **Fukushi, T., Shikata, E., and Kimura, I.,** Some morphological characters of rice dwarf virus, *Virology,* 18, 192, 1962.

45. **Reddy, D. V. R., Kimura, I., and Black, L. M.,** Co-electrophoresis of dsRNA from wound tumor and rice dwarf viruses, *Virology,* 60, 293, 1974.

46. **Nakata, M., Fukunaga, K., and Suzuku, N.,** Polypeptide components of rice dwarf virus, *Ann. Phytopathol. Soc. Jpn.,* 44, 288, 1978.

47. **Omura, T. and Inoue, H.,** Rice gall dwarf virus, *CMI/AAB Descriptions of Plant Viruses,* No. 296, 1985.

48. **Omura, T., Inoue, H., Morinaka, T., Saito, Y., Chetanachit, D., Putta, M., Parejarearn, A., and Disthaporn, S.,** Rice gall dwarf, a new virus disease, *Plant Dis.,* 64, 795, 1980.

49. **Omura, T., Morinaka, T., Inoue, H., and Saito, Y.,** Purification and some properties of rice gall dwarf virus, a new phytoreovirus, *Phytopathology,* 72, 1246, 1982.

50. **Hibi, T., Omura, T., and Saito, Y.,** Double-stranded RNA of rice gall dwarf virus, *J. Gen. Virol.,* 65, 1585, 1984.

51. **Omura, T., Minobe, Y., Matsuoka, M., Nozu, Y., Tsuchizaki, T., and Saito, Y.,** Location of structural proteins in particles of rice gall dwarf virus, *J. Gen. Virol.,* 66, 811, 1985.

52. **Milne, R. G., Boccardo, G., and Ling, K. C.,** Rice ragged stunt virus, *CMI/ABB Descriptions of Plant Viruses,* No. 248, 1982.

53. **Hibino, H., Roechan, M., Sudarisman, S., and Tantera, D. M.,** A virus disease of rice (kerdil hampa) transmitted by brown planthopper, *Nilaparvata lugens* Stal., in Indonesia, *Contrib. Centr. Res. Inst. Agric. Bogor.,* No. 35, 1977.

54. **Ling, K. C., Tiongco, E. R., and Aguiero, V. M.,** Rice ragged stunt, a new virus disease, *Plant Dis. Rep.,* 62, 701, 1978.

55. **Hibino, H., Saleh, N., and Roechan, M.,** Reovirus-like particles associated with rice ragged stunt diseased rice and insect vector cells, *Ann. Phytopathol. Soc. Jpn.,* 45, 228, 1979.

56. **Kawano, S., Uyeda, I., and Shikata, E.,** Particle structure and double-stranded RNA of rice ragged stunt virus, *J. Fac. Agric. Hokkaido Univ.,* 6, 408, 1984.

57. **Omura, T., Minobe, Y., Kimura, I., Hibino, H., Tsuchizaki, T., and Saito, Y.,** Improved purification procedure and RNA segments of rice ragged stunt virus, *Ann. Phytopathol. Soc. Jpn.,* 49, 670, 1983.

58. **Hagiwara, K., Minobe, Y., Nozu, Y., Hibino, H., Kimura, I., and Omura, T.,** Component proteins and structure of rice ragged stunt virus, *J. Gen. Virol.,* 67, 1711, 1986.

59. **Iwasaki, M., Nakano, M., and Shinkai, A.,** Detection of rice grassy stunt virus in planthopper vectors and rice plants by ELISA, *Ann. Phytopathol. Soc. Jpn.,* 51, 450, 1985.

60. **Chen, G. Y.,** The application of the ELISA for determining the percentage of active individuals transmitting rice stripe disease in a vectors population, *Acta Phytopathol. Sinica,* 11, 73, 1984 (in Chinese).

61. **Toyoda, S., Kimura, I., and Suzuku, N.,** Purification of rice dwarf virus, *Ann. Phytopathol. Soc. Jpn.,* 30, 225, 1965.

62. **Suzuku, N.,** Purification of single and double-stranded vector-borne RNA viruses, in *Viruses, Vectors, and Vegetation,* Maramorosh, K., Ed., Interscience, New York, 1969, 557.

63. **Kimura, I.,** Improved purification of rice dwarf virus by the use of polyethylene glycol, *Phytopathology,* 66, 1470, 1976.

64. Shanghai Institute of Biochemistry and Chekiang Institute of Agric. Res., The pathogens of some virus diseases of cereals in China. IV. Serological determination of the percentage of active individuals transmitting rice dwarf disease in a population of insect vectors, *Acta Biochem. Biophys. Sinica,* 10, 355, 1978 (in Chinese).

65. **Chen, H. K., Wang, G. C., and Sheng, F. J.,** Use of polyethylene glycol (PEG) in the preparation of rice dwarf virus (RDV) antiserum, *Acta Phytopathol. Sinica,* 10, 83, 1980 (in Chinese).

66. **Uyeda, I. and Shikata, E.,** Ultrastructure of rice dwarf virus, *Ann. Phytopathol. Soc. Jpn.,* 48, 295, 1982.

67. **Lin, R. F., Chen, G. Y., Gao, D. M., Jin, D. D., Chen, S. X., and Ruan, Y. L.,** Studies on the application of the reversed passive carbon agglutination test to the detection of viruliferous individuals of rice dwarf virus, *Acta Microbiol. Sinica,* 20, 173, 1980 (in Chinese).

68. **Matsuoka, M., Minobe, Y., and Omura, T.,** Reaction of antiserum against SDS-dissociated rice dwarf virus and a polypeptide of rice gall dwarf virus, *Phytopathology,* 75, 1125, 1985.

69. **Kitagawa, Y., Matsumoto, T., Okuhara, E., and Shikata, E.,** Immunogenicity of rice dwarf virus-ribonucleic acid, *Tohoku J. Exp. Med.,* 122, 337, 1977.

70. **Luisoni, E., Milne, R. G., and Roggero, P.,** Diagnosis of rice ragged stunt virus by enzyme-linked immunosorbent assay and immunosorbent electron microscopy, *Plant Dis.,* 66, 929, 1982.

71. **Parejarearn, A. and Hibino, H.,** Purification and serology of ragged stunt virus (RDV), *Int. Rice Res. Newsl.,* 12(3), 25, 1987.

72. **Hsu, H. T., Aebig, J., and Rochow, W. F.,** Differences among monoclonal antibodies to barley yellow dwarf viruses, *Phytopathology,* 74, 600, 1984.

73. **Clark, M. F. and Adams, A. N.,** Characteristics of the microplate method of enzyme-linked immunosorbent assay for the detection of plant viruses, *J. Gen. Virol.,* 34, 475, 1977.

74. **Bercks, R.,** Methodische Untersuchungen über den serologischen Nachweis pflanzenpathogener Viren mit dem Bentonite-Flockungstest, dem Latex-Test und dem Bariumsulfat-Test, *Phytopathol. Z.,* 58, 1, 1967.

75. **Cabauatan, P. Q., Hibino, H., and Hsu, H. T.,** Dot-blot immunoassay (DBI) for detecting rice grassy stunt virus (GSV), *Int. Rice Res. Newsl.,* 13(4), 34, 1988.

76. **Hibino, H., Roechan, M., and Sudarisman, S.,** Association of two types of virus particles with penyakit habang (tungro disease) of rice in Indonesia, *Phytopathology,* 68, 1412, 1978.

77. **Falk, E. L., Hsu, H. T., Aebig, J., and Franke, J.,** Production of monoclonal antibodies against three ilarviruses and alfalfa mosaic virus and their use in serotyping, *Phytopathology,* 74, 367, 1984.

78. **Kishimoto, R.,** Synoptic weather conditions inducing long-distance immigration of planthopper *Sogatella furcifera* Horvath and *Nilaparvata lugens* Stal., *Ecol. Entomol.,* 1, 95, 1976.

79. **Rosenberg, L. J. and Magor, J. I.,** A technique for examining the long-distance spread of plant virus diseases transmitted by the brown planthopper, *Nilaparvata lugens* (Homoptera:Delphacidae), and other wind-borne insect vectors, in *Plant Virus Epidemiology,* Plumb, R. T. and Thresh, J. M., Eds., Blackwell Scientific, Oxford, 1983, 228.

80. **Flores, Z. M., Hibino, H., and Perfect, J.**, Rice grassy stunt (GSV) and rice ragged stunt (RSV) carriers, *Int. Rice Res. Newsl.*, 11(4), 26, 1986.
81. **Flores, Z. M. and Hibino, H.**, Survey of rice virus carriers among brown planthopper (BPH) *Nilaparvata lugens* populations in Laguna, Philippines, *Int. Rice Res. Newsl.*, 14(5), 25, 1989.
82. **Khush, G. S. and Virmani, S. S.**, Breeding rice for disease resistance, in *Progress in Plant Breeding I.*, Russel, G. F., Ed., Blackwell Scientific, Oxford, 1985, 239.
83. **Sakurai, Y.**, Varietal resistance to stripe, dwarf, yellow dwarf, and black-streaked dwarf, in *The Virus Diseases of the Rice Plant*, Johns Hopkins Press, Baltimore, 1969, chap. 22.
84. **Parejarearn, A., Salamat, G., Jr., and Hibino, H.**, Growth retardation, yield reduction, and virus concentration in ragged stunt virus-infected rice, *J. Plant Prot. Tropics*, 6, 131, 1989.
85. **Hasanuddin, A., Daquioag, R. D., and Hibino, H.**, A method for screening resistance to tungro (RTV), *Int. Rice Res. Newsl.*, 13(6), 13, 1988.

*Insect Cell Fusion*

Chapter 10

# FUSION OF CULTURED INSECT CELLS

## Jun Mitsuhashi

## TABLE OF CONTENTS

# I. INTRODUCTION

In animal cell cultures, cell fusion became a routine technique to make hybridomas for the production of a monoclonal antibody. In plant cell cultures, cell fusion technique has been used to obtain hybrid plants. Contrary to these practical uses of cell fusion, development of cell fusion techniques is far more retarded in insect cell cultures. Earlier, insect cell fusion had been studied intensively; however, many difficult problems arose, and as far as I know, no one has worked on insect cell fusion recently. Consequently, no hybrid cell lines have existed. One of the reasons establishment of a technique of insect cell fusion has seemingly been abandoned may be lack of prospective application of insect hybrid cells.

In this review, the author introduces instances of insect cell fusion studies hitherto made.

# II. FUSION WITHIN THE SAME CELL LINE

Fusion between the cells of the same cell line species was first attempted by Becker[1] with Echalier and Ohanessian's *Drosophila melanogaster* cell line. A group of cells were labeled with [3H]thymidine and mixed with unlabeled cells. A solution of concanavalin A (Con A) was added to the cell mixture at a final concentration of 100 μg/ml in the presence of $CaCl_2$ and $MnSO_4$ at $10^{-4}$ $M$. Following treatment for 30 min, the cells were seeded in a Leighton tube and rinsed with fresh culture medium after they attached to the vessel. Cytoplasmic bridges between attached cells were observed within 1 h after the Con A treatment. During the next 10 h, heterokaryons were formed and most of them proceeded to form synkaryons within 24 h. Hybrid cells were indicated by autoradiographs showing labeled and unlabeled nuclei in a single cells; however, hybrid cells were not isolated.

Rizki et al.[2] examined effects of wheat germ agglutinin (WGA) on Schneider's *D. melanogaster* cells. They found that a high concentration of WGA (100 to 200 μg/ml) shriveled cell surface, whereas a low concentration (5 to 10 μg/ml) caused fusion of cells as observed with a scanning electron microscope. The cell fusion was evident after treatment for 3 min with WGA. Synkaryon formation was not examined.

Halfer and Petrella[3] examined effects of Con A and lysolecithin on two karyotypically different cell lines of *D. melanogaster* cells, $GM_1$ and IB5. The lysolecithin was more rapid and drastic in inducing cell fusion, and more cytotoxic to the cells compared with the former. They confirmed the formation of dikaryons by means of autoradiography. However, they did not examine the formation of synkaryons.

Bernhardt[4] demonstrated that polyethyleneglycol (PEG) caused fusion of *D. melanogaster* cells as it does in the case of vertebrate cells and plant protoplasts. He used [3H]thymidine-labeled Kc strain of a *D. melanogaster* line, and unlabeled Kc cells or dissociated imaginal disc cells isolated from the 3rd instar larvae as parent cells. The mixtures of parent cells were treated in 50 m$M$ PEG (4000 mol wt) for 10 min at 25°C. The results were checked autoradiographically. The estimated frequency of the fusion in both combination Kc × Kc and Kc × imaginal disc cells was 5%. The resultant heterokaryons, however, did not survive.

Nakajima and Miyake[5] obtained two temperature-sensitive (ts) mutants for parent cells of cell fusion. They treated the $GM_1$ cell line of *D. melanogaster* with a mutagen, ethylmethane sulfonate (500 μg/ml). Mutants that did not grow at a high temperature were selected by poisoning colonies growing at 30°C with 5-fluoro-2-deoxyuridine (5-FldU) at a concentration of 25 μg/ml. The 5-FldU treatment was repeated four times. As a result, two clones, designated as ts-15 and ts-58, respectively, were obtained. They formed colonies at 23°C about one third as well as the wild-type cells seeded at a density of $10^3$ cells per dish (60 × 15 mm glass dish), but they formed no colonies at 30°C. These mutants were used as parent cell lines for fusion. These two ts mutant cell lines were mixed and treated with 50% PEG (6000 mol wt) for 1 min. In some experiments 15% dimethyl sulfoxide was added to the PEG solution. When the treated cells were seeded at a density of $10^4$ cells per dish, a few colonies were formed at 30°C. These colony-forming cells showed plating efficiency comparable to that of the wild-type cell line at 30°C. Karyotype analysis showed that the PEG-treated and selected cell lines were nearly tetraploid, while both the parent ts mutants were nearly diploid. This result seemed to indicate that fusion of these two distinct ts mutants produced hybrids that complemented the functional deficiency, permitting growth at a high temperature. Unfortunately, the hybrid cell lines obtained have not been maintained.

From the *D. melanogaster* Kc cell line, Wyss[6] obtained a clone (MDR3) resistant to 6-methylpurine and diaminopurine. The clone was an adenine salvage-deficient variant and was sensitive to a selection medium, which was called ZH1% medium, containing thymidine, adenine, and methotrexate (TAM). He attempted to fuse this variant to either of two wild-type (TAM resistant) cell lines, Schneider's line 3 (S3) and Dübendorfer's line 1 (D1), which were unable to proliferate in the medium ZH1%. After treatment of a pellet of mixed parent cells with 45% PEG (1000 mol wt) for 30 s, hybrids were selected in the medium containing TAM. The hybrid cells were confirmed by isozyme analysis of isocitrate dehydrogenase bands. MDR3 and S3 were ecdysone sensitive and did not proliferate in the presence of 20-hydroxyecdysone (20-OH-Ecd). Some ecdysone-resistant clones, however, were obtained from MDR3 and were designed as MDER. A great majority of all hybrids formed between ecdysone-sensitive parents were also sensitive to 20-OH-Ecd, whereas most hybrids resulting from fusion of MDER to S3 proved to be resistant to 20-OH-Ecd.[7] When MDR3 was hybridized to cells derived from wild-type *D. melanogaster* embryos, the resulting hybrids could proliferate in the presence of 20-OH-Ecd.[7] On the other hand, no hybrids resulting from fusion of MDR3 to primary cells from eye-antennal discs of the 3rd instar larvae were resistant to 20-OH-Ecd.[8]

Berger and Wyss[9] studied hybridization between *D. melanogaster* S3 cell lines with contrasting phenotypes and different responses to 20-OH-Ecd. The above-mentioned S3 line cells maintained a high basal level of acetylcholinesterase (AChE), which was lost and then reinduced following exposure to 20-OH-Ecd. The MDR line cells has a lower basal AChE level, which was not elevated by 20-OH-Ecd. S3/MDR and S3/MDER hybrids were produced by treatment with PEG and selected by cloning in a soft agar medium containing ZH1% + TAM. In S3/MDR hybrids, the high basal level phenotype was extinguished and the 20-OH-Ecd-induced AChE level was modest. In S3/MDER hybrids two of the hybrid clones, F1 and F6, showed a phenotype and response to 20-OH-Ecd, similar to those of MDER parent. However, one hybrid clone, F7, was ecdysone-sensitive and was induced by 20-OH-Ecd to a specific higher level of AChE production.

Berger et al.[10] examined polypeptide synthesis of *D. melanogaster* cell lines, S3 and MDR, by labeling polypeptides with [$^{35}$S]methionine. The synthesized polypeptides were compared before and after the hormone treatment by means of two-dimensional polyacrylamide gel electrophoresis. A set of 10 of the more than 300 resolvable spots were selected as internal standards for subsequent quantitative works. Five peptide spots appeared to either increase or decrease in relative labeling intensity following hormone treatment. Significant changes occurred in the rate of synthesis of several peptides in two different *D. melanogaster* cell lines following exposure to the hormone. For peptides 7/8, which could not be excised separately because of their close proximity, the synthesis rate increased by a factor of 10 in line S3 and by a factor of 1.3 in line MDR. In hormone-sensitive hybrids S3/MDR and S3/MDER-F7, the rate increased by a factor of 3 to 4.5. For peptide 10, the synthesis rate decreased by a factor of 8 in line S3 but increased by a factor of 2 in line MDR. In hormone-insensitive cell lines, MDER, S3/MDER-F1 and S3/MDER-F6, no change in the rate of peptide (both 7/8 and 10) synthesis occurred in the presence of the hormone.

Fusion of cells in primary cultures of nerve cells of wild type *D. melanogaster* was attempted.[11] Cells of larval central nervous system were dissociated and plated at a density of $10^5$ to $10^6$ cells per square centimeter. After culturing for 6 to 24 h, 30 to 60% PEG (400 to 20,000 mol wt) in Schneider's medium, was introduced at room temperature. The treatment lasted for 60 to 90 s, and immediately after the treatment the PEG was removed by repeated washing. The best fusion efficiency and survival rate of fused cells were obtained by the treatment of 50% PEG (6000 mol wt); the fusion index was about 30% and more than 60% of fused cells were viable. The fused cells formed multinucleate cells and grew frequently for more than a week. Neurite outgrowth of the fused cells was observed. The process of the synkaryon formation did not seem to occur. Neurons from mutants defective in membrane exitability (*nap* and *para*) showed similar response to PEG.

Mitsuhashi[12] treated a cell line, NIAS-AeAl-2, derived from whole neonate larval bodies of the one striped mosquito, *Aedes albopictus* or a cell line, NIH-SaPe-4, derived from whole embryos of the fleshfly, *Sarcophaga peregrina* with 50% PEG (6000 mol wt) for 1 min, respectively. Karyotype analyses were made on the PEG-treated cells after cultivating them for a week. The parent lines of the mosquito cells were mostly tetraploid. After PEG treatment the percentage of octaploid cells increased from 30% level to 45% level. Likewise, the parent line of the fleshfly cells were predominantly diploid, and the percentage of tetraploid cells increased from 6% to 26% after PEG treatment. These increases in rate of polyploid cells seemed to indicate occurrence of cell fusion.

## III. FUSION OF CELLS BETWEEN DIFFERENT SPECIES

It is well known that chromosomal complements of lepidopterans consist of many small chromosomes, while those of dipterans have fewer large chromosomes. Therefore, if inter-order hybrids between lepidopteran and dipteran cells were obtained, they might be easily identified by karyotype analysis. With this idea, fusion of cells of such combination was attempted.[12]

A lepidopteran cell line, NIAS-PX-58, derived from pupal ovaries of the swallow tail butterfly, *Papilio xuthus,* and a dipteran cell line, NIAS-AeA1-2 (aforementioned), derived from a mosquito were used. The former has about 100 small chromosomes and the latter has 12 large chromosomes. Both of the cell lines were cultured in Mitsuhashi and Mara-morosch's medium (MM)[13] containing 3% fetal bovine serum (FBS). The suspensions of both cell lines were mixed and centrifuged at 1000 rpm for 5 min. The cell pellet was treated with 50% PEG (6000 mol wt) for 3 min. After being washed, the cells were resuspended in MM with 3% FBS. Many cells were found to attach each other. Homologous combinations, PX-58 to PX-58 or AeAl-2 to AeAl-2, were predominant suggesting occurrence of some cell-discrimination. There were also heterologous cell combinations, PX-58 to AeAl-2, although the frequency was small. The attachment of cells to each other did not indicate true cell fusion, because cell membranes were still visible between the attached cells. In some cases, however, cells which appeared to be really fused were observed. In such cases, homologous cell fusion was observed more frequently than heterologous fusion. Synkaryon formation could not be ascertained, and interorder hybrid cell lines were not obtained.

## IV. INTERPHYLUM CELL FUSION

Fusion of a mosquito cell line to a human cancer cell line, HeLa, has been attempted. The mosquito cell line was later proved to be a Grace's moth (*Antherea eucalypti*) cell line[14] which was erroneously used.[15] Zepp et al.[16] and Conover et al.[17] first labeled the HeLa cells with [$^3$H]thymidine, and then suspended the cells in UV-inactivated hemagglutinating virus of Japan (HVJ). After thorough mixing, the suspension was used to resuspend moth cells. The mixture was shaken at a low speed for 20 min in a 4°C water bath, and then for 60 min at 37°C. Autoradiography of direct smears immediately after these procedures gave evidence of HeLa-moth heterokaryon formation. This was also evident from the morphology of the cells, because the nuclei of both parents were morphologically distinguishable. The HeLa cells had been cultured in Eagle's minimum essential medium (MEM) at 37°C. They do not grow at 26 to 28°C. The moth cells could not be maintained in MEM, and they cannot withstand the high temperature of 37°C. When a mixture of equal parts of complete MEM and moth culture medium was made (EPM), HeLa cells grew in it at 37°C. The HVJ-treated cell population could be subcultured 10 times in EPM. The population contained hybrid cells and residual HeLa cells. Evidence of synkaryon formation was obtained from chromosome analysis of 36 to 40 h first-passage cultures. The presence of both HeLa and moth chromosomes, which apparently share the same spindle apparatus, was observed. However, there was no instance in which complete summation of parental chromosomes was observed. Chromosome analysis at each passage also showed both parental chromosome types present within single cells; however, it also demonstrated a predominance of the HeLa chromosomes with progressively smaller numbers of the moth chromosomes. This might be due to the fact that the culture conditions were more suited to HeLa cells than moth cells.

## V. INTERKINGDOM CELL FUSION

An attempt of electrofusion between insect cells and plant protoplasts has been made.[18] Selected parents consisted of a continuous cell line, NIAS-PX-58, from butterfly (aforementioned) and tobacco mesophyll protoplasts isolated from *Nicotiana tabacum* var. Samsun plants. The butterfly cells were maintained in MM medium with 3% FBS by a weekly subculture. The cells harvested 7 d after subculture were washed three times with 0.5% sorbitol containing 2.5 m$M$ MgCl$_2$ (chamber medium) by centrifugation at 40 $\times$ $g$ for 5 min. The cells were then mixed with tobacco protoplasts at a ratio of 10:1 or 1:1 in the same medium, and centrifuged. The mixed cell pellet was added with deionized Dispase (a bacterial protease, Godo Shusei Co., Tokyo) solution so that the final concentration of Dispase was 2000 units/ml/5 $\times$ 10$^6$ cells dissolved in the chamber medium. An aliquot of this mixture was placed in an electrofusion chamber equipped with two platinum plate electrodes. After standing for about 5 min to allow the mixed cells to settle to the bottom, a single exponentially decaying fusion pulse (initial field intesity = 0.5 to 0.7 kV/cm; time constant 0.1 to 0.2 ms) was applied to the mixture by the capacitor-discharge method. No AC electric field was applied before the fusion pulse, because the density of the mixed cells was high enough to put the cells in contact with each other.

As soon as the fusion pulse was applied, the mixed cells started to fuse. Interkingdom fusion was recognized easily, because both parent cells and protoplasts were quite distinguishable in their morphology and size. As the result, variously shaped interkingdom fusion products were observed. Round-shape or longitudinally shaped butterfly cells fused with the mesophyll protoplasts. Large cell bodies which seemed to be formed by homologous cell fusion of the butterfly cells, were observed. These large cell bodies or clumps of the cells also fused to the mesophyll protoplasts. In round-shaped interkingdom heterokaryons, the butterfly cell part generally tended to locate at the periphery of the plant protoplasts. The use of Dispase before and during electrofusion process was essential for successful interkingdom electrofusion. The effect of the enzyme is attributable to the proteolytic activity of the enzyme and may be interpreted as the modification of the surface membrane of insect cells and plant protoplasts is necessary for mutual fusion. Although interkingdom synkaryon formation could not be detected, this technique will contribute to the studies of cell membrane properties and to the transfer of genes from animal to plant cells or vice versa.

## VI. ISOLATION OF HYBRID CELLS

It is extremely important to devise a method for isolating hybrid cells efficiently from the mixed fused cell population. Without this device, practical production of hybrid cell lines will not be possible.

It is theoretically possible to isolate hybrid cells or hybrid cell clones mechanically by means of micromanipulation or dilution-colony formation method, if the hybrid cell is appropriately marked. The cloning method, such as the single cell cloning method or colony formation method is, however, applicable to a limited number of insect cell lines. Most insect cell lines are highly dependent on population density, and isolation of a single cell in either of liquid medium or soft agar medium results in death of the cell even if a conditioned medium is used. Development of the single cell cloning method is, therefore, prerequisite for this method to be used practically.

It is also difficult to mark parent cells. The marking of parent cells is indispensable, because in the above isolation method, hybrid cells are recognized only when they possess visible characteristics. Marking of cells with radioisotope cannot be used, because the cells should be isolated as living cells. The use of parent cells with quite different morphological characteristics may be applicable. In plant cells, such cell lines can be found easily; some cell lines, for example, contain plastids such as chloroplasts and chromoplasts. In insect cell lines, however, cell lines with specific morphological characteristics are rare. Then marking of cells may be considered. An easy way to mark cells is by vital staining, which has been successfully used in vertebrate cell fusion. An attempt was made to stain one parent insect cell population with neutral red and to fuse it to unstained parent cells by the use of PEG. Unfortunately, this method proved unsuccessful, because the dye was rapidly lost during the washing procedure after PEG treatment.[12] However, vital staining of a mosquito *A. albopictus* cell with truidine blue was successfully used in the fusion experiments with an arbovirus.[19] Vital staining of insect cell lines with a fluorescent dye Hoechst 33258 or ethidium bromide was also tried. Cells of one parent were stained with Hoechst 33258, and cells of the other, with ethidium bromide before fusion of cells. This technique was also unsuccessful because Hoechst 33258 stained only part of the cell population, and ethidium bromide was very toxic to the cell and its fluorescence was weak.[12] In vertebrate cells, vital staining with fluorescent dye has been successfully used.[20]

Although no insect cell has inclusion bodies such as chloroplasts in plant cells, it is possible to let insect cells acquire some characteristic particles by phagocytosis. This may be used to mark cells. Marking by phagocytosis of fluorescent beads was attempted. Parent insect cell lines were given greenish fluorescent beads (Polyscience 9847) and purplish fluorescent beads (Polyscience 7769), respectively. The cells phagocytized these beads well. However, when the cells were examined under a fluorescence microscope after PEG treatment, cell fusion was difficult to judge by this method. The fluorescence of Polyscience 7769 was weak, and free beads, derived from unphagocytized beads or from cells that disintegrated during treatment, were phagocytized after PEG treatment. In vertebrate cells, marking cells with phagocytized fluorescent beads has been used.[20,21]

There are methods to use mutant cell clone or strain, which lack some normal cell function, as parent cells to facilitate isolation of hybrid cells. As already mentioned Nakajima and Miyake[5] used ts mutants of *D. melanogaster* cells and successfully isolated hybrid cells. Miltenburger et al.[22] used Actinomycin D and a puromycin-treated *Cydia pomonella* cell line, which lacked proliferating ability, and obtained proliferating hybrid cells as is mentioned in detail in the next section. In vertebrate cells, the HGPRT⁻-mutant cells lacking hypoxanthin-guanine-phosphoribosyltransferase activity, and the TK⁻-mutant cells lacking thymidine kinase activity were commonly used. Both enzymes are members of the salvage pathway of DNA biosynthesis. These mutant cells synthesize DNA only by *de novo* synthesis and cannot survive in HAT medium (a selection medium supplemented with hypoxanthine-aminopterin-thymidine), because the *de novo* synthesis of DNA is blocked by aminopterin. However, hybrid cells which recovered their salvage pathways by complimentation of the functional deficiency can grow in the HAT medium, in which the salvage pathway of DNA is permitted to function. In insect cells, application of the same system has been examined. Becker[23] and Moiseenko and Kakpakov,[24] however demonstrated that *D. melanogaster* cells have no functional HGPRT. Then, Wyss[6] developed the TAM process, analogs to the HAT selection system. He selected a clone (MDR3) resistant to 6-methylpurine and diaminopurine from the Kc line of *D. melanogaster*. This clone, an adenine salvage-deficient variant and which was sensitive to TAM selection medium, was used as one of parent cells, and wild-type cell lines (TAM-resistant), which cannot proliferate in ZH1% medium, as the other parent cells. This allowed the selection of hybrids between MDR3 and wild-type cells in TAM selection medium after PEG treatment. By the use of this system, he obtained hybrids as already mentioned in Section II.

It may be necessary to develop more efficient methods for selecting hybrid cells. If such methods are developed, one can get hybrid cell clones easily, even if the percentage of hybridization is low.

# VII. APPLICATION OF CELL FUSION

Hybrid cells, if appropriate hybrid cells are obtained, may find their use to enhance production of insect pathogenic microorganisms for microbial control. As already stated in the preceding section, however, hybridization of different cell lines are so far not successful.

There is another practical use of hybridization. It is somewhat like hybridoma production, namely, fusion of continuously growing cells to dissociated cells from specific tissues. Usually, cells isolated from intact insects do not proliferate immediately in primary cultures. This is especially true for cells which are differentiated so as to show a specific function. On the other hand, cells in continuous cell lines generally have proliferating ability but lack specific function. It may be possible to obtain growing cells with specific function if hybridization between cells from continuous cell lines and primary cells of dissociated tissues can be made.

Wyss[7] fused MDR3 mutant cells of *D. melanogaster* (aforementioned) to cells derived from wild-type embryos of the same species by treating a pellet of the mixed cells with 45% PEG 1000 for 30 s. After selection in a selection medium, the resulting hybrids demonstrated alteration of a part of characteristics of MDR3 cell line; the hybrid became resistant to 20-OH-Ecd while MDR3 parent cells were ecdysone sensitive. However, no hybrids resulting from fusion of MDR3 to primary cells from eye-antennal discs of the third instar larvae were resistant to 20-OH-Ecd.

Miltenburger et al.[22] succeeded in obtaining a hybrid between continuously growing cells and nonproliferating primary cells of a lepidopteran, *C. pomonella*. In order to isolate hybrid cells, they used cells which were irreversibly inhibited in their growth by drugs, as parents. A cell line (IZD-Cp-2202) from hemocytes was treated with Actinomycin D (0.25 $\mu$g/ml) and puromycin (2 $\times$ 10$^{-4}$ $M$) for 3 h at 27°C. About 35% of the cells were alive at the time of fusion. However, only a small percentage (0.1%) of the cells recovered after the blocking. The cells, which survived but were unable to proliferate, were used as parent cells. Other parent cells were prepared directly from embryos of the same species. About 200 eggs, 2 to 3 days old, were crushed after surface-sterilization and passed through a steel mesh. The homogenate, containing single cells and small cell aggregates, was used for fusion after washing. The mixed cell pellet was treated with PEG for 1 min at 37°C. After washing, the cells were seeded into the wells of a hybridoma tissue culture tray. About 2 weeks later, some colonies were obtained. The cells of the colonies obtained differed in morphology from the permanent parent cells, IZD-Cp-2202. Isozyme analysis for esterases was particularly indicative of the hybrid nature of the resulting cell, because several enzymes with esterase activity, found either in the primary cell line or in IZD-Cp-2202 cells, were present together in the resulting cells. The hybrid cells were also found to be much less susceptible to *C. murinana* nuclear polyhedrosis virus than was IZD-Cp-2202.

# VIII. VIRUSES INDUCING CELL FUSION

Since demonstration of cell fusion by a virus, hemagglutinating virus of Japan (HVJ),[25] the HVJ has been used widely to fuse vertebrate cells. As already mentioned, the HVJ was successfully used to fuse a human cell, HeLa, to an insect cell,[16,17] whereas fusion of insect cells to insect cells by the HVJ was not successful.[12] In the former case, the HeLa cells were treated with UV-inactivated HVJ first, and then insect cells were added. The first step might change the nature of the HeLa cell membrane to facilitate fusion. The membrane of insect cells seems to lack receptors for paramyxoviruses,[5] and this may be the reason for unsuccessful fusion within insect cells. Besides the HVJ, there are many other viruses including insect viruses and plant viruses which induce cell fusion.[12,26]

It is well known that alphaviruses and flaviviruses can replicate in mosquito cell lines, the former showing no cytopathic effect in Singh's *Aedes albopictus* cell line and the latter forming syncytia.

Paul et al.[27] reported that the West Nile virus, the Japanese encephalitis virus and dengue 2 virus caused formation of multinucleate giant cells and syncytia when inoculated to cultures of Singh's *A. albopictus* cell line.

Stollar and Thomas[28] found a cell-fusing agent (CFA) in cultures of Peleg's *Aedes aegypti* cell line. The CFA caused marked syncytium formation of Singh's *A. albopictus* cells, but not of BHK, KB, or Vero cells. Based upon examination with the electron microscope, plaque assay, and other procedures, the CFA was supposed to be a virus belonging to ungrouped togaviruses which contaminated Peleg's *A. aegypti* cells.[29]

Späth and Koblet[30] reported syncytium formation of a subline of Singh's *A. albopictus* cell line after inoculation with Semliki forest virus (SFV). The fusion of cells was found to occur in at least two steps, namely, a fast initial step which is pH dependent and temperature independent, and a second slower process which is pH independent and temperature dependent.[31] The initiation step was induced by low pH (around 6.0) exposure, and resulted in an irreversible conformational change of protein located at the cell surface.[32,33] The shortest effective exposure time to low pH was less than 15 s, suggesting that low pH functions as a trigger to initiate the fusion process. The fusogenic protein was later identified as $E_1$ viral glycoprotein.[19,34] It is, therefore, evident that the above mechanism by which SFV induces fusion of cells was a "fusion from within",[34,35] which is caused by replication of the inoculated virus in the cells.

Low pH condition was also required for the fusion of a mosquito *A. albopictus* cell line with Sindbis virus.[36] Fusion of the mosquito cells cultured in Eagle's MEM medium by Sindbis virus occurred optimally at pH 4.6 (a pH which results in destruction of the virus particle). This fusion of cells seemed to be culture medium-dependent, because the same mosquito cells cultured in MM medium did not fuse at any pH (as low as 4.0), although the cells were infected by the virus with equal efficiency.

Syncytia were also formed when a cell line from nonvector mosquito *Toxorhynchites amboinensis* was infected with dengue viruses (DEN 1-4).[37] Typical syncytia appeared from 5 to 10 d postinoculation and progressively grew larger to cover the entire flask surface in 12 d. The dengue viruses also caused syncytium formation in another mosquito cell culture from *Aedes pseudoscutellaris*.[37] Singh's *A. albopictus* cells did not show syncytium formation upon infection with dengue viruses,[37] although Paul et al.[27] reported that they did so with dengue virus 2 as mentioned above.

Inoculation of a mosquito cell line from Singh's *A. aegypti* with an insect virus *Chilo iridescent virus* (CIV) induced massive formation of syncytia at 2 to 4 h postinoculation.[38] Polykaryocytosis did not require viral genome expression. Thus cell fusion appeared in nonpermissive insect cell lines such as Singh's *A. albopictus* or Quiot's gypsy moth, *Lymantria dispar*, cell line or in a vertebrate cell line, $CV_1$, from monkey. Polykaryocytosis occurred faster at 28°C than at 20°C, whereas no cell fusion was detected at 4°C. Ability to fuse cells was lost when CIV suspension was heated at 53°C for 30 min. CIV suspension rendered noninfectious by UV-irradiation still retained the ability to induce cell fusion. However, the viruses neutralized with antiserum lost the ability to induce polykaryocytosis. Another iridovirus *Tipula* iridescent virus (TIV) also showed formation of multinucleate cells with as many as 10 nuclei when it infected cultures of the cell line from hemocytes of the salt marsh caterpillar, *Estigmene acrea*. This cytopathic effect was observed from 7 to 29 h postinoculation.[39]

Densonucleosis virus (DNV) of greater wax moth, *Galleria mellonella*, also has the ability to produce syncytia. When the DNV was inoculated to mouse L cells, the multinucleate cells appeared from the 5th d postinoculation and increased their number with the time. Syncytia containing 4 to 10 nuclei were common.[40,41]

Some plant virus can also be an inducer of cell fusion. According to Hsu,[42] potato yellow dwarf virus (PYDV) induced extensive cell fusion when it infected a leafhopper vector cell line from embryos of *Aceratagallia sanguinolenta*. Similar cell fusion was also observed in a nonvector cell line from a leafhopper, *Dalbulus elimatus*, when the cell-monolayers were inoculated with concentrated virus inoculum. However, vertebrate cells such as L cells and BHK-21 cells did not form polykaryocytes by the inoculation with PYDV. Virus inoculum rendered noninfectious by exposing it to UV light still caused cell fusion, and the longer the UV irradiation the greater the capacity of the preparation to cause cell fusion.

As cited above, there are many reports concerning fusion of insect cells by a virus. However, no one reported synkaryon formation and isolation of hybrid cell clones, except in the case of the human-insect cell hybrid.[16,17]

# REFERENCES

1. **Becker, J.-L.,** Fusion *in vitro* de cellules somatiques en culture de *Drosophila melanogaster*, induites par la Concanavaline A, *C.R. Acad. Sci. Paris Ser. D*, 275, 2969, 1972.
2. **Rizki, R. M., Rizki, T. M., and Andrews, C. A.,** *Drosophila* cell fusion induced by wheat germ agglutinin, *J. Cell Sci.*, 18, 113, 1975.
3. **Halfer, C. and Petrella, L.,** Cell fusion induced by lysolecithin and concanavalin A in *Drosophila melanogaster* somatic cells cultured *in vitro*, *Exp. Cell Res.*, 100, 399, 1976.
4. **Bernhard, H. P.,** *Drosophila* cells: fusion of somatic cells by polyethylene glycol, *Experientia*, 32, 786, 1976.
5. **Nakajima, S. and Miyake, T.,** Cell fusion between temperature-sensitive mutants of a *Drosophila melanogaster* cell line, *Somatic Cell Genet.*, 4, 131, 1978.
6. **Wyss, C.,** TAM selection of *Drosophila* somatic cell hybrids, *Somatic Cell Genet.*, 5, 29, 1979.
7. **Wyss, C.,** Loss of ecdysone sensitivity of a *Drosophila* cell line after hybridization with embryonic cells?, *Exp. Cell Res.*, 125, 121, 1980.
8. **Wyss, C.,** Cell hybrid analysis of ecdysone sensitivity and resistance in *Drosophila* cell lines, in *Invertebrate Systems In Vitro*, Kurstak, E., Maramorosch, K., and Dübendorfer, A., Eds., Elsevier/North Holland, Amsterdam, 1980, 279.
9. **Berger, E. and Wyss, C.,** Acetylcholinesterase induction by β-ecdysone in *Drosophila* cell lines and their hybrids, *Somatic Cell Genet.*, 6, 631, 1980.
10. **Berger, E., Ireland, R., and Wyss, C.,** Pattern of peptide synthesis in *Drosophila* cell lines and their hybrids, *Somatic Cell Genet.*, 6, 719, 1980.
11. **Suzuki, N. and Wu, Chun-Fang,** Fusion of dissociated *Drosophila* neurons in culture, *Neurosci. Res.*, 1, 437, 1984.
12. **Mitsuhashi, J.,** Fusion of insect cells, in *Biotechnology in Invertebrate Pathology and Cell Culture*, Maramorosch, K., Ed., Academic Press, San Diego, 1987, 387.
13. **Mitsuhashi, J.,** Media for insect cell cultures, in *Advances in Cell Culture Vol. 2*, Maramorosch, K., Ed., Academic Press, New York, 1982, 133.

14. **Grace, T. D. C.,** Establishment of four strains of cells from insect tissues grown *in vitro, Nature,* 195, 788, 1962.

15. **Greene, A. E., Charny, J., Nickol, W. W., and Coriell, L.,** Species identity of insect cell lines, *In Vitro,* 7, 313, 1972.

16. **Zepp, H. D., Conover, J. H., Hirschhorn, K., and Hodes, H. L.,** Human-mosquito somatic cell hybrids induced by ultraviolet-inactivated Sendai virus, *Nature (London) New Biol.,* 229, 119, 1971.

17. **Conover, J. H., Zepp, H. D., Hirschhorn, K., and Hodes, H. L.,** Production of human-mosquito somatic cell hybrids and their response to virus infection, *Curr. Topics Microbiol. Immunol.,* 55, 85, 1971.

18. **Morikawa, H., Mitsuhashi, J., and Yamada, Y.,** Interkingdom electrofusion between tobacco mesophyll protoplasts and cultured butterfly cells, *Plant Tissue Culture Lett.,* 5, 90, 1988.

19. **Omar, A. and Koblet, H.,** Application of mosquito cell culture and toga virus for studying the mechanism of membrane fusion, in *Invertebrate Cell System Applications,* Mitsuhashi, J., Ed., CRC Press, Boca Raton, FL, 1989, 151.

20. **Schaap, G. H., Verkerk, A., van der Kamp, A. W. M., and Jongkind, J. F.,** Selection of proliferating cybrid cells by dual laser flow sorting. Isolation of teratocarcinoma × neuroblastoma and teratocarcinoma × endoderm cybrids, *Exp. Cell Res.,* 140, 299, 1982.

21. **Jongkind, J. F., Verkerk, A., Schaap, G. H., and Galjaard, H.,** Non-selective isolation of fibroblast-cybrids by flow sorting, *Exp. Cell Res.,* 130, 481, 1980.

22. **Miltenburger, H. G., Naser, W. L., and Schliermann, M. G.,** Establishment of a lepidopteran hybrid cell line by use of a biochemical blocking method, *In Vitro Cell. Develop. Biol.,* 21, 433, 1985.

23. **Becker, J. L.,** Purine metabolism pathway in *Drosophila* cells grown *in vitro:* Phosphoribosyl transferase activities, *Biochimie,* 56, 779, 1974.

24. **Moiseenko, E. V. and Kakpakov, V. T.,** The absence of hypoxanthine-guanine phosphoribosyltransferase in extracts of *Drosophila melanogaster* flies and established embryonic diploid cell line, *Drosophila Inf. Serv.,* 51, 44, 1974.

25. **Okada, Y., Suzuki, T., and Husaka, Y.,** Interaction between influenza virus and Ehrlich's tumor cells. II. Fusion phenomenon of Ehrlich's tumor cells by the action of HVJ Z strain, *Med. J. Osaka Univ.,* 7, 709, 1957.

26. **Mazzone, H. M.,** Cell fusion studies on invertebrate cells *in vitro,* in *Invertebrate Cell System Applications,* Vol. 1, Mitsuhashi, J., Ed., CRC Press, Boca Raton, FL, 1989, 135.

27. **Paul, S. D., Singh, K. R. P., and Bhat, U. K. M.,** A study on the cytopathic effect of arboviruses on cultures from *Aedes albopictus* cell line, *Ind. J. Med. Res.,* 57, 339, 1969.

28. **Stollar, V. and Thomas, V. L.,** An agent in the *Aedes aegypti* cell line (Peleg) which causes fusion of *Aedes albopictus* cells, *Virology,* 64, 367, 1975.

29. **Igarashi, A. M., Harrap, K. A., Casals, J., and Stollar, V.,** Morphological, biochemical, and serological studies on a viral agent (CFA) which replicates in and causes fusion of *Aedes albopictus, Virology,* 74, 174, 1976.

30. **Spāth, P. J. and Koblet, H.,** *Aedes albopictus* (mosquito) cells showing strong response to Semliki forest virus (SFV) infection, *Experientia,* 34, 1664, 1978.

31. **Koblet, H., Kempf, C., Kohler, U., and Omar, A.,** Conformational change at pH 6 on the cell surface of Semliki forest virus-induced *Aedes albopictus* cells, *Virology,* 143, 334, 1985.

32. **Spāth, P. J. and Koblet, H.,** Alphavirus induced syncytium formation in *Aedes albopictus* cell cultures, in *Invertebrate Systems In Vitro,* Kurstak, E., Maramorosch, K. and Dübendorfer, A., Eds., Elsevier/North-Holland, Amsterdam, 1980, 375.

33. **Reigel, F., Stalder, J., and Koblet, H.,** Semliki forest virus assembly in *Aedes albopictus* cells is inhibited at low pH, *Experientia,* 38, 1369, 1982.

34. **Koblet, H., Omar, A., Kohler, U., and Kemph, Ch.,** Investigation of cell-cell fusion in Semliki forest virus (SFV) infected C6/36 (mosquito) cells, in *Invertebrate and Fish Tissue Culture,* Kuroda, Y., Kurstak, E., and Maramorosch, K., Eds., Japan Science Society Press/Springer-Verlag, Tokyo, 1987, 140.

35. **Omar, A., Flaviano, A., Reigel, F., Kohler, U., and Koblet, H.,** Syncytium formation and inhibition in Semliki-forest virus-induced *Aedes albopictus* cells at low pH, in *Invertebrate Cell System Applications,* Mitsuhashi, J., Ed., CRC Press, Boca Raton, FL, 1989, 147.

36. **Edwards, J. and Brown, D. T.,** Sindbis virus induced fusion of tissue cultured *Aedes albopictus* (mosquito) cells, *Virus Res.,* 1, 705, 1984.

37. **Kuno, G.,** Replication of dengue, yellow fever, St. Louis encephalitis and vesicular stomatitis viruses in a cell line (TRA-171) derived from *Toxorhynchites amboinensis, In Vitro,* 17, 1011, 1981.

38. **Cerutti, M. and Devauchelle, G.,** Cell fusion induced by an invertebrate virus, Brief report, *Arch. Virol.,* 61, 149, 1979.

39. **Mathieson, W. B. and Lee, P. E.,** Cytology and autoradiography of *Tipula* irridescent virus infection of insect suspension cell cultures, *J. Ultrastructure Res.,* 74, 59, 1981.

40. **Kurstak, E., Chagnon, A., Hudon, C., and Trudel, M.,** Formation de cellules polynucléees dans un système monocellulaire de la souche L de tissu sous-cutané de souris en contact avec le virus de la densonucléose, 2nd Int. Colloq. Invertebr. Tissue Culture, Como, 1967, 264.
41. **Kurstak, E., Belloncik, S., and Brailovsky, C.,** Transformation de cellules L de souris par un virus d'invertébrés: le virus de la densonucléose(VDN), *C.R. Acad. Sci. Paris Ser. D,* 269, 1716, 1969.
42. **Hsu, H. T.,** Cell fusion induced by a plant virus, *Virology,* 84, 9, 1978.

*Medical Aspects*

Chapter 11

# CONTROL OF MOSQUITO VECTORS BY GENETICALLY ENGINEERED *BACILLUS THURINGIENSIS* AND *B. SPHAERICUS* IN THE TROPICS

## Somsak Pantuwatana

## TABLE OF CONTENTS

# I. INTRODUCTION

Malaria, dengue hemorrhagic fever, encephalitis, and filariasis are relatively common mosquito-borne diseases that cause serious problems in many countries in the tropics, especially in Thailand. Since we still lack practical immunization or therapeutic measures for preventing or treating these diseases and the development of drug-resistant malarial organisms in some areas, control of these diseases relies entirely on the control of mosquito vector, on the early identification of cases, and, if possible, on the isolation of cases from contact with mosquito vectors.

Chemical insecticides are conventionally used to control mosquitos in many countries. However, concerns over the environmental consequences, the resulting legal constraints regarding the use of nonselective and persistent pesticides, and the development of insecticide resistance by target species have brought about intensified efforts to search for alternative vector control strategies. One option is to exploit organisms that are natural enemies of vectors and naturally regulate vector population growth. Biological agents had for some time been successfully used for agricultural pest control, and the research community was beginning to take an interest in their possible application to the control of vectors of human disease. A number of viruses, bacteria, and fungi have been identified as potential vector control agents and are being tested in the laboratories of the few research groups involved in this work. Biological agents tend to affect only a narrow range of organisms and are therefore ecologically more acceptable than most chemical pesticides. In the search for safer and more lasting methods, biologists have turned their attention to the possibility of using organisms as biological control agents, and microbiologists are contributing to the development of microbial control agents.

# II. PROBLEMS IN MOSQUITO CONTROL

Several measures which have been put into practice to control mosquito vectors are

1. Measures designed to prevent or reduce the breeding of mosquitoes by eliminating the collection of water or by altering the environment
2. Measures designed to destroy the larvae of mosquitoes
3. Measures designed to destroy adult mosquitoes

Since the immature stages of mosquito are permanent residents of an aquatic environment, source reduction, in which the elimination of mosquito-producing sites by environmental modification and environmental manipulation includes all procedures that specifically modify the environment in which mosquitoes breed so that it is no longer suitable for that purpose, has been used as the basic approach to the control of mosquito larvae for quite sometime. However, this approach is often difficult to achieve over large areas or is exceedingly expensive, especially for the developing countries located in the tropics. In addition to this, manipulating the environment for mosquito control may result in new problems, i.e., draining and filling marshes may eliminate fish and wildlife habitats. Thus, source-reduction programs may not eliminate the source of mosquitoes or disease.

Area-wide spraying of insecticides i.e., "space spraying", provides an important means for reducing or eliminating adult mosquito populations on an emergency basis. Control of adult mosquitoes by space spraying of any kind is only temporary, since mosquitoes from nonsprayed areas can move rapidly into the sprayed area following spray applications, and there is usually little or no effect on the aquatic stages, so that emergence of adults will continue.

In general, space sprays may be applied as thermal fogs, or ultra-low volume (ULV) cold fogs, either with ground-based or aerial equipment. However, the use of pesticides for mosquito control requires considerable caution to assure the safety of the public and the spray team. The greatest potential exposure of the public is encountered when larvicides are used. In some cases, if there is an accidental spill of the concentrated insecticide, the spray crew comes into contact with technical grade insecticides regularly and may experience skin contamination, as well as exposure to aerosols. At high risk are the laborers who fill the ULV machines, the men operating the machines, and the field men who perform larviciding. The dangers of accidentally ingesting the insecticide through eating and drinking practices associated with working conditions should be considered. Adulticiding operations, especially by aircraft, can present a hazard to certain nontarget animal species. Honey bees are especially susceptible to some insecticides applied when bees are active.

In summary, there are some limitations in the use of pesticides. First, chemical control is temporary and must be repeated frequently. Second, pesticides may present toxic hazards to man and the environment. Third, some pesticides may cause unsightly strains or damage automobile finishes and other surfaces. Fourth, specialized equipment and skilled personnel are required for the proper application of insecticides. Fifth, insecticides are expensive and costs continue to rise; and finally, insecticide resistance may develop within mosquito populations after prolonged usage.

## III. PROSPECTS OF USING MICROBIAL CONTROL AGENTS IN THE FIELD

In the tropics, especially in Southeast Asia, the four major mosquito-borne diseases are dengue hemorrhagic fever, which is transmitted by *Aedes aegypti* and *Ae. albopictus;* Japanese encephalitis, which is transmitted by *Culex tritaeniorhynchus* and *Cx. gelidus;* malaria, which is transmitted by anopheline mosquitoes, especially *An. minimus* and *An. dirus;* and filariasis, which is transmitted by *Culex* and *Mansonia* species. All of these mosquito vectors occur in different and wide varieties of water habitats that make efforts to control mosquitoes more difficult.

Chemical agents had been selected and used as effective weapons to control mosquitoes since 1938. However, this prophecy was not long in fulfilment when evidence of resistance to chemical insecticides in mosquito species including anopheline species[1] was increasingly encountered. This gradual and relentless spread of resistance to insecticides promoted an increase in research to find new insecticides to replace those to which resistance had developed, and to find alternative ways to control mosquitoes, which in turn will slow down the rate of development of resistance.

Recently, two spore-forming bacterial agents have been considered as promising agents in mosquito control programs. These two bacilli are *Bacillus thuringiensis* subsp. *israelensis* (*Bti*) and *B. sphaericus*. They produce endotoxins and can be cultured relatively easily. The endotoxins act as stomach poisons when ingested by insect larvae. These products can be formulated, stored, and shipped like other chemical agents, which is a great advantage over living organisms that must be handled much more carefully. Several large companies are now manufacturing and formulating *Bti* and *B. sphaericus* on a large scale for public health uses, and field tests and operational trials of these agents are under investigation. Therefore, there is a prospect for the operational use of these two bacterial agents to control mosquitoes in the tropics.

*Bti* was first isolated by Goldberg and Margalit.[2] After the isolation, this strain was used in numerous laboratory studies on its activity against mosquitoes and blackfly larvae. In general, *Aedes* species were most susceptible, followed by *Culex* and *Anopheles,* in that order. It was also documented that younger instars were more susceptible to a given preparation or formulation than older instars. In every case, susceptibility decreased with increasing instars.[3]

In Thailand, similar results were obtained and it was demonstrated that *Ae. aegypti* larvae were the most susceptible to *Bti*.[4] Since *Ae. aegypti* larvae are found mainly in jars or artificial water container indoors, thus detailed studies on field efficacy of *Bti* formulations are limited. Field trials conducted consisted of simulated field tests, and indoor conditions in jars using field water, and *Ae. aegypti* larvae from colonies or field-collected specimens.

Preliminary small-scale field trials were conducted in Ban Plu village, Song Khla province, which was known as an endemic area of dengue hemmorhagic fever and malaria. This small village was well isolated from other villages and surrounded by rubber plantations. Technical powder of *Bti* obtained from Abbott Laboratories was used to evaluate against *Ae. aegypti* larvae, in a water jar containing 50 l of well water. The application was made by adding *Bti* powder in the amount to obtain a given final concentration, i.e., 400, 100, 10, 4, 1, 0.4, and 0.1 mg/l. The assessment of larvicidal activity was done by using 50 4th instar larvae of laboratory-reared *Ae. aegypti* at weekly intervals and results were observed after 48 h exposure. It was shown that *Bti* gave good activity against *Ae. aegypti* larvae in about 41 d at a concentration of 0.4 mg/l.[5] In another case of simulated indoor conditions in Bangkok, 100% reduction was obtained in *Ae. aegypti* and *Cx. quinquefasciatus* at the rate of $1.33 \times 10^4$ spores per milliliter of product made from Abbott wettable powder formulation. The level of control persisted up to 100 and 70 d for *Ae. aegypti* and *Cx. quinquefasciatus*, respectively.[6] Poor control of *Cx. quinquefasciatus* was demonstrated in polluted waste water in Bangkok at the rate of application of 16 l/1600 m$^2$ of the concentrated liquid formulation made locally ($2 \times 10^{10}$ viable spores per milliliter). Field studies reported by others[7] clearly demonstrated that currently available formulations of *Bti* produce an immediate high level of control of larvae, with the extent of control declining rapidly after 7 d. Thus, repeated applications (weekly treatment) are needed in situations where continuous breeding and recruitment are taking place.

*Bacillus sphaericus* was isolated and its larvicidal activity was also elucidated for quite a length of time. Recently, two potent strains (1593 and 2362) of *B. sphaericus* have been isolated and evaluated against mosquitoes under laboratory conditions resulting in a satisfactory result.[3,8,9] In most of these studies, it was determined that *Aedes* species were less susceptible or unaffected by *B. sphaericus*. It was also documented that younger instars were more susceptible to a given preparation or formulation than older instars, and susceptibility decreased with increasing instars similar to those found with *Bti*.[10]

Trials with *B. sphaericus* on *Cx. quinquefasciatus* larvae breeding in polluted and sewage water produced a good reduction in larval populations. This evidence was based on results of studies in Bangkok, Thailand.[11] Preparations of *B. sphaericus* 1593 containing $3.53 \times 10^{10}$ viable spores per milliliter were employed against *Cx. quinquefasciatus* in artificial pools. The treated sites varied from small pools to 1600 m$^2$. In one trial, the bacterial preparation (cell count $3.53 \times 10^{10}$ viable spores per milliliter) at the concentration of $10^{-8}$ dilutions was evaluated in 1 m$^2$ ponds containing polluted sewage water against *Cx. quinquefasciatus* in Bangkok. The efficacy was assessed by placing 100 *Cx. quinquefasciatus* larvae consisting of second-, third-, and fourth-instar larvae from a colony or from field-collected larvae in sentinel cages. The sentinel cages were made from plastic boxes of 7 cm × 7.5 cm × 10 cm in size that had been cut on each side and then covered with nylon screen on all sides. The cage allowed the larvae to have access to the water surface. The mortality of the sentinel larvae was observed and recorded 48 h after being placed in the test pond. The dosage applied ranged from $3.0 \times 10^5$ to $1.8 \times 10^8$ cells, per milliliter. *B. sphaericus* gave excellent larvicidal activity against sentinel larvae for 60 d and the percent mortality that was higher than 50% lasted up to 190 d in the first pool. In the second pool, *B. sphaericus* also gave excellent activity against sentinel larvae for 53 d. The larvicidal activity fluctuated above and below the 50% mortality from 60 d onward. The populations of bacteria fluctuated in both pools. The sentinel technique, at the concentration used, produced 80 to 100% mortality when exposed at 48 h posttreatment intervals, and the treatment was effective for at least 190 d.

In another trial, a concentrated liquid formulation of *B. sphaericus* 1593 consisting of $1.0 \times 10^{10}$ to $3.0 \times 10^{10}$ viable spores per milliliter was applied at 16 l/1600 m² to polluted sewage ponds supporting population of *Cx. quinquefasciatus* in Bangkok during September and November 1985. The bacterial agent yielded 80 to 100% control of treated larvae in all four trials, with the level of control persisting up to 10 d posttreatment.

In similar fashion, a concentrated liquid formulation of *B. sphaericus* 1593 consisting of $2.4 \times 10^{7}$ to $5.7 \times 10^{8}$ viable cells per milliliter was also evaluated against *Anopheles minimus* larvae in a clear-water slow-running stream in the Nakorn Ratchasima province in Thailand. This stream was supporting populations of *An. minimus* and *An. maculatus* with the majority of 99% being *An. minimus*. This small stream was about 9 km long, starting from the foothill. During the peak season of *An. minimus,* the water level was very low and the average discharge of water was measured at 2,228.5 cm³/min in the area of slow current, and 0.61 m³/min at the fastest running section of the stream. The width of the stream varied from 30 cm to 2 m. The average temperature of the water during the course of the study was $22 \pm 2°C$ and the pH of water was 7.5. The average temperature during the course of the study was 24.8°C. The application of *B. sphaericus* 1593 was made twice at 1-week intervals. The first application was made by dropping through the faucet from the container into the stream at the rate of 15 l/2 h, the second application was made by water sprinkling along the margin of the stream for 450 m long in distance. The assessment of mosquito larval populations was made by using a dipping technique and sentinel larvae. The application yielded 60 to 100% control of treated larvae with the level of control persisting for up to 9 weeks posttreatment. *B. sphaericus* gave excellent larvicidal activity against sentinel *An. minimus* larvae for at least 10 weeks.

The residual activity of *B. sphaericus* was studied in different biotypes to which the pathogen was applied and the bacterial agent was found to be surviving in the polluted water of the treated habitats for at least 6 months after treatment. In tap water and clear water the organism was also found to be surviving for at least 9 and 3 months after treatment, respectively.[6] It was shown that *B. sphaericus* 1593 remained active for at least 6 months against *Cx. quinquefasciatus* larvae in artificial pools containing sewage water.[11]

In summary, the residual activity of these two bacterial agents is still selective against certain mosquito species and varied according to mosquito species and environmental conditions.

## IV. RESTRAINTS IN THE USE OF EXISTING MICROBIAL CONTROL AGENTS IN THE FIELD

Following preliminary investigations including small- and large-scale field trials and in laboratory conditions, the use of these bacteria for practical vector control has both advantages and disadvantages. The main advantage is the production of stable spore-crystal mixtures. However, the evidence of epizootics is very rare and the bacterial preparations are rapidly inactivated after application, so only short-term protection from the vector population is obtained from a single treatment.

*Bti* is found to be effective against *Aedes, Culex,* and to some extent, *Anopheles* and *Mansonia.* However, this biocontrol agent in its present formulation tends to disappear quickly from the feeding zone of vector mosquito larvae, particularly anophelines. It appears to be unable to propagate itself effectively in the biotope, and thus there is a need to apply it repeatedly to obtain good protection from the vector population. The information on the biology and ecology of *Bti* and on its reproduction in nature is very limited at the present time. It has been used effectively against *Simulium* in large-scale operations,[12,13] since only short exposures in specific formulation are required.

*B. sphaericus* is effective against mosquitoes, but not against *Simulium*. Among the mosquitoes, it is effective particularly against species of *Culex,* and generally less effective against *Anopheles, Mansonia,* and *Aedes* in that order. Although it demonstrates a recycling potential in certain environmental conditions,[14] it is yet to be confirmed how long *B. sphaericus* remains effective in the biotope. If it is proven that it is able to propagate in all kinds of biotope, this would reduce the rate of application. There is also a little information on the biology and ecology of *B. sphaericus*. It has been shown that UV rays affect its efficacy in clear water.[15]

One drawback concerning the application of these two bacterial agents is the dosage of application. However, the technical material from which bacterial larvicides are formulated is comparable in vector toxicity to many organochlorine and organophosphate insecticides. It has been estimated that the $LC_{50}$ of powders made by drying fermentation slurries of *Bti* and *B. sphaericus* is usually in the range of 0.1 to 0.2 mg per dose against 4th instar larvae of *Cx. quinquefasciatus*.[16,17] When applied in the field, these powders, applied at rates of 100 to 400 g/ha, yield mortality levels above 90% for 3rd and 4th instars of many mosquito species.[7] However, somewhat higher application rates are needed to ensure effective control. In *Bti,* the toxins are believed to represent 20 to 25% of the cell dry weight, and in *B. sphaericus* the toxin is thought to represent only 5% of the dry weight. Therefore, because these toxins are proteins, improvement of the toxic potential, using the tools of genetic engineering, particularly in the case of *B. sphaericus,* is a realistic possibility.

## V. ROLES OF MOLECULAR BIOLOGY AND GENETIC ENGINEERING IN THE DEVELOPMENT OF BIOCONTROL OF DISEASE VECTORS

Since the discovery of *Bti,* it has become one of the most promising potential bioinsecticides against larvae of mosquitoes. A number of field trial studies are in progress to utilize this bacterium for biological control of mosquito vectors. There are three properties which are foreseen as limitations to the effectiveness of *Bti* for mosquito control. The first is its genetic instability from the point of view of toxin production; the second is its limited persistence in natural environments, and finally, it has a narrow host range in which the dosage required for control of malaria-carrying mosquitoes (*Anopheles* spp.) is rather high. On the other hand, *B. sphaericus,* another famous biological control agent, demonstrates a good potential on recycling and persistence in certain conditions.[6,14,18,19] However, recycling does not always occur, but seems to be favored by organic pollution. In general, *Anopheles* are more susceptible to *B. sphaericus* than they are to *Bti,* but this varies according to species and even strains of the same species. Particulate matter, strong UV radiation, low temperatures, and alkalinity decrease larvicidal activity.[20,21] It is more interesting to find a good strain of these two bacteria that can persist better in the environment and is more effective against *Anopheles* larvae than the existing strains. One disadvantage in using the existing strains of *B. sphaericus* is that *B. sphaericus* requires media rich in amino acid for growth. Thus, the cost of production is a little more expensive than *Bti* which requires media rich in carbohydrates. Attempts have been made to engineer strains of *B. sphaericus* that will grow on media rich in carbohydrates. Consequently, improvement in larvicidal activity reduces the cost of production, and broadening of the host range is a more appropriate alternative against malaria-carrying mosquitoes.

Gene manipulation techniques have brought a new dimension to applied genetics. The initial step in developing a biotechnological process is generally a search for suitable organisms. Such organisms will be expected to create a product or service that will generate a financial return to that industry. Selection of improved organisms is tedious and time-demanding and most of the methods available to the geneticists, up until recently, have involved trial and error. However, new genetic technologies, i.e., protoplast fusion, conjugation, and the use of recombinant DNA techniques, allow new approaches by which useful genetic traits can be inserted directly into the chosen organisms. Thus, totally new capabilities can be engineered and microorganisms in particular may be made to produce substances beyond their naturally endowed genetic capabilities.

The new methods of genetic engineering give the biologist direct access to the genome. A gene can be inserted, deleted, altered, or duplicated. For example, a gene from the organism can be transferred into the genome of another, and such a transfer of genetic information can be accomplished even across the evolutionary distance that separates a human being from a bacterium. An artificial gene, encoding instructions for making a molecule unknown in nature, can be assembled by chemical means and introduced into a microorganism. It has been shown that the activity of *Bti* and *B. sphaericus* is a result of the activity of toxins produced by the bacteria. These toxins are coded by genes found inside the bacterial cell. Therefore, it is possible to improve the toxicity of these two bacteria by using the new methods of genetic engineering.

Among the practical considerations to be dealt with are persistence in the larval feeding zone and the inevitable problem of larval resistance. Learning more about the regulation of protoxin synthesis and the number of factors involved should be helpful in appreciating why these organisms may or may not persist in a particular environment. Protoxin gene stability (i.e., plasmid replication properties), germination rates, nutritional requirements, etc. will all play a role. Furthermore, due to restraints in the use of existing microbial control agents in the field, it is feasible to use recombinant DNA methods to engineer strains providing higher toxin production, or to produce toxins in easily handled organisms that can propagate freely or persist long enough in the feeding zone of mosquito larvae, particularly of *Anopheles* and *Culex* species.

## VI. BASIC METHODOLOGY OF GENETIC ENGINEERING

Genetic engineering has currently brought a revolution in the study of gene structure and function. It is now possible to isolate genes and to determine their nucleotide sequence. This includes both the coding region and the regulatory region which determine the expression. It is also possible to modify the coding region by directed mutagenesis or by gene fusion. This allows a detailed study of the structure and function relationship of a protein. The expression can also be influenced by modifying the regulatory region. Finally, it is possible to transfer and express genes into new hosts. In this manner, desired traits can be exchanged between unrelated organisms.

The basis of this new technology, which is most typically referred to as recombinant DNA technology or genetic engineering, rests on our ability to cleave DNA at specific sites using restriction enzymes, and then transfer, select, and study genes of interest and their protein products in pure form in, or as obtained from, other hosts. To date, the most commonly used alternate host is the bacterium *Escherichia coli (E. coli)*, although a variety of other hosts, including yeasts, algae, and the cells of higher animals and plants, can now be used to express and study foreign proteins.

With regard to microbial control agents for vectors, the development of recombinant DNA technology has provided the opportunity to study and improve the efficacy of microbial larvicides, particularly those based on the bacteria *Bti* and *B. sphaericus*, whose insecticidal properties are due to the highly selective protein toxins they produce.

## VII. STATE OF GENETIC ENGINEERING OF
## *B. THURINGIENSIS* H-14

Information obtained from thorough analysis of protein crystal indicates that the toxicity of *Bti* is due to one or more proteins in the parasporal body.[22,23] Some evidence suggests that the high toxicity characteristic of the parasporal body may not be due to a single protein, but rather to an interaction, perhaps synergistic, of two or more proteins.[24]

Attempts have been made by using recombinant DNA techniques as a tool to improve toxicity of *Bti*. Efforts are emphasized on transferring genes encoding toxic proteins that have relative molecular weight of about 28,000 and 130,000 Da into several selected species of microorganisms. These efforts rely on the facts that strain of *Bti* harbor several resident plasmids, the size of which range from 4.5 to 180 kilobase pairs (kb). From plasmid-curing experiments and by transfer of plasmids between strains, it has been shown that a 112-kb plasmid is associated with crystalline toxin production in *Bti*.[25,26]

After the sequencing of genes encoding the 28,000-Da toxin had been done, it was shown that the *E. coli* clones harboring the recombinant plasmid show a hemolytic activity and have a slight toxicity to mosquito larvae.[27,28] The corresponding insert DNA were introduced into *B. subtilis* competent cells using shuttle vectors.[29] *B. subtilis* clones harboring the recombinant plasmid produced hemolytic activity only in the sporulation phase. The synthesis of the toxin was dramatically decreased in mutants blocked at early stages of sporulation. The toxin was deposited as refractile inclusions when it was produced in sufficient amounts. The inclusions were purified and consisted solely of the 28,000-Da polypeptides. It was also demonstrated that only the native insoluble inclusions were toxic to *Ae. aegypti*, whereas solubilized extracts were not toxic. Analysis of the hydropathy plot of the 28,000-Da protein showed that it consisted of high hydrophobic regions interrupted by short hydrophylic stretches.

It appeared that two polypeptide bands were revealed when the 130,000-Da polypeptide was resolved by SDS-PAGE. This observation suggested the existence of two genes. One of these genes has been cloned successfully into *B. megaterium* by ligating DNA fragments of the 112-kb plasmid with the plasmid pBC 16 and transformed into the recipient cell.[26] The resulting recombinant strain produced irregular phase refractile crystalline inclusions during sporulation and its toxicity to mosquito larvae was similar to that of the original *B. thuringiensis* strain. The recombinant plasmid induced the synthesis of a 130-kDa polypeptide which reacted with antibodies directed against the *Bti* crystal proteins. Two genes encoding the 130-kDa polypeptide were also cloned successfully into selected organisms by several investigators. So far, the gene encoding the 28-kDa and 130-kDa protein has been successfully cloned into selected organisms, i.e., *E. coli*, *B. megaterium*, *B. subtilis*, *B. sphaericus*, and cyanobacteria.[26,27,29-37]

# VIII. STATE OF GENETIC ENGINEERING OF *B. SPHAERICUS*

*B. sphaericus* is also one of the spore-forming bacteria which had been recommended as promising biological control agents for mosquito vectors. Strain improvement for better and broad-spectrum larvicidal activity of this organism has been emphasized.

It has been demonstrated that polypeptides of 110 and 43 kDa synthesized by *B. sphaericus* are toxic to *Cx. pipiens* larvae. However, when a tissue culture system is used, only 43-kDa protein is found to be toxic to tissue-cultured cells of *Cx. quinquefasciatus*.[38] Some evidence suggests that in *B. sphaericus* there are two sequential pathways of toxin activation: a conversion of 125-kDa to 43-kDa proteins during sporulation and a conversion of 43-kDa to 40-kDa proteins in the mosquito larval gut.[38-41]

Genes encoding for mosquito toxins of *B. sphaericus* had been cloned successfully into selected recipient organisms, i.e., *E. coli* and cyanobacterium, *Anacystis nidulans* R2.[38-41] Progress has been made to clone mosquito toxins into selected organisms that can propagate freely in aquatic environments and survive in the larval feeding zone.

Recently, it has been demonstrated that *B. sphaericus* transformants carrying the *Bti* toxin gene (3.7 kb) have good toxicity against *Ae. aegypti*, *An. dirus*, and *Cx. quinquefasciatus* larvae in which toxicity against *Ae. aegypti* was increased about 100-fold when compared with the standard strains 1593 and 2362. These transformants were found to be quite stable. It was found that the stability of the plasmid on media without the presence of selective pressure remained for at least 4 weeks.[37] Thus, the biological control of mosquito vectors by using genetically engineered microorganisms will probably occur in the near future.

## IX. ASSESSMENT OF RISKS OF DELIBERATE RELEASE INTO THE ENVIRONMENT OF MICROORGANISMS CONTAINING RECOMBINANT DNA AND INTENDED FOR VECTOR CONTROL

One of the ultimate goals of strain improvement using molecular biology and genetic engineering of biological control agents is to obtain recombinant strains that can propagate freely in the environment and exist in the feeding zone long enough to provide a good control of mosquito vectors. It has been extrapolated that the fate of natural and manipulated genetic material is dependent on the survival, establishment, and growth of the microbial hosts which house the genetic material, in the natural habitats into which the hosts are introduced. Survival, establishment, and growth are, in turn, dependent on the genetic constitution of the host and on the physical (e.g., temperature, pressure, electromagnetic radiation, surfaces, spatial relations), chemical (e.g., carbonaceous substrates, inorganic nutrients, growth factors, ionic composition, toxicants), and biological (e.g., characteristics of and positive and negative interactions between microbes) factors of the various habitats.[42,43] The ability of an organism to survive, grow, and colonize new habitats will influence its ability to successfully transfer genetic information.

The transfer of genes between bacteria, either by conjugation, transformation, or transduction, and the splicing of genes to create novel DNA sequences, either by transposons or insertion sequences, are probably natural events that occur in the biosphere. Although the transfer of genes between bacteria can occur across interspecific and intergeneric boundaries, and gene splicings may result in novel nucleotide sequences, the genetic constitution of bacteria that have been isolated from natural habitats appears to have remained basically unchanged. Under certain conditions, anthropogenic activities have modified the bacterial gene pool. This is perhaps best illustrated by the increased incidence of bacteria containing multiple antibiotic resistance encoded on conjugative plasmids, as this increase appears to be correlated with the increased use of antibiotics, both as chemotherapeutic agents and as supplements to animal feeds. The possible adverse influence of genetically engineered bacteria, whether those that have escaped inadvertently or those released purposefully, on the homeostasis of the biosphere, needs to be studied.

Understanding of the risk of the either intentional or accidental release of genetically engineered microbes to the biosphere has been hindered by the lack of information concerning the survival, establishment, and growth of, and on the transfer of genetic information by, genetically engineered microbes in natural environments. One good example is that plasmid pBR322 has been extensively used as a vector for inserting engineered genes into bacteria. Since this plasmid is known to be both nonconjugative and poorly mobilizable, it is considered to be safe. However, it has been demonstrated that this plasmid can be transferred to bacteria indigenous to the human gastrointestinal tract by means of triparental matings.[44] Furthermore, many of the data on gene transfer between bacteria have been obtained with laboratory strains and with the genetic recombination studies performed under optimal laboratory conditions for gene transfer. Such data, however, may not be directly applicable to genetic transfer *in situ*, e.g., in natural aquatic and terrestrial environments. There is insuffficient information on the survival and genetic exchange, neither *in vivo* nor *in situ*, of bacteria other than *E coli*. It is probable that bacterial species engineered for a specific function (e.g., pesticides and pest control) will be compatible with the environment into which they will be introduced. Thus, they will have a vastly greater potential for survival *in situ* than enteric bacteria, and therefore, they will constitute a potentially greater risk to the biosphere.

One aspect concerning safety considerations is the release of genetically engineered microbes. Because of the uncertainty of the potential ecological impacts of the release into natural environments of these microbes, thus, the initial releases should perhaps be into isolated and insulated environments. Islands, considerably distant from the mainland, should be considered for such test releases, not only because of their isolation and insulation, but also because the relative simplicity of insular biotas should enable such potential impacts to be evaluated more easily than impacts on more complex mainland biotas. However, because of the apparent rapid airborne (as well as possible waterborne) spread of microbes between continents, testing on islands should not be considered as an absolute guarantee for the distributional restriction or containment of released genetically engineered microorganisms. Nevertheless, the potential biospheric dangers of such releases may be reduced if initial releases are to insular environments.

Another aspect is the potential effects of the release of genetically engineered microbes on the ecological structure and function of natural environments. It is difficult to design experiments to test the potential perturbations that an engineered microbe might have on the multitude of ecological events. The acquisition of a plasmid that may result in a spectrum of unrelated, unanticipated, and nonpredicted biochemical and other physiological alterations in recipient bacteria suggest that studies designed to evaluate the survival of, and gene transfer by genetically engineered bacteria in natural habitats be alert for such unanticipated and nonpredicted alterations.

## X. DEVELOPMENT OF METHODS TO ASSESS AND REDUCE THE RISK OF INTRODUCING GENETICALLY ENGINEERED LARVICIDAL ORGANISMS INTO THE ENVIRONMENT

For safety considerations, in order to test and evaluate any risk associated with genetically engineered larvicidal organisms, there is a need to identify a set of general factors which may contribute to risk. The following items need to be considered:

1.  Since these organisms possess biological activity, they may control mosquito larvae by a variety of modes of action: lethality, pathogenicity, or deterrence of feeding. The organisms are designed for environmental application for vector control in which they are close to homes. Therefore, exposure to nontarget organisms and humans may be widespread. These microbes are expected to survive and replicate freely in nature. Once released into the environment, absolute containment is not possible. Thus, prior to any release of genetically engineered organisms, their capacity to flourish and find new ecological niches or to compete successfully against other organisms in the environment must be carefully evaluated. Any assessment of environmental impact must take into consideration expected levels of the organisms in the environment but must include reliable predictions of the effect of unusual environmental conditions and their influence on localized increases or blooms in the microbial population. In addition, genetically engineered organisms that are able to persist in the environment in new niches in sufficient numbers may lead to adverse effects on nontarget species. Increased environmental persistence may be a problem under certain circumstances. For example, the *Bti* toxin gene can be inserted into a long-lasting or readily reproducing recipient or host, such as blue green algae. This can be desirable for enhancing the useful lifetime and availability of the *B. thuringiensis* toxin to mosquito larvae, but may also lead to unforeseen problems, especially if the engineered strain is not narrowly focused in its target insect specificity.

2.  Any testing scheme for characterizing genetically engineered microbes or for screening their effects on or exposure to nontarget organisms provides data for characterizing risk. However, placing an engineered organism into a new habitat can lead to unforeseen consequences. Thus, the ecological scenario for any proposed application of engineered organisms must be carefully evaluated, using approaches from a variety of scientific disciplines.

3.  It has been shown that certain types of engineered vectors have a high probability of genetic transfer. This can lead to two problems: inserted genes might be expressed in other species of microorganisms in the environment and, in addition, multiple copies of the gene might possibly be produced in an individual species of microorganisms, thereby increasing the activity. It is desirable to eliminate these kinds of genetic segments from the vectors used for genetic engineering.

4.  Whenever antibiotic resistance traits are used as markers for selection, antibiotic resistance may be a significant problem in the environment. Therefore, it is desirable, when possible in research and development, to use antibiotics which are not important in the treatment of human or animal disease.

5.  Certain genetically engineered microbial pesticides may have enhanced efficacy against the target organism or may incorporate two toxins in one organism. Even though the host range of the nonengineered organism and its effect on beneficial insects may be known, it is worthwhile to reevaluate the effects of the engineered microorganism by screening pathogenicity and infectivity in nontarget and beneficial insects.

6.  With widespread use, after an engineered microbial agent passes all safety screens, the target insect may develop resistance to microbial toxins to the extent that they persist for long periods or continuously in the environment at high levels.

# REFERENCES

1. **Brown, A. W. A.,** Insecticide resistance in mosquitoes: a pragmatic review, *J. Am. Mosq. Control Assoc.,* 2, 123, 1986.

2. **Goldberg, L. J. and Margalit, J.,** Bacterial spore demonstrating rapid larvicidal activity against *Anopheles sergentii, Uranotaenia unguiculata, Culex univittatus, Aedes aegypti,* and *Culex pipiens, Mosq. News,,* 37, 355, 1977.

3. **Anon.,** Biological control agent data sheet: *Bacillus thuringiensis* serotype H-14, WHO Mimeographed Document, WHO/VBC/79.750 Rev. 1, WHO, Geneva, 1979, 46p.

4. **Pantuwatana, S. and Youngvanitsed, A.,** Preliminary evaluation of *Bacillus thuringiensis* serotype H-14 and *Bacillus sphaericus* strain 1593 for toxicity against mosquito larvae in Thailand, *J. Sci. Soc. Thailand,* 10, 101, 1984.

5. **Pantuwatana, S., Silapanuntakul, S., Samasanti, W., and Santinanalert, P.,** Small field trials of mosquito larvicidal *Bacillus thuringiensis* and *Bacillus sphaericus* in Thailand, *Microb. Util. Renewable Resour.,* 4, 366, 1985.

6. **Silapanuntakul, S., Pantuwatana, S., Bhumiratana, A., and Charoensiri, K.,** The comparative persistence of toxicity of *Bacillus sphaericus* strain 1593 and *Bacillus thuringiensis* serotype H-14 against mosquito larvae in different kinds of environments, *J. Invertebr. Pathol.,* 42, 387, 1983.

7. **Davidson, E. W., Sweeney, A. W., and Cooper, R.,** Comparative field trials of *Bacillus sphaericus* strain 1593 and *B. thuringiensis* var. *israelensis* commercial powder formulations, *J. Econ. Entomol.,* 74, 350, 1981.

8. **Davidson, E. W., Singer, S., and Briggs, J.,** Pathogenesis of *Bacillus sphaericus* SSII-1 infection in *Culex pipiens quinquefasciatus* ( = *C. pipiens fatigans*) larvae, *J. Invertebr. Pathol.,* 25, 179, 1975.

9. **Singer, S.,** Insecticidal activity of recent bacterial isolates and their toxins against mosquito larvae, *Nature,* 224, 110, 1973.

10. **Anon.,** Biological control agent data sheet: *Bacillus sphaericus* strain 1593-4, WHO Mimeographed Document, WHO/VBC/80.777, WHO, Geneva, 1980, 16p.

11. **Pantuwatana, S., Maneeroj, R., and Upatham, E. S.,** Long residual activity of *Bacillus sphaericus* 1593 against *Culex quinquefasciatus* larvae in artificial pools, *Southeast Asian J. Trop. Med. Public Health*, 20, 421, 1989.

12. **Undeen, A. H. and Colbo, M. H.,** The efficacy of *Bacillus thuringiensis* against blackfly larvae (Diptera: Simuliidae) in their natural habitat, *Mosq. News*, 40, 181, 1980.

13. **Lacey, L., Escaffre, H., Philippon, B., Seketeli, A., and Guillet, P.,** Large river treatment with *Bacillus thuringiensis* (H-14) for the control of *Simulium damnosum* S.1. in the Onchocerciasis control programme, *Tropenmed. Parasitol.*, 33, 97, 1982.

14. **Hertlein, B. C., Levy, R., and Miller, T. W., Jr.,** Recycling potential and selective retrieval of *Bacillus sphaericus* from soil in a mosquito habitat, *J. Invertebr. Pathol.*, 33, 217, 1979.

15. **Burke, W. F., Jr., McDonald, K. O., and Davidson, E. W.,** Effect of UV light on spore viability and mosquito larvicidal activity of *Bacillus sphaericus* 1593, *Appl. Environ. Microbiol.*, 46, 954, 1983.

16. **Singer, S.,** *Bacillus sphaericus* (bacteria), in *Biological Control of Mosquitoes*, Bull. No. 6, American Mosquito Control Association, Lake Charles, LA, Chapman, H. C., Ed., March 1985, 132.

17. **Lacey, L.,** *Bacillus thuringiensis* serotype H-14, in *Biological Control of Mosquitoes*, Bull. No. 6, American Mosquito Control Association, Lake Charles, LA, Chapman, H. C., Ed., March 1985, 132.

18. **Hornby, J. A., Hertlein, B. C., Levy, R., and Miller, T. W., Jr.,** Persistent activity of mosquito larvicidal *Bacillus sphaericus* 1593 in fresh water and sewage, WHO Mimeograph Document, WHO/VBC/ 81.830, WHO, Geneva, 1981.

19. **Davidson, E. W., Urbina, M., Payne, J., Mulla, M., Darwazeh, H., Dulmage, H., and Correa, J.,** Fate of *Bacillus sphaericus* 1593 and 2362 spores used as larvicides in the aquatic environment, *Appl. Environ. Microbiol.*, 47, 125, 1984.

20. **Mulligan, F., Schaefer, C., and Wilder, W.,** Efficacy and persistence of *Bacillus sphaericus* and *B. thuringiensis* H-14 against mosquitoes under laboratory and field conditions, *J. Econ. Entomol.*, 73, 684, 1980.

21. **Wright, W. P., Molloy, D., Jamnback, H., and McCoy, P.,** Effects of temperature and instar on the efficacy of *Bacillus thuringiensis* var. *israelensis* and *Bacillus sphaericus* strain 1593 against *Aedes stimulans* larvae, *J. Invertebr. Pathol.*, 38, 78, 1981.

22. **Whiteley, H. R., and Schnepf, H. E.,** The molecular biology of parasporal crystal body formation in *Bacillus thuringiensis, Annu. Rev. Microbiol.*, 40, 549, 1986.

23. **Höfte, H. and Whiteley, H. R.,** Insecticidal crystal proteins of *Bacillus thuringiensis, Microbiol. Rev.*, 53, 242, 1989.

24. **Wu, D., and Chang, F. N.,** Synergism in mosquitocidal activity of 26- and 65-kDa proteins from *B. thuringiensis* sub. *israelensis* crystal, *FEBS Lett.*, 190, 232, 1985.

25. **Gonzalez, J. M. and Carlton, B. C.,** A large transmissible plasmid is required for crystal toxin production in *Bacillus thuringiensis* var. *israelensis, Plasmid*, 11, 28, 1984.

26. **Sekar, V. and Carlton, B. C.,** Molecular cloning of the delta-endotoxin gene of *Bacillus thuringiensis* var. *israelensis, Gene*, 33, 151, 1985.

27. **Waalwijk, C., Dullemans, A. M., Van Workum, M. E. S., and Visser, B.,** Molecular cloning and the nucleotide sequence of the Mr 28,000 crystal protein gene of *Bacillus thuringiensis* subsp. *israelensis, Nucleic Acid Res.*, 13, 8207, 1985.

28. **Bourgouin, C., Klier, A., and Rapaport, G.,** Characterization of the genes encoding the hemolytic toxin and the mosquitocidal delta-endotoxin of *Bacillus thuringiensis israelensis, Mol. Gen. Genet.*, 205, 390, 1986.

29. **Ward, E. S., Ridley, A. R., Ellar, D. J., and Todd, J. A.,** *Bacillus thuringiensis* var. *israelensis* delta-endotoxin: encoding and expression of the toxin in sporogenic and asporogenic strains of *Bacillus subtilis, J. Mol. Biol.*, 191, 13, 1986.

30. **Ward, E. S., Ellar, D. J., and Todd, J. A.,** Cloning and expression in *E. coli* of insecticidal delta-endotoxin gene of *Bacillus thuringiensis* var. *israelensis, FEBS Lett.*, 175, 377, 1984.

31. **Ward, E. S. and Ellar, D. J.,** *Bacillus thuringiensis* var. *israelensis* delta-endotoxin nucleotide sequence and characterization of the transcripts in *Bacillus thuringiensis* and *E. coli, J. Mol. Biol.*, 191, 1, 1986.

32. **Pantuwatana, S., Bhumiratana, A., Panyim, S., and Wilairat, P.,** Improvement of Bacterial Agents for Control of Mosquito Vectors, in Proc. His Majesty's 5th Cycle Commemorative Conf. USAID Sci. Res. Grantees, held in Nakorn Pathom, Thailand, July 24 to 26 1987, 65.

33. **Angsuthanasombat, C., Chungjatupornchai, W., Kertbundit, P., Luxananil, C., Settasatian, C., Wilairat, P., and Panyim, S.,** Cloning and expression of 130-kDa mosquito-larvicidal delta-endotoxin gene of *Bacillus thuringiensis* var. *israelensis* in *E. coli, Mol. Gen. Genet.*, 208, 384, 1987.

34. **McLean, K. M. and Whiteley, H. R.,** Expression in *E. coli* of a cloned crystal protein gene of *Bacillus thuringiensis* subsp. *israelensis, J. Bacteriol.*, 169, 1017, 1987.

35. **Ward, E. S. and Ellar, D. J.,** Cloning and expression of two homologous genes of *Bacillus thuringiensis* subsp. *israelensis* which encode 130-kDa mosquitocidal proteins, *J. Bacteriol.*, 170, 727, 1988.

36. **Chungjatupornchai, W.,** Characterization and Expression of Genes Encoding Mosquitocidal Proteins of *Bacillus thuringiensis* in *E. coli, Bacilli,* and Cyanobacteria, Ph.D. thesis, Free University of Brussels, Brussels, Belgium, 1989.

37. **Trisrisook, M., Pantuwatana, S., Bhumiratana, A., and Panbangred, W.,** Molecular cloning of the 130-kDa mosquitocidal delta-endotoxin gene of *Bacillus thuringiensis* subsp. *israelensis* in *Bacillus sphaericus, Appl. Environ. Microbiol.,* submitted.

38. **Davidson, E. W.,** Effects of *Bacillus sphaericus* 1593 and 2362 spore/crystal toxin on cultured mosquito cells, *J. Invertebr. Pathol.,* 47, 21, 1986.

39. **Ganesan, S., Kamdar, H., Jayaraman, K., and Szulmajster, J.,** Cloning and expression in *E. coli* of a DNA fragment from *Bacillus sphaericus* coding for biocidal activity against mosquito larvae, *Mol. Gen. Genet.,* 189, 181, 1983.

40. **Louis, J., Jayaraman, K., and Szulmajster, J.,** Biocide gene(s) and biocidal activity in different strains of *B. sphaericus.* Expression of the gene(s) in *E. coli* maxicells, *Mol. Gen. Genet.,* 195, 23, 1984.

41. **Tandeau de Marsac, N., de la Torre, F., and Szulmajster, J.,** Expression of the larvicidal gene of *Bacillus sphaericus* 1593 M in the cyanobacterium *Anacystis nidulans* R2, *Mol. Gen. Genet.,* 209, 396, 1987.

42. **Stotzky, G.,** in *Microbial Ecology,* Laskin, A. I. and Lechevalier, H., Eds., CRC Press, Boca Raton, FL, 1974, 57.

43. **Stotzky, G. and Krasovsky, V. N.,** In *Molecular Biology, Pathogenicity and Ecology of Bacterial Plasmids,* Levy, S. B., Clowes, R. C., and Koenig, K. L., Eds., Plenum Press, New York, 1981, 31.

44. **Levine, M. M., Kaper, J. B., Lockman, H., Black, R. E., Clements, N. L., and Falkow, S.,** Recombinant DNA risk assessment studies in humans. Efficacy of poorly mobilizable plasmids in biologic containment, *J. Infect. Dis.,* 148, 699, 1983.

Chapter 12

# BIOMEDICAL APPLICATIONS OF MONOCLONAL ANTIBODIES (McAb) AGAINST *ENTAMOEBA HISTOLYTICA*

N. Thammapalerd and S. Tharavanij

## TABLE OF CONTENTS

# I. INTRODUCTION

Amebiasis, a disease caused by *Entamoeba histolytica,* with or without clinical symptoms, occurs throughout the world, especially in tropical and subtropical climates and in places with poor sanitary conditions due to fecal contamination of water and food or direct fecal-oral contact. *E. histolytica* is the third leading parasitic cause of death in the world after malaria and schistosomiasis and infects about 10% of the world's population, 10% again of whom (approximately 48 million cases) develop invasive colitis or liver abscess with at least 40,000 deaths per year, while the other 90% are asymptomatic.[1] The high percentage of asymptomatic cases suggests that differences in invasiveness and/or pathogenicity occur among two different isolates of amebae, pathogenic and nonpathogenic strains, as first mentioned by Brumpt in 1925.[2] *In vitro* and *in vivo* experimental studies have demonstrated that *E. histolytica* adherence and cytolytic mechanisms are absolutely required for the pathogenesis of amebiasis.[3-5] Intestinal amebiasis occurs after ingestion of contaminating mature cysts. Excystation occurs in the lower ileum followed by intestinal colonization via adherence of ameba trophozoites to gut epithelial cells throughout the colon. Encystation takes place in the lower colon and the cysts leave the host via feces. Pathogenic amebae invade the intestinal wall, causing lesions and the symptoms of amebic colitis, and if they spread via the blood stream they may reach and invade other organs of the body, especially the liver, lung, and brain, causing extraintestinal amebiasis. Amebic brain infection causes changes in the CSF that mimic meningitis. Therefore, development of vaccines effective in preventing invasive amebiasis and even intestinal colonization by amebae would be the most cost-effective approach for prevention.[4] To develop a vaccine against amebiasis or other parasitic infections using recombinant DNA or synthetic peptide approaches, it is important to identify protective antigens, which may be facilitated by the use of monoclonal antibodies (McAb). *E. histolytica*-specific McAb have been produced in many laboratories[6-8] and have been used in a McAb-based ELISA for detection of amebic antigen in clinical specimens,[9] for typing of pathogenic amebae,[10] for prevention of parasite-mediated cytotoxicity of target cells,[7] for characterization of proteins involved in adherence to target cells,[11] for amebic antigen purification,[8] and hopefully in the future for the development of vaccine against invasive amebiasis.

# II. PRODUCTION, CONCENTRATION, AND PURIFICATION OF McAb

## A. PRODUCTION OF McAb

Murine McAb to pathogenic strains of *E. histolytica* were produced essentially according to techniques modified from that of Köhler and Milstein[12] (Table 1).[6-8,13,14] The procedure used in our laboratory was based on that of Galfre and Milstein[15] and can be summarized as follows: BALB/c mice were immunized intraperitoneally at least 3 times with $10^6$ cells of ultrasonically disrupted amebic trophozoites incorporated in Freund's complete adjuvant initially, and incomplete adjuvant subsequently at 3-week intervals with the same amount of antigen. Ten days after the last immunizing dose, the mouse was bled from the retroorbital plexus and tested by indirect fluorescent antibody (IFA) and enzyme-linked immunosorbent assay (ELISA) for the presence of antibodies against *E. histolytica.* Mice with the highest antibody titers were immunized intravenously with ultrasonically disrupted amebae equivalent to $2 \times 10^5$ trophozoites. Three days after the last immunizing doses, the spleen cells were fused with Sp2/O mouse myeloma cells at a ratio of 4:1 to 10:1 using polyethylene glycol 4000 (PEG 4000, Sigma Chemical, St. Louis) as the fusing agent. Cells were then suspended in RPMI 1640 (Gibco, Grand Island, New York) containing 20% fetal bovine serum, kanamycin (30 μg/ml), hypoxanthine (50 μ$M$), and azaserine (10 μ$M$), placed in

## TABLE 1
### Reported Procedures for Raising Monoclonal Antibodies against *E. histolytica*

| Antigen | Immunization schedule | Myeloma cell | Spleen cell/ myeloma cell | Fusant PEG (mol wt) | Methods of screening of hybrid | Purification | Ref. |
|---|---|---|---|---|---|---|---|
| Live T, NIH strain | i.p. of 8 × 10⁵ amebae followed by i.v. booster of 8 × 10⁵ amebae on day 11 | NS-1 | 4:1 and 7:1 | 1000 | ELISA | IgG2b protein-A affinity chromatography | 8 |
| Sonicated T, HM-1 strain | i.m. of 5 × 10⁵ amebae in CFA followed by two i.p. boosters of 10⁵ in week 3 and week 5; sacrifice of mouse on day 4 after i.v. injection of 10⁴ amebae | P3 × 63-Ag8 | 5:1—7:1 | 1500 | ELISA, IFA | IgG McAb by affinity chromatography on protein A-Sepharose CL-4B, IgM McAb by precipitation in distilled H₂O at 4°C, followed by gel filtration on Sephacryl S-300 | 7 |
| Sonicated T, HK-9 strain | i.p. of 10⁶ amebae in CFA initially followed by two i.p. boosters of 10⁵ amebae in week 3 and 5; sacrifice of the mouse on day 4 after i.v. of 10⁴ amebae | Sp2/0 | 4:1—10:1 | 4000 | ELISA, IFA | (NH₄)₂SO₄ precipitation followed by protein A affinity chromatography | 6 |
| Sonicated T, HM-1, HK-9, and Laredo strains | Four doses of 100 μg at 1 week intervals with CFA initially and IFA subsequently | X63-Ag8.653 | 6:1 | 1000 | ELISA | (NH₄)₂SO₄ precipitation | 13 |
| Live T and their membranes | i.p. of 1 × 10⁶ amebae followed 1 month later with 100 μg of membrane | X63-Ag8.653 | 5:1—10:1 | 1000 | ELISA | 50% (NH₄)₂SO₄ precipitation followed by protein A-Sepharose 4B chromatography | 14 |

*Note:* i.p. = intraperitoneal injection. i.m. = intramuscular injection. i.v. = intravenous injection. T = trophozoites. CFA = complete Freund's adjuvant. ICFA = incomplete Freund's adjuvant.

96-well microtiter plate (Nunc, Kamstrup, Denmark) and incubated at 37°C in a 5% $CO_2$ incubator. On the 4th to 5th day after fusion, fresh medium supplemented with 20% FCS and hypoxanthine (100 $\mu M$) was added to the hybridoma culture without removal of old spent medium. The spent media obtained at 3 to 6 weeks from wells with proliferating clusters of cells were screened for antiamebic antibodies by IFA test and ELISA. Antibody-secreting hybrids were subcloned by limiting dilution. Large amounts of McAb were prepared as ascites fluid by inoculating the clones intraperitoneally in the pristane-primed BALB/c mice or as spent medium, by growing them in large quantities of tissue culture medium.

## B. CONCENTRATION AND PURIFICATION OF McAb

The globulin fraction from the antibody-producing hybrids was usually concentrated to 20 times less than the original volume by 50% ammonium sulfate precipitation followed by exhaustive dialysis against PBS pH 7.2.[16] By this method, antibodies are not pure and are still contaminated with other high molecular weight precipitates. If pure antibody preparations are needed, additional steps are necessary. IgG McAb can be purified by affinity chromatography on a Protein A- or antisubclass-specific antibody Sepharose CL-4B column.[16] McAb of IgM class from ascites fluid can be purified simply by passing through Sephadex G150 to G200 followed by $(NH_4)_2SO_4$ precipitation. The presence of IgM can be visualized by SDS-PAGE showing migration with relative mobility of 70 to 80 kDa.

# III. CHARACTERIZATION OF McAb

Conventional characterization of anti-*E. histolytica* McAb involves identification of isotypes and subisotypes of immunoglobulins, identification of molecules recognized by McAb by using immunoprecipitation or Western blot analysis, and determination of their functional activities. In this review, characterization of anti-*E. histolytica* McAb will be restricted to three methods comprising (1) indirect fluorescent antibody (IFA)-staining patterns, (2) agglutination with live amebic trophozoites, and (3) inhibition of *in vitro* adherence of amebic trophozoites to target cells.

## A. INDIRECT FLUORESCENT ANTIBODY (IFA)-STAINING PATTERNS

There have been only a few reports on IFA patterns of McAb against *E. histolytica*. We have therefore attempted to establish an IFA-based characterization of anti-*E. histolytica* McAb. Antigen-coated slides used in our laboratory were prepared according to the method described by Garcia et al.,[17] with slight modification. Trophozoites from three strains of axenically grown amebae (HM-1, HK-9, and Laredo) and one monoxenically-grown VN strain were used. The cells were harvested, washed, resuspended in PBS pH 7.2 to give $2 \times 10^6$ cells per milliliter and spotted on clean and dry microscopic slides. The slides were dried, fixed in absolute ethanol for 10 min, wrapped with tissue paper, kept with silica gel as desiccant in sealed polyethylene bags, and stored at $-20°C$ until used. Culture supernatant or ascites fluid to be tested as well as unrelated McAb, PBS, and positive and negative control sera were applied to the slides and incubated in a humidified box at 37°C for 2 h. The slides were washed, air-dried, and then reacted with FITC-labeled IgG rabbit antimouse IgG conjugate, followed by incubation at 37°C for another hour, washed, dried, mounted in freshly prepared Tris-buffered glycerol containing 5% *n*-propyl gallate[18] and examined by fluorescence microscopy. The fluorescence intensity was recorded as negative or positive from one to four plus. Photographs were taken with Kodak VR 400 color film. There were at least five IFA staining patterns which were subsequently used to classify McAb into 5 groups. Group I McAb stained the granules and plasmalemma with or without staining of released products. Group II McAb stained only the granules. Group III McAb stained the cytosol only. Group IV McAb stained the membrane and cytosol, and group V McAb stained the granules and cytosol (Figure 1, Table 2). A literature survey of reported anti-*E. histolytica*

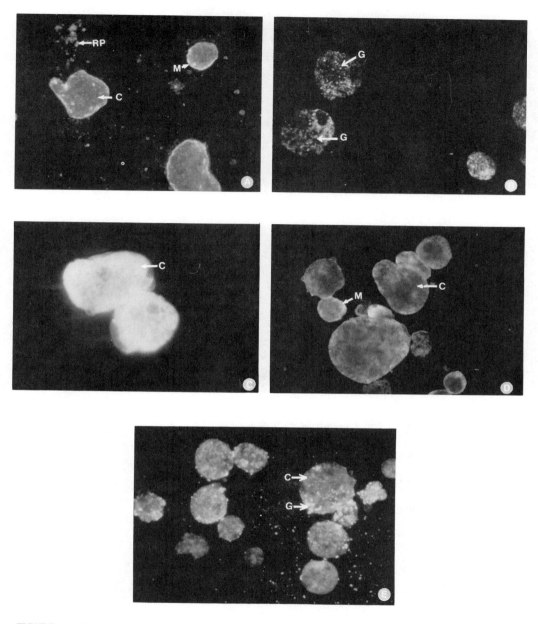

FIGURE 1. Five groups of IFA-staining patterns of McAb against *E. histolytica*. (A) Group I McAb stained the granules and plasmalemma with or without staining of released products; (B) group II McAb stained only the granules; (C) group III McAb stained the cytosol only; (D) group IV McAb stained the membrane and cytosol; (E) group V McAb stained the granules and cytosol. C = cytosol, G = granules, M = membrane, RP = released products.

McAb revealed that they were mostly reactive against the surface membrane and were involved in either agglutination of the parasites or inhibition of cytoadherence of amebae to target cells (Table 3).[7,10,14,19,20] With uncertainty about their reactivities with other amebic organelles, it is therefore difficult to categorize those McAb in our scheme. However, McAb reported by Strachan et al.,[10] will fit best with group II.

**TABLE 2**

**Classification of Anti-*E. histolytica* McAb Based on IFA
Staining Patterns**

| Group | Staining pattern | McAb | Isotype |
|---|---|---|---|
| I | Generalized staining of amebae including the granules, cytosol, plasmalemma, internal membrane, and with or without released products | 208C2-2<br>35C1-3<br>12/6[a]<br>9/17<br>20/1 | IgG1<br>IgG1<br>IgM<br>IgM<br>IgM |
| II | Cytoplasmic granules | 208C2-3<br>19C6<br>65B8 | IgG1<br>IgG1<br>IgG1 |
| III | Cytosol | 90<br>11/8<br>22/8<br>26/2* | IgG1<br>IgG1<br>IgG1<br>IgG1 |
| IV | Membrane and cytosol | 23C17 | IgG1 |
| V | Granules and cytosol | 34/2<br>116C1 | IgG1<br>IgG1 |

[a]   Agglutinated with live *E. histolytica* trophozoites.

**TABLE 3**

**Possible IFA-Based Categorization of Some Anti-*E. histolytica* McAb Reported in the
Literature**

| McAb | IFA reactivities | $M_r$ (kDa) | Isotype | Functional activity or use | Possible group | Ref. |
|---|---|---|---|---|---|---|
| 22.3 | Internal granule | Not reported | Not reported | Typing of pathogenic zymodemes | II | 10 |
| 22.5 | Internal vacuole | Not reported | Not reported | Typing of pathogenic zymodemes | | |
| MC004 | Surface membrane | Not determined | IgG1 | Agglutination of live amebic trophozoites | ? | 17 |
| F14 | Nonreactive with acetone-fixed, surface staining of live amebae | 170 kDa | IgM | Inhibition of cytoadherence to target cells | ? | 7,19, 20 |
| H8-5 | No detailed description | 170 kDa | IgG2b | Inhibition of cytoadherence to target cells | ? | |
| Adh-1 | Surface membrane | 112 kDa (approximate) | IgG | Inhibition of cytoadherence | ? | 28 |
| Adh-2 | Surface membrane | 112 kDa (approximate) | IgG | Inhibition of cytoadherence | ? | |
| EH403 | Surface membrane | 96 kDa | IgM | Agglutination | ? | 8 |
| EH106 | Surface membrane | Non-precipitable | IgM | Agglutination | ? | |
| EH335 | Surface membrane | 95 kDa | IgG2b | Nonagglutination of live amebae | ? | |

## B. AGGLUTINATION OF LIVE AMEBIC TROPHOZOITES

This technique is used in general to demonstrate the presence of the membrane-associated antigen usually in conjunction with demonstration of antibody capping by IFA of live trophozoites. Ortiz-Ortiz et al. were the first to demonstrate that some McAb could agglutinate live amebic trophozoites.[14] This was done by mixing McAb with packed amebic trophozoite in a tube followed by incubation at 4°C. Thereafter, the amebae were resuspended in FITC-labeled rabbit antimouse globulin, incubated at 4°C, centrifuged, and the packed trophozoites were examined with a fluorescence and phase contrast microscope. Purified membrane antigens reacting with an agglutinating MC004 McAb of IgG1 isotype in an enzyme-linked immunoelectrotransfer blot (EITB) showed three polypeptides, the molecular weights of which were not well defined. It was not known whether the three molecules had precursor-product relationship or whether they represented enzymatic degradation products from a single polypeptide.[14] In our laboratory, one McAb of group I McAb (12/6, IgM isotype) and three other McAb of group III (26/2, 26/3 and 26/4, all IgG1 isotypes), which showed surface membrane fluorescence of live HK-9 strain also caused agglutination (Thammapalerd et al.[66]). A simpler agglutination technique was developed by Torian et al.[8] by mixing a McAb with a suspension of $10^3$ trophozoites in a Microtiter plate followed by incubation at 37°C for 10 min. Three *E. histolytica*-specific McAb (EH106, EH282, and EH403) of IgM isotype were shown to agglutinate live trophozoites. One of these McAb (EH403) recognized a polypeptide of 96 kDa on the surface membrane, whereas the molecules recognized by the remaining two McAb (EH106 and EH282) could not be demonstrated.

## C. INHIBITION OF *IN VITRO* ADHERENCE OF AMEBIC TROPHOZOITES TO TARGET CELLS

Adherence of *E. histolytica* to cells or tissues is a prerequisite for amebic cytolytic activity.[21,22] *In vitro* and *in vivo* intestinal models of amebiasis indicated that amebae adhere to host inflammatory cells and intestinal mucosa before invasion.[23-25] Since amebic adherence to a human intestinal epithelial cell line, Chinese hamster ovary (CHO) cells, and human neutrophils is inhibited by *N*-acetyl-glucosamine (GlcNAc) and *N*-acetyl-galactosamine (GalNAc) or galactose,[26-27] McAb causing amebic adherence inhibition could be selected on the basis of their binding to a target cell line, e.g., CHO cells preexposed to a partially purified preparation of the amebic GalNAc inhibitable lectin.[7] McAb positive by this selection will not necessarily inhibit amebic adherence, since antibodies from only four of nine positive subclones inhibited adherence of *E. histolytica* to CHO cells.[7] Conversely, antibodies from hybrids with negative binding to lectin-exposed CHO cells could inhibit amebic adherence to target cells. Though the molecules recognized by adherence-inhibitory McAb could not be identified in the initial report,[7] subsequent Western blot analysis showed that one McAb (F14) bound to a 170-kDa protein purified from an amebic homogenate by either galactose-silica bead or adherence-inhibitory McAb-conjugated protein A-agarose column chromatography.[19]

*In vitro* McAb-mediated inhibition of trophozoite adherence, erythrophagocytosis and cytopathic effect was also reported by Arroyo et al.[28] McAb Adh-1 and Adh-2 of IgG isotype reacting with membrane antigen of approximately 112 kDa inhibited trophozoite adherence and phagocytosis of RBC as well as inhibition of cytolysis of Madin Darby canine kidney (MDCK) cells, whereas two other McAb (MAb-3 and MAb-24) did not. MAb-24 did not react to amebic membrane antigen, and thus failure to inhibit cytoadherence was understandable. Nevertheless, ability to react with the membrane antigen does not necessarily endow such McAb with the ability to inhibit cytoadherence, as is the case with MAb-3 reacting with a membrane antigen. Neither does cytoadherence inhibitory activity indicate protective activity, since Adh-1 and Adh-2 McAb did not confer protection in experimental animals.[28] More research is needed to determine whether other cytoadherence-inhibitory McAb will have corresponding activities *in vivo* or have any protective role. If they do, then these McAb will be useful in the assessment of immune mechanisms in amebiasis and possibly in future vaccine development.

# IV. BIOMEDICAL APPLICATIONS OF McAb

## A. TYPING OF PATHOGENIC *E. HISTOLYTICA*

It has been suggested by Brumpt that *E. histolytica* is comprised of two morphologically identical organisms, one pathogenic and invasive (*E. dysenteriae*) and the other a harmless commensal (*E. dispar*).[2] These two types of amebae have subsequently been shown by several investigators to differ in a number of characteristics including zymodemes. Sargeaunt and co-workers demonstrated that pathogenic properties of amebae, as judged by clinical criteria, correlated with the presence of a β-band and the absence of an α-band in phosphoglucomutase (PGM) and by the presence of fast-running hexokinase (HK) bands on starch-gel electrophoresis of the culture of freshly isolated *E. histolytica*.[29-31] Up to now, 22 distinct zymodemes have been recognized, of which only 10 have been associated with clinical evidence of tissue invasiveness.[32] Zymodeme type has been considered to be a stable characteristic, with the conclusion that pathogenic *E. histolytica* can be identified by zymodeme typing.[32] However, this conclusion is not consistent with data from the Weizmann Institute, Israel, showing that zymodeme characteristics are not sufficiently stable to be considered as a marker for pathogenicity. Zymodeme type is influenced by the microenvironment conditions including those provided by intestinal bacteria. Cloned or uncloned cultures of *E. histolytica* with nonpathogenic zymodemes adapted to grow axenically in the presence of lethally irradiated bacteria became pathogenic with concomitant change from nonpathogenic to pathogenic zymodemes.[33,34] The validity of the finding of Mirelman's group was challenged by Sargeaunt[35] on the ground that the number of experiments was too small and the results thus obtained could not be compared with thousands of proven "clinical isolates" from various places in the world. Though the stability of zymodeme is still debatable, techniques for detection of pathogenic zymodemes including McAb-based IFA,[10] and DNA probes (P145 and B133) have been reported.[36] In a recent review, McKerrow[37] cited the work of Reed et al.,[38] who showed that secreted thiol proteases could be used to distinguish between pathogenic and nonpathogenic *E. histolytica* strains isolated directly from patient stool specimens. Cloned mutants of virulent HM-1 amebae which lack protease expression do not produce a cytopathic effect in vitro.[39] Room is now open to study whether McAb to thiol proteases can be used to differentiate between pathogenic and nonpathogenic *E. histolytica*.

McAb-based IFA typing of pathogenic zymodemes has been reported recently.[10] Two McAb (22.3 and 22.5) producing a granular internal fluorescence and an internal vacuolar staining pattern, respectively, bound exclusively to isolates of *E. histolytica* with fast hexokinase bands. The test was shown to be specific for pathogenic amebae, since all 11 *E. histolytica* including four reference strains (NIH 200, HK-9, HM-1:IMSS, and Loon Lake) and seven cultured isolates with fast moving HK-bands were positive, and all 23 isolates with slow-moving HK-bands were negative. These two McAb did not react directly with *E. histolytica* cysts and preliminary culture is still required. The time needed to obtain sufficient organisms for IFA testing was only 2 d, in contrast to the 7 d needed to prepare the lysate for isoenzyme analysis.[10]

In our laboratory, 36 McAb were tested with IFA against 36 ameba isolates comprising 25 pathogenic zymodeme II, 7 nonpathogenic zymodeme I, 1 *E. histolytica*-like ameba (Laredo strain), and 3 non-*E. histolytica*.[40] The result (Table 4) showed that there were four promising McAb (Eh19C7, Eh19C8, Eh35C1, and Eh35C5) which reacted to more than 75% of pathogenic zymodemes and less than 20% of nonpathogenic amebae, including the Laredo strain. The Eh35C1 and Eh35C5 McAb recognized the same proteins of 183 and 170 kDa, whereas the molecules recognized by the Eh19C7 and Eh19C8 McAb could not be demonstrated by Western blot analysis. Our results was not as clear-cut as that of Strachan et al.,[10] and further development is needed. It is however apparent that McAb-based IFA

## TABLE 4
### IFA Reactivities of Monoclonal Antibodies to *E. histolytica* Isolates from Thailand

| McAb | Zymodeme | | *E. histolytica*-like (Laredo) and non-*E. histolytica* amebae |
|------|-----|-----|-----|
| | I | II | |
| Eh6C1 | 4/7 | 18/24# | 2/4 |
| Eh6C2 | 3/7 | 16/25 | 0/3 |
| Eh6C2/8 | 0/6 | 10/21 | 1/3 |
| Eh6C12 | 2/7 | 18/25 | 1/4 |
| Eh6C13 | 1/7 | 10/24 | 0/3 |
| Eh6C14 | 2/7 | 10/25 | 1/4 |
| Eh6C18 | 2/7 | 16/24 | 0/3 |
| Eh6C19 | 2/7 | 15/23 | 0/4 |
| Eh6C20 | 4/7 | 19/25 | 1/4 |
| Eh6C22 | 1/7 | 14/24 | 0/3 |
| Eh6C24 | 1/7 | 14/24 | 1/4 |
| Eh6C26 | 1/7 | 14/23 | 0/4 |
| Eh16C1 | 1/7 | 12/24 | 0/3 |
| Eh16C2 | 0/7 | 7/24 | 0/4 |
| Eh16C3 | 1/7 | 10/24 | 0/3 |
| Eh16C7 | 0/7 | 7/25 | 0/4 |
| Eh16C8 | 0/7 | 11/24 | 0/4 |
| Eh19C1 | 0/7 | 11/23 | 0/4 |
| Eh19C4 | 1/7 | 11/22 | 1/4 |
| Eh19C5 | 1/7 | 11/24 | 0/4 |
| Eh19C6 | 0/7 | 4/24 | 0/4 |
| Eh19C7* | 1/7 | 19/25 | 0/4 |
| Eh19C8* | 1/7 | 19/24 | 0/4 |
| Eh19C9 | 1/7 | 8/24 | 0/4 |
| Eh19C11 | 1/7 | 8/23 | 1/3 |
| Eh19C13 | 1/7 | 11/24 | 1/4 |
| Eh19C14 | 2/7 | 14/24 | 0/4 |
| Eh35C1* | 2/7 | 19/23 | 0/4 |
| Eh35C1/4* | 1/6 | 19/25 | 0/3 |
| Eh35C3 | 1/7 | 12/19 | 1/4 |
| Eh35C4 | 0/7 | 17/25 | 1/4 |
| Eh35C5* | 2/7 | 20/25 | 0/4 |
| Eh106C1 | 0/1 | 9/24 | 0/3 |
| Eh208C1 | 4/6 | 9/14 | 1/4 |
| Eh208C2 | 5/6 | 16/22 | 0/3 |
| Eh208C3 | 5/6 | 21/22 | 0/4 |

\# Number with positive IFA/number tested.

\* McAb showing >75% IFA positivity against pathogenic zymodemes and less than 20% IFA positivity against the combined nonpathogenic zymodemes and *E. histolytica*-like (Laredo) or non-*E. histolytica* amebae.

could possibly provide a relatively cheap and rapid method for determining the invasive potential of clinical isolates of *E. histolytica*.

Very recently McAb to the amebic Gal/GalNac-inhibitable lectin with molecular mass of 260 kDa have been tested by a radioimmunoassay (RIA) against 48 pathogenic and nonpathogenic strains for the presence of Gal/GalNac lectin.[41] The assay was positive against all 16 pathogenic zymodemes from several countries, but negative in all 32 nonpathogenic zymodemes. The authors suggested that lectin RIA was a simple and rapid method to distinguish pathogenic from non-pathogenic strains of amebae in culture.

## B. DETECTION OF AMEBIC ANTIGEN IN CLINICAL SPECIMENS

The most commonly used diagnostic methods for amebiasis are directed to the detection of antibodies in the circulation.[42-46] However, long persistence of antibodies makes it difficult to differentiate between present and past infections in individuals living in tropical countries, especially in deciding whether antiamebic treatment should be given to serologically positive patients with symptoms not compatible with classical amebiasis. Specific diagnosis could be reliably made by demonstration of the parasites (cysts or trophozoites) in clinical specimens by microscopic examination. During the past decade, there have been a number of reports on the development of diagnostic tests based on the detection of amebic antigens. In 1978, Root and co-workers reported polyclonal antibody (PAb)-based sandwich ELISA on microporous membranes fixed on polystyrene slides which formed a basis of the Millipore immunoenzyme kit.[47] In 1982, Grundy et al. developed a double antibody sandwich ELISA for *E. histolytica* antigen detection in stool samples containing significant numbers of trophozoites, but the assay was not sensitive enough to detect amebic antigens in cyst passers.[48] The assay was specific for *E. histolytica* and did not give a cross-reaction with *E. coli* and *E. nana*. The assay was subsequently improved with triple antibody sandwich ELISA whereby the plate was coated with sheep anti-*E. histolytica* IgG, washed, reacted with antigen in clinical samples, washed, reacted with rabbit anti-*E. histolytica* IgG, washed, reacted with peroxidase-conjugated sheep antirabbit IgG, and finally the substrate added.[49] The assay was shown to be specific for amebiasis and did not give cross-reaction with *E. hartmanni*, *E. nana*, *I. buetschlii*, *H. nana*, *G. lamblia*, *Trichomonas*, and *Ascaris*. The test was more sensitive than the previous assay from the same laboratory in that as many as 70.6% of patients with *E. histolytica* cysts were positive.

The supply of reagents of consistent quality may be a constraint for extensive use of PAb-based ELISA, and thus substitution of PAb with McAb is desirable. Ungar et al.[9] developed a double antibody sandwich ELISA to detect *E. histolytica* antigens whereby a commercial source of McAb (Synbiotex Corporation, San Marcos, California) was used for plate coating, followed by stepwide addition with washing after each step with lysate of *E. histolytica*, rabbit anti-*E. histolytica*, alkaline phosphatase-labeled goat antirabbit IgG and *p*-nitrophenyl phosphate substrate. The assay detected 1 to 57 trophozoites from 6 *E. histolytica* strains in 25 μl of PBS. Stool specimens were positive by ELISA in 18 of 22 (82%) patients with *E. histolytica* and in 2 of 186 (2%) of patients without demonstrable *E. histolytica* in their stools. Of the latter, one sample was from a child living near an asymptomatic cyst carrier and another was from a traveler with giardiasis who had recently taken antibiotics. The overall sensitivity of the ELISA was 82%, specificity 98%. The false positive rate was 2% and the false negative rate was 18%. The predictive value of a positive ELISA was 86% and the predictive value of a negative ELISA was 98%.

A McAb-based ELISA similar to that developed by Ungar et al.[9] has recently been established in our laboratory (Wonsit et al.[67]). The McAb (208C2-2) used recognized four proteins of 122, 115, 111 and 65 kDa, respectively. It was not clear whether these four proteins were precursor-product-related or whether they represented enzymatic degradation products from a single polypeptide. In our assay, the 96-well plates were coated with 100 μl of the McAb (20 μg IgG/ml), followed by incubation at 4°C overnight. The plates were washed and blocked for 30 min with 100 μl of 5% nonfat dry milk in PBS pH 7.2 and incubated at 37°C. After washing, samples were added to the wells and the plates incubated at 37°C for 2 h. The plates were again washed, and 100 μl of rabbit anti-*E. histolytica* (HM-1:IMSS strain) IgG diluted in PBS-0.5% Tween 20 (10 μg/ml) added, followed by incubation at 37°C for 1 h. After washing, 100 μl of alkaline phosphatase-conjugated goat antirabbit IgG was added and the plates were incubated at 37°C for 1 h. The plates were washed and *p*-nitrophenyl phosphate substrate was added, followed by incubation at room temperature for 45 min. The result was read at 405 nm with a Titertek Multiskan MCC 340

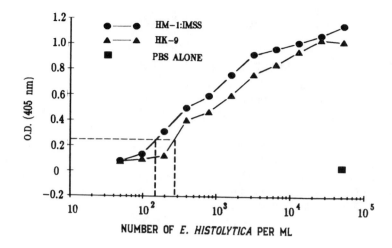

FIGURE 2. Sensitivity of ELISA in the detection of parasite antigen in PBS lysates of HM-1:IMSS and HK-9 strains. With the O.D. of 0.25 cutoff, sensitivity of the assay was 11 and 28 parasites in 100 μl of PBS, respectively.

ELISA reader. Based on mean O.D. + 3SD from assay of parasitologically negative stool samples (1:10 dilution) from 30 healthy individuals, the cutoff O.D. was 0.25. The sensitivity limits of our assay were approximately 110 and 280 amebae per milliliter for the HM-1:IMSS and HK-9 strains, respectively, which would be equivalent to 11 and 28 amebae per well (Figure 2). This level of sensitivity was similar to that reported by Ungar et al.[9] The test was positive in 31 of 40 stool samples (77.5%) with *E. histolytica* cysts and/or trophozoites, whereas only 1 of 48 (2.1%) stool samples with other protozoa and/or helminths were positive. One of 36 (2.7%) parasitologically negative individuals was positive. Though our assay did not appear as sensitive as the one reported by Ungar,[9] the results of these two assays could not be accurately compared in view of the uncertainty of the actual parasite load in the samples tested.

Tissue invasion by virulent pathogenic *E. histolytica* has been shown to be accompanied by formation and liberation of electron-dense granules (EDG). SDS-PAGE and silver staining of EDG showed 15 polypeptides, one of which was a 40-kDa protein recognized by a McAb (B-4). McAb B-4 reacted with an extract from pathogenic strains of *E. histolytica* but not with nonpathogenic species including *E. moshkovskii* and *E. invadens* or with *G. lamblia*.[50] The authors suggested that McAb B-4 might be useful as a marker to make a differential diagnosis of enteric amebiasis. This suggestion can be extended to include the possibility of using this McAb for the detection of *E. histolytica*-specific antigen in clinical specimens.

It should be stressed that McAb-based ELISA for specific diagnosis of pathogenic *E. histolytica* or cyst antigens has not been established. Since McAb reactive only to pathogenic zymodemes have been produced, it is possible, therefore that ELISA for detection of specific antigens of pathogenic zymodemes could be developed in the near future. Specific detection of cyst antigens which are intimately associated with amebic cyst passers could also be made possible in view of the successful production of cyst-specific McAb (JAM3 and JAM5) to epitopes on the 250 kDa sialoglycoproteins of *E. invadens*.[51] Though cyst-specific antigens of *E. histolytica* have not been demonstrated, their existence was suggested by the observation of Chayen et al.[52] that during encystation of *E. histolytica*-bacterium cultures in hypoosmotic medium depleted of nutrients, there was a *de novo* synthesis of two glycoproteins with apparent molecular sizes of 100 and 150 kDa containing sialic acid which is absent in trophozoites of *Entamoeba* species. These sialoglycoproteins might probably be cyst-specific antigens of *Entamoeba* parasites. Monoclonal antibodies raised against these two stage-specific antigens would, therefore, be useful for diagnosis of cyst antigens in human stool.

## C. PURIFICATION OF DESIRED ANTIGENIC MOLECULES

Purification of the surface antigen of 96 kDa by using McAb has been accomplished by Torian et al.,[8] using a two-step procedure of immunoprecipitation and elution. Unlabeled whole trophozoite antigens were first immunoprecipitated with either IgM (EH403) or IgG2b (EH335) McAb, followed by SDS-PAGE. A gel slice containing the 96-kDa polypeptide was excised and subjected to electroelution and electrophoretic concentration. With this technique, a purified 96-kDa protein was obtained and should be further studied. Petri et al. purified a 170-kDa galactose-binding lectin of *E. histolytica* by passing detergent-solubilized lysate of *E. histolytica* through a protein A-agarose column conjugated with an adherence-inhibitory McAb (H8.5) followed by elution with glycine.[19] The purified protein competitively inhibited amebic adherence to target CHO cells in a Gal-inhibitable manner.[19]

## D. VACCINE DEVELOPMENT

Substantial information has been accumulated for the existence of partial protective immunity to amebiasis in experimental animals receiving whole amebic trophozoites followed by drug cure,[53] crude amebic cell lysate,[54] partially purified fraction of high molecular weight fractions of the crude amebic extract,[55] and lysosome fractions.[56] Yet little is known about the protective immune mechanisms and the nature of protective molecules which could be broadly categorized into two groups comprising membrane-associated and nonmembrane-associated antigens. The protective function of membrane-associated antigens was evident from the work of Purnima, who showed that membrane protein antigens entrapped in multilamellar phosphatidyl choline liposomes conferred 100% protection against a subsequent intrahepatic amebic challenge in the hamsters, while intradermal injections of live amebic trophozoites failed.[57] Membrane antigens could play important roles in adhesion which has been shown to precede the contact-mediated cytotoxic damage and cytolysis of target cells.[21,22] Amebic adhesins consist of at least two soluble lectins, one of which is inhibited by GalNAc or galactose residue,[22] and the other by GlcNAc.[26] McAb against amebic adhesins have been produced[7,19,20,28] and at least two of them (F14 and H8-5) have been shown to react with the 170-kDa galactose-binding lectin.[19,20] Since the 170-kDa proteins was shown to react not only with McAb H8-5 but also with sera from patients in several countries in the world who had recovered from amebic liver abscess, it was therefore postulated that this protein was highly conserved,[20] and hence could be a vaccine candidate. The other adherence-inhibitory McAb have been shown to recognize a 112-kDa protein, alteration of which in adhesion-deficient mutants was associated with loss of amebic cytopathic effects in target cells *in vitro*.[28] Another candidate membrane antigen involved in agglutination of *E. histolytica* is the 96-kDa molecule recognized by McAb Eh403.[8] It is not clear whether the 112-kDa or 96-kDa proteins are in any way related to galactose-, GalNAc-, or GlcNAc-inhibitable lectins. Nevertheless, molecular cloning of the genes encoding the amebic adhesins is needed if a vaccine aiming at prevention of amebic adhesion is to be developed. McAb against amebic adhesins will be useful for screening of the recombinant DNA products for use in such vaccine development.

Nonmembrane-associated antigens which may play a major role in cytotoxic damage of target cells include amebic proteinases (endopeptidase)[58] and amebapore.[59,60] Cathepsin B (a cysteine proteinase), collagenase (metalloproteinase), and probably cathepsin D (aspartic proteinase) are major proteinases with cytotoxic activity.[58] Since protease has been shown in hookworm to function as an anticoagulant,[61] and a vaccine based on purified or recombinant hookworm antigen is being developed,[37] it is therefore possible that a proteinase vaccine could be developed for amebiasis, especially when purification of 16-kDa cathepsin B has been made possible.[62] If McAb to thiol proteinases could be raised, they could be used to screen recombinant clones expressing the products relevant to vaccine development.

The presence of a 30-kDa amebapore or amebic pore-forming material (PFM) in *E. histolytica* was independently reported in 1982 by Young et al.,[59] and Lynch et al.[60] The protein is secreted by *E. histolytica* after exposure to calcium ionophore, concanavalin A and lipopolysaccharide, but not by unstimulated *E. histolytica*. Amebapores have been purified[63,64] but blocking antibodies to amebapore could not be produced.[64] Recently, Aucott et al. succeeded in producing McAb against an ion channel protein (poretoxin) (PT) prepared by detergent extraction of ultracentrifuged particulate fraction of amebic cell lysate.[65] These McAb (EH-C-7, EH-C-8, EH-C-9) inhibited [51]Cr release from Chang liver cells after amebic contact. It is likely but not yet proved that PT is identical to or is a variant of amebapores. Availability of anti-PT McAb would help in further examination of the role of ion channel forming activity in the cytopathogenicity of *E. histolytica*. They will also help in the screening of recombinant clones secreting proteins with vaccine potential.

# V. CONCLUSION

McAb against *E. histolytica* can be characterized by several techniques including IFA-staining patterns, agglutination of live amebic trophozoites, and inhibition of adherence of amebic trophozoites to target cells. Based on IFA-staining patterns, McAb could be divided arbitrarily into 5 groups. Group I McAb stained the granules and plasmalemma with or without staining of released products. Group II McAb stained only the granules. Group III McAb stained by cytosol only. Group IV McAb stained the membrane and cytosol and group V McAb stained the granules and cytosol. Biomedical applications of McAb include typing of pathogenic zymodemes, detection of *Entamoeba histolytica* antigens in clinical specimens, purification of antigenic molecules relevant to vaccine development. McAb-based IFA identification of pathogenic zymodemes could replace the cumbersome procedure of zymodeme typing by electrophoresis. McAb-based ELISA for detection of *E. histolytica* can potentially be applied to clinical diagnosis of amebiasis especially when large numbers of specimens are to be processed. Purification of amebic antigens of functional interest can be achieved by affinity chromatography. McAb recognizing important functional and highly conserved antigens will be useful for screening of recombinant DNA products for vaccine development.

# ACKNOWLEDGMENTS

We thank Dr. L. S. Diamond, Laboratory of Parasitic Diseases, National Institutes of Health, Bethesda, for the HK-9 strain of *E. histolytica,* and Dr. Gordon Bailey, Department of Biochemistry, Morehouse School of Medicine, Atlanta, Georgia, for the HM-1:IMSS strain for use in our study. We are grateful to P. G. Sargeaunt, Department of Medical Protozoology, London School of Hygiene and Tropical Medicine, for the technology transfer of zymodeme typing to the senior author. Manuscript typing by Miss Thanya Nirantasook is acknowledged.

The authors received financial support from USAID, Grant Number 936-5542-G-00-6029-00 and from the National Research Council of Thailand.

# REFERENCES

1. **Walsh, J. A.**, Problems in recognition and diagnosis of amebiasis: estimation of the global magnitude of morbidity and mortality, *Rev. Infect. Dis.*, 8, 228, 1986.
2. **Brumpt, E.**, Étude sommaire de l'*Entamoeba dispar* n.sp. amibe à kystes quadrinucléés parasite de l'home, *Bull. Acad. Med. (Paris)*, 94, 943, 1925.
3. **Ravdin, J. I.**, Pathogenesis of disease caused by *Entamoeba histolytica:* studies of adherence, secreted toxins and contact dependent cytolysis, *Rev. Infect. Dis.*, 8, 247, 1986.
4. **Ravdin, J. I.**, *Entamoeba histolytica:* from adherence to enteropathy, *J. Infect.*, 159, 420, 1989.
5. **Chadee, K., Johnson, M. L., Orozco, E., Petri, W. A., Jr., and Ravdin, J. I.**, Binding and internalization of rat colinic mucins by the galactose/*N*-acetyl-D-galactosamine adherence lectin of *Entamoeba histolytica, J. Infect. Dis.*, 158, 398, 1988.
6. **Kotimanasvanij, D., Thammapalerd, N., Khusmith, S., and Tharavanij, S.**, Production of monoclonal antibodies against *Entamoeba histolytica* antigen extract, Abstr., in Current Research Works of the Faculty of Tropical Medicine, Mahidol University, Bangkok, August 7, 1984, 3.
7. **Ravdin, J. I., Petri, W. A., Jr., Murphy, C. F., and Smith, R. D.**, Production of mouse monoclonal antibodies which inhibit *in vitro* adherence of *Entamoeba histolytica* trophozoites, *Infect. Immun.*, 53, 1, 1986.
8. **Torian, B. E., Lukehart, S. A., and Stamm, W. E.**, Use of monoclonal antibodies to identify, characterize and purify a 96,000-dalton surface antigen of pathogenic *Entamoeba histolytica, J. Infect. Dis.*, 156, 334, 1987.
9. **Ungar, B. L. P., Yolken, R. H., and Quinn, T. C.**, Use of monoclonal antibody in an enzyme immunoassay for the detection of *Entamoeba histolytica* in fecal specimens, *Am. J. Trop. Med. Hyg.*, 34, 465, 1985.
10. **Strachan, W. D., Chiodini, P. L., Spice, W. M., Moody, A. H., and Ackers, J. P.**, Immunological differentiation of pathogenic and non-pathogenic isolates of *Entamoeba histolytica, Lancet*, 1, 561, 1988.
11. **Meza, I., Cazares, F., Rosales-Encina, J. L., Talamas-Rohana, P., and Rojkind, M.**, Use of antibodies to characterize a 220-kilodalton surface protein from *Entamoeba histolytica, J. Infect. Dis.*, 156, 798, 1987.
12. **Köhler, G. and Milstein, C.**, Continuous cultures of fused cells secreting antibody of predefined specificity, *Nature*, 256, 495, 1975.
13. **Lopez, J. S., Jensen, F. J., Mendoza, F., and Ortiz-Ortiz, L.**, Anticuerpos monoclonales contra *Entamoeba histolytica, Arch. Invest. Med. (Mex).*, 13 (Suppl. 3), 291, 1982.
14. **Ortiz-Ortiz, L., Ximenez, C., Mendoza, F., Michalak, C., Melendro, E. I., and Oliva, A.**, *Entamoeba histolytica:* specific antigen recognized by a monoclonal antibody, *Exp. Parasitol.*, 61, 390, 1986.
15. **Galfre, G. and Milstein, C.**, Preparation of monoclonal antibodies: Strategies and procedure, in Properties of the Monoclonal Antibodies Produced by Hybridoma Technology and their Application to the Study of Diseases, UNDP/World Bank/WHO Special Program for Research and Training in Tropical Diseases, 1982, chap. 1.
16. **Goding, J. W.**, Purification, fragmentation and isotopic labeling of monoclonal antibodies, in *Monoclonal Antibodies: Principles and Practice*, 2nd ed., Goding, J. W. Ed., Academic Press, New York, 1986, chap. 4.
17. **Garcia, L., Bruckner, D. A., Brewer, T. C., and Shimizu, R.**, Comparison of indirect fluorescent-antibody amebic serology with counterimmune-electrophoresis and indirect hemagglutination amebic serologies, *J. Clin. Microbiol.*, 15, 603, 1982.
18. **Giloh, H. and Sedat, J. W.**, Fluorescent microscopy: reduced photobleaching of rhodamine and fluorescein protein conjugates by *n*-propyl gallate, *Science*, 217, 1252, 1982.
19. **Petri, W. A., Jr., Broman, J., Healy, G., Quinn, T., and Ravdin, J. I.**, Antigenic stability and immunodominance of the Gal/GalNAc adherence lectin of *Entamoeba histolytica, Am. J. Med. Sci.*, 297, 163, 1989.
20. **Petri, W. A., Jr., Smith, R. D., Schlesinger, P. H., Murphy, C. F., and Ravdin, J. I.**, Isolation of the galactose-binding lectin that mediates the *in vitro* adherence of *Entamoeba histolytica, J. Clin. Invest.*, 80, 1238, 1987.
21. **Ravdin, J. I., Croft, B. Y., and Guerrant, R. L.**, Cytopathogenic mechanisms of *Entamoeba histolytica, J. Exp. Med.*, 152, 377, 1980.
22. **Ravdin, J. I. and Guerrant, R. L.**, The role of adherence in cytopathogenic mechanisms of *Entamoeba histolytica.* Study with mammalian tissue culture cells and human red blood cells, *J. Clin. Invest.*, 68, 1305, 1981.
23. **Galindo, J. M., Martinez-Palomo, A., and Chavez, B.**, Interaccion entre *Entamoeba histolytica* y el epitelio cecal del cobayo, *Arch. Invest. Med. (Mex)*, 9 (Suppl. 1), 261, 1978.
24. **Orozco, E., Martinez-Palomo, A., Gonzales-Robles, A., Guarneros, G., and Mora-Galindo, J.**, Las interacciones lectina-receptor median la adhesion de *E. histolytica* a celulas epiteliales. Relacion de la adhesion con la virulencia de las cepas, *Arch. Invest. Med. (Mex)*, 13 (Suppl. 3), 159, 1982.

25. **Chadee, K. and Meerovitch, E.**, *Entamoeba histolytica:* early progressive pathology in the cecum of the gerbil *(Meriones unguiculatus), Am. J. Trop. Med. Hyg.*, 34, 283, 1985.
26. **Kobiler, D. and Mirelman, D.**, Adhesion of *Entamoeba histolytica* trophozoites to monolayer of human cells, *J. Infect. Dis.*, 144, 539, 1981.
27. **Ravdin, J. I., Murphy, C. F., Salata, R. A., Guerrant, R. L., and Hewlett, E. L.**, *N*-acetyl-D-galactosamine-inhibitable adherence lectin of *Entamoeba histolytica.* I. Partial purification and relation to amebic virulence *in vitro, J. Infect. Dis.*, 151, 804, 1985.
28. **Arroyo, R. and Orozco, E.**, Localization and identification of an *Entamoeba histolytica* adhesin, *Mol. Biochem. Parasitol.*, 23, 151, 1987.
29. **Sargeaunt, P. G., Williams, J. E., and Grene, J. D.**, The differentiation of invasive and non-invasive *Entamoeba histolytica by isoenzyme electrophoresis, Trans. R. Soc. Trop. Med. Hyg.*, 72, 519, 1978.
30. **Sargeaunt, P. G. and Williams, J. E.**, Electrophoretic isoenzyme patterns of the pathogenic and non-pathogenic intestinal amoebae of man, *Trans. R. Soc. Trop. Med. Hyg.*, 73, 225, 1979.
31. **Sargeaunt, P. G., Williams, J. E., Jackson, T. F. H. G., and Simjee, A.**, Biochemical homogeneity of *Entamoeba histolytica* isolates, especially those from liver abscess, *Lancet,* 1, 1386, 1982.
32. **Sargeaunt, P. G.**, The reliability of *Entamoeba histolytica* zymodemes in clinical diagnosis, *Parasitology Today*, 3, 40, 1987.
33. **Mirelman, D., Bracha, R., Chayen, A., Aust-Kettis, A., and Diamond, L. S.**, *Entamoeba histolytica:* effect of growth conditions and bacterial associates on isoenzyme patterns and virulence, *Exp. Parasitol.*, 62, 142, 1986.
34. **Mirelman, D., Bracha, R., Wexler, A., and Chayen, A.**, Changes in isoenzyme patterns of a cloned culture of a non-pathogenic *Entamoeba histolytica* during axenization, *Infect. Immun.*, 54, 827, 1986.
35. **Sargeaunt, P. G.**, A reply to letters, *Parasitology Today*, 3, 353, 1987.
36. **Garfinkel, L. I., Giladi, M., Huber, M., Gitler, C., Mirelman, D., Revel, M., and Rozenblatt, S.**, DNA probes specific for *Entamoeba histolytica* having pathogenic and non-pathogenic zymodemes, *Infect. Immun.*, 57, 926, 1989.
37. **McKerrow, J. H.**, Minireview, parasite proteases, *Exp. Parasitol.*, 68, 111, 1989.
38. **Reed, S. L., Keene, W. E., and McKerrow, J. H.**, Cysteine proteinase expression and pathogenicity of *E. histolytica,* as cited in **McKerrow, J. H.**, *Exp. Parasitol.*, 68, 111, 1989.
39. **Keene, W. E., Hidalgo, M., Orozco, E., and McKerrow, J. H.**, Evidence that a secreted thiol proteinase mediates the cytopathic effects of *E. histolytica,* Joint ASCB/AJBMB Meeting, San Francisco, as sited in **McKerrow, J. H.**, *Exp. Parasitol.*, 68, 111, 1989.
40. **Thammapalerd, N., Tharavanij, S., Charoenpol, M., Techasathirakul, P., Lamom, C., Khusmith, S., Wonsit, R., Maskerewong, C., Bunnag, D., and Radomyos, P.**, Application of monoclonal antibodies against *Entamoeba histolytica* in tropical medicine research, Workshop on AID/SCI Funded Research in Immunology in Thailand, Chiengmai, December 4 to 6, 1989.
41. **Petri, W. A., Jr., Jackson, T. F. H. G., and Mirelman, D.**, The galactose and *N*-acetyl-D-galactosamine (GAL/GALNAc) adherence lectin of *Entamoeba histolytica* is detected by RIA only in pathogenic strains, 38th Annu. Meet. Am. Soc. Trop. Med. Hyg., Abstr. No. 378, Honolulu, Hawaii, December 10 to 14, 1989.
42. **Kessel, J. F., Lewis, W. P., Pasquel, C. M., and Turner, J. A.**, Indirect hemagglutination and complement fixation tests in amebiasis, *Am. J. Trop. Med. Hyg.*, 14, 540, 1965.
43. **Jeanes, A. L.**, Indirect fluorescent antibody tests in diagnosis of hepatic amebiasis, *Br. Med. J.*, 1, 1464, 1966.
44. **Savanat, T. and Chaicumpa, W.**, Immunoelectrophoresis for amebiasis, *Bull. WHO*, 40, 343, 1969.
45. **Morris, M. N., Powell, S. J., and Elsdon-Dew, R.**, Latex agglutination test for invasive amebiasis, *Lancet*, 1, 1362, 1970.
46. **Samrejrongroj, P. and Tharavanij, S.**, Assessment of validity of counterimmunoelectrophoresis and ELISA in the routine diagnosis of amebiasis, *Southeast Asian J. Trop. Med. Public Health*, 16, 365, 1985.
47. **Root, D. M., Cole, F. X., and Williamson, J. A.**, The development and standardization of an ELISA method for the detection of *Entamoeba histolytica* antigens in fecal samples, *Arch. Invest. Med. (Mex.)*, 9 (Suppl. 1), 203, 1978.
48. **Grundy, M. S.**, Preliminary observations using a multi-layer ELISA method for the detection of *Entamoeba histolytica* trophozoite antigen in stool samples, *Trans. R. Soc. Trop. Med. Hyg.*, 76, 396, 1982.
49. **Grundy, M. S., Voller, A., and Warhurst, D.**, An enzyme-linked immunosorbent assay for the detection of *Entamoeba histolytica* antigens in fecal material, *Trans. R. Soc. Trop. Med. Hyg.*, 81, 627, 1987.
50. **Munoz, M. L., Perez Garcia, J. N., de la Torre, M., Tovar, G. R. and Leon, G.**, A monoclonal antibody specific for electron-dense granules secreted by *Entamoeba histolytica,* in 38th Annu. Meet. Am. Soc. Trop. Med. Hyg., Abstr. No. 374, Honolulu, Hawaii, December 10 to 14, 1989.
51. **Avron, B., Chayen, A., Stolarsky, T., Schauer, R., Reuter, G., and Mirelman, D.**, A stage specific sialoglycoprotein in encysting cells of *Entamoeba invadens, Mol. Biochem. Parasitol.*, 25, 257, 1987.

52. **Chayen, A., Avron, B., Nuchamowitz, Y., and Mirelman, D.,** Appearance of sialoglycoproteins in encysting cells of *Entamoeba histolytica, Infect. Immun.,* 56, 673, 1988.
53. **Swartzwelder, J. C. and Avant, W. H.,** Immunity to amebic infection in dogs, *Am. J. Trop. Med. Hyg.,* 1, 567, 1952.
54. **Swartzwelder, J. C. and Muller, G. R.,** A comparison of infection rate and gross pathology of amebic infection in normal and antigen-infected rats, *Am. J. Trop. Med.,* 30, 181, 1950.
55. **Krupp, I. M.,** Protective immunity to amebic infection demonstrated in guinea pigs, *Am. J. Trop. Med. Hyg.,* 23, 355, 1974.
56. **Sepulveda, B.,** Induccion de immunidad protectora antiamibiana con ''nuevos'' antigenos en el hamster lactante f. Comentarios y conclusiones, *Arch. Invest. Med. (Mex),* 9 (Suppl. 1), 327, 1978.
57. **Purnima, C., Nain, K., and Vinayak, V. K.,** Elicitation of protective immunity to *Entamoeba histolytica*— an experimental study, *Immunol. Cell Biol.,* 65, 217, 1987.
58. **Lushbaugh, W. B.,** Proteinases of *Entamoeba histolytica,* in *Amebiasis, Human Infection by Entamoeba histolytica,* Ravdin, J. I., Ed., Churchill Livingstone, New York, 1988, chap. 14.
59. **Young, J. D. E., Young, T. M., Lu, L. P., Unkeless, J. C., and Cohn, Z. A.,** Characterization of a membrane pore-forming protein from *Entamoeba histolytica, J. Exp. Med.,* 156, 1677, 1982.
60. **Lynch, E. C., Rosenberg, I., and Gitler, C.,** An ion channel forming produced by *Entamoeba histolytica, EMBO J.,* 1, 801, 1982.
61. **Hotez, P. J. and Cerami, A.,** Secretion of a proteolytic anticoagulant by *Ancylostoma* hookworms, *J. Exp. Med.,* 157, 1594, 1983.
62. **Lushbaugh, W. B., Hofbauer, A. F., and Pittman, F. E.,** Purification of cathepsin B activity of *Entamoeba histolytica* toxin, *Exp. Parasitol.,* 59, 328, 1985.
63. **Rosenberg, I., Bach, D., Loew, L. M., and Gitler, C.,** Isolation, characterization and partial purification of a transferable membrane channel (amebapore) produced by *Entamoeba histolytica, Mol. Biochem. Parasitol.,* 33, 237, 1989.
64. **Rosenberg, I. M., Bach, D., Loew, L. M., and Gitler, C.,** Ionophore activity (amebapore) on *Entamoeba histolytica,* in *Amebiasis, Human Infection by Entamoeba histolytica,* Ravdin, I., Ed., Churchill Livingstone, New York, 1988, chap. 18.
65. **Aucott, J. N., Maholtra, I. J., and Salata, R. A.,** Inhibition of *in vitro* cytotoxicity and ion channel forming activity of *Entamoeba histolytica* by murine monoclonal antibodies, in 38th Annu. Meet. Am. Soc. Trop. Med. Hyg. Abstr. No. 375, Honolulu, Hawaii, December 10 to 14, 1989.
66. **Thammapalerd, N. et al.,** unpublished.
67. **Wonsit, R. et al.,** unpublished.

Chapter 13

# HUMAN ONCHOCERCIASIS: NEW IMMUNODIAGNOSTIC ASSAYS AND CONTROL MEASURES

**Ricardo Luján and Eddie W. Cupp**

## TABLE OF CONTENTS

# I. INTRODUCTION

## A. BIOLOGY OF THE PARASITE

*Onchocerca volvulus,* the causative agent of "river blindness" in Africa and Robles' disease in the New World, follows a pattern of morphogenesis and development generally similar to related taxa within the nematode superfamily Filarioidea.[1] Hematophagous arthropods (in this case members of the family Simuliidae — "blackflies") serve as intermediate hosts for development of the initial larval stages ($L_1$-$L_3$) after ingesting microfilariae (these embryonic forms of *O. volvulus* are found in the skin) from an infected human. Following development in the thoracic musculature, *O. volvulus* $L_3$s (infective stage larvae) migrate to the head where, during the course of bloodfeeding by the female, they escape into the skin by breaking through the terminal portions of the mouthparts. Because blackflies are telmophagous, i.e., the rasping action of their mouthparts causes the formation of tiny subdermal hemorrhages, infective larvae escape into an environment of pooled blood and hemolymph from the ruptured mouthparts of the vector. This method of transmission and infection is clearly distinct from that of mosquito-associated filariases.

Following infection of the human host, the $L_3$ molts to the fourth larval stage ($L_4$); prior to ecdysis, the worm enters a state of lethargus for 24 to 48 h and then escapes from the old cuticle. *In vitro* studies,[1] as well as those using infective stage larvae implanted into laboratory mice,[2] indicate that the $L_3$ to $L_4$ molt by *O. volvulus* occurs within 2 to 4 d. The $L_4$ is highly motile and presumably is the migratory stage, allowing the parasite to move to its final anatomical site for a final molt ($L_4$-juvenile) and subsequent growth to reproductive competence. Estimates for the length of time in which the parasite remains in the $L_4$ stage vary, ranging from 1 to $2^1/_2$ months.[1,3]

The prepatent period appears to be 12 to 15 months. Adult worms are usually encapsulated in connective tissue that forms subcutaneous nodules or that are located deeply within the body or along the pelvic region of the skeleton.[4] The nodule not only protects the adults but nourishes the parasite through a vascularization process which brings blood vessels adjacent to the worm's cuticle and into a central fibrin lake.[5] Females are multiply-inseminated by itinerant males.[6] Following fertilization, embryonation begins, ultimately producing unsheathed microfilarie which congregate in the uterus and from which they are expelled. Larvating eggs of *O. volvulus* readily bind a variety of lectins; however, neither *in utero* derived microfilariae nor those found in the skin do so,[7] suggesting a loss of surface antigenicity. Once inside the blackfly vector, the cuticular surface of the developing larva continues to change in an apparent attempt to also evade the arthropod's immune system.[8]

Based on classical taxonomic,[9,10] parasitological/entomological[1] and molecular evidence,[11] *O. volvulus* appears to be a relative "new" parasite of humans, having probably evolved from an antecedent bovid species. This assumption is based on the high degree of morphological and vectorial similarity among those species, the recent human cultural adaptation of domesticating and tending cattle, as well as the fact that no other species of *Onchocerca* naturally parasitizes primates. This unique evolutionary aspect has important ramification for control in that the reservoir for *O. volvulus* is limited to human hosts. It also poses problems in making accurate identifications of the parasite in the vector population where closely related *Onchocera* species of ungulates may occur and blackfly taxa are zoophilic.

## B. DISTRIBUTION AND VECTORS

The Simuliidae, and in particular, certain members of the *Simulium damnosum* sibling species complex serve as vectors of *O. volvulus.* This parasite, which is the second leading cause of infectious blindness in the world, infects an estimated 11 million persons living in 32 countries; at least 663,000 persons are irreversibly blind as a result of chronic onchocerciases, and some 85,583,780 persons are at risk of infection.[12] Thus, this is a vector-associated disease of major proportions.

The distribution of the parasite in Africa is brokered over a large part of its range through a well-balanced coevolutionary relationship within the *S. damnosum* complex, i.e., major savanna zone vectors are composed primarily of two chromosomally distinct cytospecies (*S. damnosum* sensu stricto; *S. sirbanum*) with three others (*S. yahense, S. soubrense,* and *S. squamosum*) serving as primary vectors in the rain forest.[13] Vector competency is partitioned, i.e., forest vectors are incompetent for the savanna form of *O. volvulus* and savanna vectors are incompetent for the forest form of the parasite.[14] Accurate identification of wild-caught adult flies is frequently difficult due to the lack of consistently distinct morphological characters of the various sibling taxa. Consequently, approaches involving the use of DNA probes,[15] cuticular hydrocarbons,[16] Malpighian tubule chromosomes,[17] and allozymes[18] have been suggested for differentiation; however, only the latter two techniques appear promising under field conditions in Africa. Several alpha-level species belonging to the *S. neavei* group serve as vectors in the highlands of East Africa.

Important New World vector taxa (where *O. volvulus* was probably introduced several hundred years ago) include *S. ochraceum* (Guatemala, Mexico), *S. exiguum* (Colombia, Ecuador), *S. metallicum* (Venezuela, Guatemala) and *S. oyapockense* s.l. (Brazil, Venezuela).[19] Microfilarial intake and subsequent infection of the hemocoel in *S. ochraceum* is regulated by a bucco-pharyngeal armature which shreds most of the ingested worms;[20] however, once microfilarie successfully bypass this barrier, infection of the flight muscles and development to the third larval stage is readily achieved.

Experimental cross-infections using *Simulium* spp. and *O. volvulus* from Africa and the New World (Guatemala, Venezuela) revealed that a nonreciprocal pattern exists intercontinentally between the vectors and the parasite,[21,22] with none of the vectors (*S. ochraceum, S. metallicum, S. damnosum* sensu lato) able to serve as competent intermediate hosts. However, both the Guatemalan and Venezuelan strains of *O. volvulus* were equally compatible with *S. metallicum* from either Guatemala (where this blackfly is a secondary vector) or Venezuela.[23] Thus partitioning of vector competence by this New World species complex does not appear to have taken place.

The pathology of human onchocerciasis varies considerably between the various endemic areas in Africa and the New World. For example, in West Africa, the prevalence of blindness and skin disease is more intense in the Sudan and Guinea savannas than in the rainforest and transition zones. Thus the "savanna strain" of *O. volvulus* is considered to be the most intrinsically pathogenic;[24] however the magnitude of annual parasite transmission[25] and intensity of skin infection[26] are important epidemiological factors as well. In Latin America, particularly Guatemala, ocular pathology is variable but is associated primarily with intensity of skin infection,[27] i.e., the number of microfilariae per milligram of skin.

## II. IMMUNODIAGNOSTIC ASSAYS

### A. IMMUNODIAGNOSIS

Diagnosis of human onchocerciasis is important for determining the prevalence and intensity of infection, identifying individuals who may require drug treatment, evaluating the success in the application of a given treatment, and assessing the impact of control efforts. Diagnostic procedures used so far have relied mainly upon the identification of characteristic microfilariae in skin biopsies from infected individuals[28] or by palpation of subcutaneous nodules (onchocercomata) that develop around the adult worm.[29,30] Neither of these methods however, can detect early prepatent infections, nor the presence of adult worms that may exist occult or free of nodules.[30] Thus, present parasitological methods may not detect parasites because of the low sensitivity of these techniques, leading to an inaccurate diagnosis.

Immunodiagnostic assays, which are more sensitive and specific for detecting parasite products (antigenic fractions) rather than the organisms themselves in body fluids of infected individuals, have been developed as an alternative to parasitological diagnosis.[31] Polyvalent antisera,[32] as well as several monoclonal antibodies (McAb) have been raised against different antigenic determinants of filarial parasites, including *O. volvulus,* and have been used in isotopic and nonisotopic assays for the detection of circulating parasite antigens.[33,34] The induction and selection of McAb has been hampered, however, by the use of heterogeneous immunogens and by the cross-reactivity observed with phosphorylcholine, which is present in many preparations of helminth antigens. Nevertheless, serological reactivity [by McAb] has been identified in low molecular weight (14 to 18 kDa) polypeptides of *O. volvulus;*[35] *O. volvulus* antigens have been detected also in other body fluids, such as breast milk.[36] In spite of the diverse applications of McAb, " . . . the practical utility of these reagents [McAb] to solve the diagnostic dilemmas [in filariasis including human onchocerciasis] has not yet been demonstrated".[37]

Filarial infections in humans induce both humoral and cellular immune response.[38] Detection of host antibodies, however, has been used more frequently because of the amplificative effect of this component of the immune system, and the easier application in serological studies on population-based surveys. A variety of immunodiagnostic methods, which have been applied for diagnosis of onchocerciasis, include the complement fixation,[39] indirect hemagglutination,[40] immunofluorescence,[41] ELISA,[42,43] and radio-allergosorbent[44] tests. The sensitivity and specificity for the diagnosis of onchocerciasis infections has been significantly improved over the traditional parasitological methods. Many problems still remain unsolved, especially in the high degree of cross-reactivity observed between onchocerciasis and other filarial and helminthic infections.[45] Most of the serologic assays have employed crude extracts of adult *O. volvulus* worms as antigens or extracts from related filarial parasites (heterologous antigens), counting on the high degree of cross-reactivity observed between these parasites.[46] Evidence, however, for species- and stage-specific reactions in human onchocerciasis has not yet been described.[47]

Attempts have been made to improve the quality of *O. volvulus* antigens used in isotopic tests, for example, by applying purification procedures such as affinity chromatography or isoelectric focusing to remove host globulins and other antigenic contaminants.[48] However, the real value of this method for increasing the specificity of the ELISA test remains to be evaluated in a large scale survey. Recently, gel filtration for purification of *O. volvulus* adult worm antigen yielded a 20 kDa surface antigenic fraction which, when radiolabeled, provided an assay with a sensitivity of 92% and a specificity of 98% in a seroepidemiological study of human onchocerciasis in Mexico.[49] This antigen, however, was not strictly species-specific since it detected high levels of antibodies in sera from Indians exposed to *Wuchereria bancrofti* in the New Delhi endemic area. Nevertheless, the specificity of serologic tests for onchocerciasis may be increased by focusing on the characterization of surface, somatic, and excreted antigens, and by identifying class or subclass-specific antibody response.

Modern methods for the characterization of surface antigens are based on the combination of sodium dodecyl sulfate-polyacrylamide gel electrophoresis (SDS-PAGE), with the most commonly used buffer system being that of Laemmli,[50] with the subsequent electrophoretical transfer of proteins onto nitrocellulose filter sheets, according to Towbin et al.[51] These procedures together with immunoassays employing isotopic and nonisotopic markers, have been used to characterize the reactive specificities of antigens and antibodies in several parasitic infections, including human onchocerciasis.[51,52-58]

Low molecular weight (LMW) antigens (ranging from 20 to 43 kDa) of adult *O. volvulus* worms have been shown by two-dimensional SDS-PAGE to be more species-specific than high molecular weight (HMW) antigens, which cross-react with sera from patients with other filariases.[53] Similar findings were obtained on immunoblotted antigens, where *Onch-*

*ocera*-specific bands were observed in the 12- to 25-kDa range of phosphate-buffered saline (PBS)- and detergent-solubilized worm extracts.[54] Interestingly, comparable LMW antigens to adult worm surface proteins,[49,54-58] have been found to be present in other parasite stages, such as skin microfilariae.[55] Based on these results, a LMW (12.5 to 35.5 kDa) antigenic fraction was recently prepared by gel filtration of crude extracts of soluble, adult *O. volvulus* and used in an ELISA test to determine the prevalence of onchocerciasis in African villages. The specificity of the test was significantly improved when the LMW fractions were used as antigens, in comparison to the crude adult worm extracts. The assay could be used to evaluate onchocerciasis prevalence, especially when testing individuals 5 to 15 years of age.[59]

The specificity of serological tests for onchocerciasis may also be increased by focusing on the detection of class or subclass-specific antibodies. It appears that IgE antibodies are more species-specific than IgG antibodies in both human onchocerciasis and lymphatic filariasis.[44,60] In addition, IgG4 type antibodies are preferentially induced during filarial infections and can recognize *O. volvulus*[58-61] and lymphatic filariasis antigens[62-64] more specifically than any of the other classes or subclasses of antibodies. Clinico-immunological relations have been suggested between signs of the disease and class or subclass-specific responses, . . . "but their broad applicability is unproven still".[37] Nevertheless, the specificity of ELISA and immunoblot tests for serological diagnosis of human onchocerciasis is improved when IgG4 or IgE responses are examined, even when whole worm soluble antigenic preparations are used.

## B. MOLECULAR BIOLOGY

It appears that one of the major problems in immunodiagnosis of human onchocerciasis (the specificity of the test) can be partly solved when antigens of low molecular mass are utilized instead of whole worm soluble antigenic preparations. Stage-specific surface, somatic, and excreted antigens may be of particular value for accurate serodiagnosis. A problem still remains, however, in providing a constant, reliable source of adult worms and key larval stages to prepare large enough quantities of these antigens. Currently, obtaining antigens is dependent on preparation from live or frozen parasites, but worldwide there is a limited availability of *O. volvulus* parasites since there is currently no suitable laboratory model for human onchocerciasis. Therefore, alternative antigen sources have to be developed, either from other *Onchocerca* species which display a high degree of cross-reactivity or by cloning DNA sequences for specific *O. volvulus* antigens.

Recombinant DNA technology can be extremely useful for filariasis research since purified reagents can be produced in large quantities.[65] Recent applications of this technology have been aimed at detecting *O. volvulus* infections in human beings and blackflies, differentiating *O. volvulus* from other helminths or filarial parasites,[66,67] distinguishing forest from savanna strains of *O. volvulus*,[68] expressing stage-specific antigens (L₃),[69] and producing proteins that may serve as specific antigens in serodiagnosis or that might be involved in developing a vaccine against *O. volvulus*. Recently, a recombinant cDNA clone derived from microfilariae-producing female *O. volvulus* and expressing a released product of 16 to 17 kDa, has been identified to be specific for the serodiagnosis of onchocerciasis, since sera of patients with other filariases did not recognize this antigen.[70]

This aspect of the review has focused primarily on immunodiagnosis of *O. volvulus* infections in humans by detecting antigen products of the parasite or host antibodies in body fluids. Other human filarial agents such as *Wuchereria bancrofti, Brugia malayi, or Loa loa,* or nonhuman filarial pathogens, such as *Dirofilaria immitis, B. pahangi, Acantho-cheilonema viteae,* and others, have not been included due to space limitations. Much information, however, has been derived from these other systems and have contributed significantly to the development of accurate immunodiagnosis for human onchocerciasis. Modern immunodiagnostic assays which rely on isotopic (e.g., RIA, IRMA, RIPEGA) or nonisotopic (e.g., ELISA, IFA) markers have more than an acceptable sensitivity. The major problem for diagnosis of human onchocerciasis has been the extensive cross-reactive anti-genicity observed between *Onchocerca* spp., other filarial parasites, and other helminths. For this reason, each serological test has an inadequate specificity which is not suitable for epidemiological studies, follow-up of infected or treated patients, or for surveillance of control areas. The introduction of immunoblotting techniques and determination of class and subclass-specific antibodies has provided a significant improvement on the specificity of immunodiagnosis over the preceding assays using crude soluble antigens. The limited source and complexity of *O. volvulus* antigens, however, is a serious constraint to the development of sensitive and specific assays. This problem may be overcome by recombinant cDNA expression products, which can be a substitute for the traditional serological assays.

## III. CONTROL MEASURES

### A. VECTOR CONTROL

Systematic attempts at controlling human onchocerciasis date back to the 1940s when DDT became available as an effective blackfly larvicide.[71] By instituting successful control of blackfly vectors using this organochlorine, transmission of *O. volvulus* was temporarily curtailed or eliminated in several parts of Africa[12,72] and Guatemala.[73] More recently, a multinational program aimed at interrupting transmission of *O. volvulus* by members of the *Simulium damnosum* sibling species complex was instituted in a large geographic zone of the savannahs of West Africa.[74] The Onchocerciasis Control Program (OCP), funded by the World Bank and administered by the World Health Organization, is scheduled for a 20-year operational period (1974 to 1994).

At its inception, because of the lack of an appropriate mass chemotherapeutic agent, the strategy of the OCP was oriented exclusively to vector control using extensive applications of organophosphate larvicides.[75] After a 10-year period of highly successful blackfly control using temephos and chlorphoxim, transmission of *O. volvulus* had been reduced below the accepted levels of 100 *O. volvulus* infective stage larvae ($L_3$)/person/year in 90% of the treatment area. In so doing, a 97% reduction in infections among children was noted, indicating that a cohort of approximately three million children were free from risk of ocular onchocerciasis.

However, attendant with the success of this program, several entomological and epidemiological problems emerged which have threatened its overall operational well-being. Chief among these are insecticide resistance[76,77] and reinvasion of areas previously sanitized by larviciding;[78] others include lack of suitable means for precise identification of the vector stages of *O. volvulus* "strains" and related bovine *Onchocerca* species, as well as the need to accurately separate savanna vector species adults (*S. damnosum* s.s., *S. sirbanum*) from forest forms at the interface of these major geobotanical zones. In response, the original research-operational goals of the OCP were expanded to include a drug discovery component, development of alternative *in vivo* and *in vitro* systems for experimental drug screening, and ancillary research focusing on modern identification methods for both vector and parasite. A territorial extension of the vector control program was also instituted to provide adequate insecticide coverage in westward areas believed to be the source of invading flies.[79] Molecular biologic approaches, including the evaluation of current methodologies for vaccine research,[65] were also independently formulated to diversify control possibilities.

In response to the occurrence of resistance in larval populations of important vector cytospecies (*S. sirbanum*, *S. damnosum* s.s.),[80] attention was directed toward the discovery of alternative larvicides which had little or no deleterious impact on aquatic nontarget organisms.[81] Because of cross-resistance to organophosphorus (OP) insecticides, permethrin (a pyrethroid) and carbosulfan (a carbamate) received strong consideration as potential replacements, even though the prospects of resistance, nontarget toxicity, and carry (distance below application with 100% effect) remain as potential or real operational problems.

Because of its specific mode of action and rapidity of toxicity for medically important, nematocerous Diptera[82] and the low potential for selection of resistance,[83] the H-14 serotype of *Bacillus thuringiensis*[84] has been used extensively in combatting the spread of OP-resistance in both forest and savanna populations of *S. damnosum* sensu lato.[85] For economical and operational reasons, the initial use of *Bacillus thuringiensis* H-14 was restricted primarily to the dry season. Implementation during this time was based on decreased river discharge, hence less material was required to treat larval sites. A flowable concentrate formulation was applied weekly by air to large rivers and was alternated with other insecticides (usually chlorphoxim) in the wet season in an attempt to curtail expansion of temephos resistance.[86] Smaller water courses, i.e., with flow rates $\leq 50$ m$^3$/s, in areas populated by resistant forms were treated throughout the year. In implementing *Bacillus thuringiensis* H-14 as a response to resistance, over one million liters of material was used by the OCP between 1981 and 1985.[87]

The *Bacillus thuringiensis* H-14 serotype is the first commercially available biological control agent for blackflies.[88] As such, it has reached a relatively wide market, in part, because of overlap with mosquito control. In contrast, the operational control potential of other simuliid pathogens appears to be relatively low, due to the inability to effectively mass produce and deliver promising entomopathogens.[89,90] Indeed, operational, wide-scale application of *Bacillus thuringiensis* H-14 is still less cost effective than traditional insecticides because of excess bulk, formulations requirements and variability in concentration of active ingredient, a biotechnological problem currently being addressed.[91]

OP-resistance gene flow within *Simulium soubrense/sanctipauli* populations in the Ivory Coast has been followed using cytogenetic techniques.[92] A chromosomal rearrangement (inversion A) on the long arm of chromosome II (IIL-A) served as a marker to trace the spread of resistance eastward from the Bandama river. The use of this inversion marker was also later verified for populations of *S. sanctipauli* from the Ivory Coast, Ghana, and Mali.[93] However, while serving as an excellent phenotypic marker, the actual genetic/biochemical mechanism(s) for temephos/chlorphoxim detoxification is still under study. For example, total esterase activity seen in larvae collected from areas where resistance occurs appeared to be predictive although the high degree of enzyme heterogeneity among individual larvae in both resistant and susceptible populations made this interpretation difficult.[94]

The occurrence of OP resistance has proven costly to the overall fiscal, operational and epidemiological aspects of the OCP.[95] In developing a resistance management scheme, current approaches involve not only the use of *Bacillus thuringiensis* H-14 to reduce the frequency of resistant genotypes but the selection of chemical larvicides to which there is less likelihood of cross-resistance or that exhibit negative cross-resistance.[80] However, the latter approach is not without difficulty due to possible "omnipotent" detoxification mechanisms controlled at other genetic loci.[96]

## B. CHEMOTHERAPY

The chemotherapy of human onchocerciasis has seen a dramatic change in the past 5 years as a result of the introduction of ivermectin, a potent microfilaricidal drug.[97] Of particular importance is the fact that ivermectin produces substantially fewer clinical reactions than drugs used previously.[98] Prior to the advent of this drug, treatments were problematic due to the nephrotoxicity of the only available macrofilaricide (Suramin) and the adverse side-reactions associated with microfilarial-killing by diethylcarbamazine, a piperazine derivative.[99] As a broad spectrum antiparasitic agent, ivermectin is free of these major problems: reported side effects including edema of the limbs,[100,101] and face,[101] and occasional headache, fever,[102] and postural hypotension.[100] Because of its safety, ivermectin provides the first opportunity for implementation of chemotherapy programs integrating mass treatment with vector control. Indeed, as of January 1989, more than 120,000 individual doses had been administered without severe adverse reactions in various WHO-sponsored field trails.[97] With its recent registration (as Mectizan) for human consumption, ivermectin is currently being dispensed in the OCP area of West Africa.[103] In addition to its treatment value for *O. volvulus*, a single dose of ivermectin is highly effective against *Ascaris lumbricoides*, *Strongyloides stercoralis* and *Trichuris trichiura* with relatively good activity against hookworm (probably *Nectar americanus*).[104] Thus the clinical implications of a chemotherapy-based program for human onchocerciasis control can also be expected to have a major impact on reducing enteric roundworm burdens as well.

The demonstration that a single treatment with ivermectin substantially reduced the uptake of *O. volvulus* microfilariae for periods of 4 to 10 months by both forest[105,106] and savanna[107] vectors in West Africa suggested that in addition to its favorable clinical properties this drug could be used to interrupt transmission as well. Because of its ability to prevent vector infection by ridding the skin of microfilariae for epidemiologically important periods of time, it has been suggested that ivermectin would be particularly effective where parasite transmission is limited by seasonality, discrete ecological boundaries, or vector inefficiency.[105] This latter possibility was explored in Guatemala where it was shown that two treatments, given at 7-month intervals to a highly infectious group of volunteers, resulted in almost complete suppression of developing or infective larvae in *S. ochraceum* for a 6-month period.[108] Also, the overall decline of vector infection was an order of magnitude lower than pretreatment levels.

Studies have begun in Africa and Guatemala to measure the effects of community-wide treatment on the annual transmission potential[12] of *O. volvulus*. Preliminary results from investigations using one[109] or two[110] community-wide treatments are encouraging and demonstrate a downturn in *S. damnosum* s.l. infection with developing larvae, hence $L_3$ production. Data from Guatemala, where four 6-monthly treatments have been completed in three communities,[111] indicate that the basic reproductive rate ($R_0$) of the parasite can be drastically lowered with sequential ivermectin treatment, thereby possibly reaching the threshold for continued transmission using chemotherapy alone.[107] These findings have broad ramifications for the planning and implementation of future control programs, particularly in the African savannas. With the availability of modern molecular techniques to specifically identify *O. volvulus* $L_3$s in zoophilic blackfly vectors,[113] it should now be possible to critically evaluate similar studies in the savanna when multiple treatments are given on a scheduled basis. In doing so, the role of vector control in concert with ivermectin chemotherapy can be determined so that cost-effective protection can be established at the community level.

## C. VACCINE DEVELOPMENT

The development of an *O. volvulus* synthetic vaccine is an emerging area which will require careful evaluation. Factors which must be considered include the social, political, and economic realities of vaccine delivery after identifying and cloning genes for protective immunogens,[114] as well as possible difficulties in demonstrating the efficacy of a prophylactic vaccine because of the chronic nature of this parasitism.[115] The spread of AIDS throughout Africa and Latin America also poses potential problems when considering a vaccine for a chronic helminth infection such as human onchocerciasis[116] because of the former's effect on immunoresponsiveness.

Nevertheless, techniques in molecular genetics and immunology have been successfully implemented in surveying the *O. volvulus* genome for potentially useful antigens. For example, several cDNA fusion products that are recognized by human sera from endemic areas have been isolated and characterized.[69,117] One of these, paramyosin, is a major immunogen in filarial infections, exhibiting a high degree of amino acid homology between *O. volvulus* and *D. immitis*.[118] As a corollary to the search for protective antigens, recent field observations in Guatemala of persons living in an endemic zone suggest that some level of natural immunity to *O. volvulus* occurs within a subset of that population; the protective mechanism appears to be associated with greater lymphocyte responsiveness involving production of interleukin-2.[119]

The recent development of several field and laboratory techniques should also prove complementary in further identifying potentially useful clones/fusion products. These include the use of *in situ* DNA-mRNA hybridization to determine the tissue location of genes encoding specific antigens,[120] greater access to specific stages of the parasite's life cycle, including infective stage larvae,[121] *in vitro*-cultured $L_4$s[122] and adult worms.[123] These techniques, when coupled with the usefulness of cryopreservation of live worms,[124] should provide added means to pursue the vaccine discovery process. However, as reviewed by Gutteridge,[125] while vaccines are potentially safer, cheaper, and more prophylactically efficacious than drugs, the latter will continue to be needed for some time to come.

# REFERENCES

1. **Cupp, E. W.**, Human onchocerciasis: developmental biology of the parasite, in *Proc. UpJohn WHO/OCP Symp. Onchocerciasis/Filariasis*, Conder, G. A. and Williams, J. F., Eds., Kalamazoo, MI, 1986, 1.
2. **Bianco, A. E., Mustafa, M. B., and Ham. P. J.**, Fate of developing larvae of *Onchocerca lienalis* and *O. volvulus* in micropore chambers implanted into laboratory hosts, *J. Helminthol.*, 63, 218, 1989.
3. **Bianco, A. E. and Muller, R.**, Experimental transmission of *Onchocerca lienalis* to calves, *Parasitology* (Suppl.), 1982, 349.
4. **Duke, B. O. L.**, Onchocerciasis: deep worm-bundles close to hip joints, *Trans. R. Soc. Trop. Med. Hyg.*, 64, 791, 1970.
5. **George, G. H., Palmieri, J. R., and Connor, D. H.**, The onchocercal nodule: interrelationships of adult worms and blood vessels, *Am. J. Trop. Med. Hyg.*, 34, 1144, 1985.
6. **Schulz-Key, H. and Karam, M.**, Periodic reproduction of *Onchocerca volvulus*, *Parasitology Today*, 2, 284, 1986.
7. **Paulson, C. W., Jacobson, R. H., and Cupp, E. W.**, Microfilarial surface carbohydrates as a function of developmental stage and ensheathment status in six species of filariids, *J. Parasitol.*, 74, 743, 1988.
8. **Ham, P. J., Smail, A. J., and Groeger, B. K.**, Surface carbohydrate changes on *Onchocerca lienalis* larvae as they develop from microfilariae to the infective third-stage in *Simulium*, *J. Helminthol.*, 62, 195, 1988.
9. **Muller, R.**, Identification of *Onchocerca*, in *Problems in the Identification of Parasites and Their Vectors*, Vol. 17, Taylor, A. E. R. and Muller, R., Eds., Blackwell Scientific, Oxford, 1979, 175.
10. **Bain, O.**, Le genre *Onchocerca*: hypothèses sur son évolution et de dichotomique des espèces, *Ann. Parasitol.*, 56, 503, 1981.
11. **Meredith, S. E. O., Unnasch, T. R., Karam, M., Piessens, W. F., and Wirth, D. F.**, Cloning and characterization of an *Onchocerca volvulus* specific DNA sequence, *Mol. Biochem. Parasitol.*, 36, 1, 1989.
12. **World Health Organization**, WHO Expert Committee on Onchocerciasis, Tech. Rep. Ser., 752, Geneva, 1987.
13. **Dunbar, R. W. and Vajime, C. G.**, Cytotaxonomy of the *Simulium damnosum* complex, in *Black Flies— The Future for Biological Methods in Integrated Control*, Laird, M., Ed., Academic Press, London, 1981, 31.
14. **Duke, B. O. L., Lewis, D. J., and Moore, P. J.**, *Onchocerca-Simulium* complexes. I. Transmission of forest and Sudan-savanna strains of *Onchocerca volvulus*, from Cameroon, by *Simulium damnosum* from various West African bioclimatic zones, *Ann. Trop. Med. Parasitol.*, 60, 318, 1966.
15. **Post, R. J.**, DNA probes for vector identification, *Parasitology Today*, 1, 89, 1985.
16. **Philips, A., Walsh, J. F., Garms, R., Molyneux, D. H., Milligan, P., and Ibrahim, G.**, Identification of adults of the *Simulium damnosum* complex using hydrocarbon analysis, *Tropenmed. Parasitol.*, 36, 97, 1985.
17. **Procunier, W. S. and Post, R. J.**, Development of a method for the cytological idientification of man-biting sibling species within the *Simulium damnosum* complex, *Trop. Med. Parasitol.*, 37, 49, 1986.

18. **Thomson, M. C., Davies, J. B., and Wilson, M. D.,** A portable allozyme electrophoresis kit used to identify members of the *Simulium damnosum* Theobald complex (Diptera:Simuliidae) in the field, *Bull. Entomol. Res.,* 79, 685, 1989.

19. **Shelley, A. J.,** Vector aspects of the epidemiology of onchocerciasis in Latin America, *Annu. Rev. Entomol.,* 33, 337, 1988.

20. **Omar, M. S. and Garms, R.,** The fate and migration of microfilariae of a Guatemalan strain of *Onchocerca volvulus* in *Simulium ochraceum* and *S. metallicum* and the role of the buccopharyngeal armature in the destruction of microfilariae, *Tropenmed. Parasitol.,* 26, 183, 1975.

21. **DeLeon, R. and Duke, B. O. L.,** Experimental studies on the transmission of Guatemalan and West African strains of *Onchocerca volvulus* by *Simulium ochraceum, S. metallicum* and *S. callidum, Trans. R. Soc. Trop. Med. Hyg.,* 60, 735, 1966.

22. **Duke, B. O. L.,** *Onchocerca-Simulium* complexes. VI. Experimental studies on the transmission of Venezuelan and West African strains of *Onchocerca volvulus* by *Simulium metallicum* and *S. exiguum* in Venezuela, *Ann. Trop. Med. Parasitol.,* 64, 421, 1970.

23. **Takaoka, H., Tada, I., Hashiguchi, Y., Baba, M., Kokenaga, M., Onofre, J., Convit, J., and Yarzabal, L.,** A cross compatibility study of Guatemalan and North Venezuelan *Onchocerca volvulus* to *Simulium metallicum* from two countries, *Jpn. J. Parasitol.,* 35, 35, 1986.

24. **Budden, F. H.,** Comparative study of ocular onchocerciasis in savanna and rainforest, *Trans. R. Soc. Trop. Med. Hyg.,* 57, 64, 1963.

25. **Thylefors, B., Philippon, B., and Prost, A.,** Transmission potentials of *Onchocerca volvulus* and the associated intensity of onchocerciasis in a Sudan-savanna area, *Tropenmed. Parasitol.,* 29, 346, 1978.

26. **Remme, J., Dadzie, K. Y., Rolland, A., and Thylefors, B.,** Ocular onchocericasis and intensity of infection in the community. I. West African savanna, *Trop. Med. Parasitol.,* 40, 340, 1989.

27. **Brandling-Bennett, A. D., Anderson, J., Fuglasang, H., and Collins, R.,** Onchocerciasis in Guatemala. Epidemiology in fincas with various intensities of infection, *Am. J. Trop. Med. Hyg.,* 30, 970, 1981.

28. **Taylor, H. R., Munoz, B., Keyvan-Larijani, E., and Greene, B. M.,** Reliability of detection of microfilariae in skin snips in the diagnosis of onchocerciasis, *Am. J. Trop. Med. Hyg.,* 41, 467, 1989.

29. **Shulz-Key, H. and Karam, M.,** Quantitative assessment of microfilariae and adults of *Onchocerca volvulus* in ethanol-fixed biopsies and nodules, *Trans. R. Soc. Trop. Med. Hyg.,* 78, 157, 1984.

30. **Nnochiri, E.,** Observations on onchocercal lesions in autopsy specimens in Western Nigeria, *Ann. Trop. Med. Parasitol.,* 58, 89, 1964.

31. **Hamilton, R. G.,** Application of immunoassay methods in the serodiagnosis of human filariasis, *Rev. Infect. Dis.,* 7, 837, 1985.

32. **Ouaissi, A., Koumemeni, L.-E., Haque, A., Ridel, P.-R., Saint André, P., and Capron, A.,** Detection of circulating antigens in onchocerciasis, *Am. J. Trop. Med. Hyg.,* 30, 1211, 1981.

33. **Weiss, N.,** Monoclonal antibodies as investigative tools in onchocerciasis, *Rev. Infect. Dis.,* 7, 826, 1985.

34. **Des Moutis, I., Ouaissi, A., Grzych, J. M., Yarzábal, L., Haque, A., and Capron, A.,** *Onchocerca volvulus*: Detection of circulating antigen by monoclonal antibodies in human onchocerciasis, *Am. J. Trop. Med. Hyg.,* 32, 533, 1983.

35. **Cabrera, A. and Parkhouse, R. M. E.,** Isolation of an antigenic fraction for diagnosis of onchocerciasis, *Parasite Immunol.,* 9, 39, 1987.

36. **Petralanda, I., Yarzábal, L., and Piessens, W. F.,** Parasite antigens are present in breast milk of women infected with *Onchocerca volvulus, Am. J. Trop. Med. Hyg.,* 38, 372, 1988.

37. **World Health Organization,** Report of the Steering Committee of the Scientific Working Group on Filariasis, WHO TDR/FIL/SC-SWG(83-88)/88.3. 1, 1988.

38. **Greene, B. M., Gbakima, A. A., Albiez, E. J., and Taylor, H. R.,** Humoral and cellular immune responses to *Onchocerca volvulus* infection in humans, *Rev. Infect. Dis.,* 7, 789, 1985.

39. **Tanaka, H., Fujita, K., Sasa, M., Tagawa, M., Naito, M., and Kurokawa, K.,** Cross-reactions in complement fixation test among filarial species, *Jpn. J. Exp. Med.,* 40, 47, 1970.

40. **Tada, I. and Ikeda, T.,** Indirect hemagglutination test in onchocerciasis with special reference to the utility of dried blood smears taken on filter papers, *J. Kanazawa Med. Univ.,* 1, 79, 1976.

41. **Collins, W. E., Campbell, C. C., Collins, R. C., and Skinner, J. C.,** Serologic studies on onchocerciasis in Guatemala using fixed-tissue sections of adult *Onchocerca volvulus, Am. J. Trop. Med. Hyg.,* 29, 1220, 1980.

42. **Bartlett, A., Bidwell, D. E., and Voller, A.,** Preliminary studies on the application of the enzyme immunoassay in the detection of antibodies in onchocerciasis, *Tropenmed. Parasitol.,* 26, 370, 1975.

43. **Luján, R., Collins, W. E., Stanfill, P. S., Campbell, C. C., Collins, R. C., Brogdon, W., and Huong, A. Y.,** Enzyme-linked immunosorbent assay (ELISA) for serodiagnosis of Guatemalan onchocerciasis: comparison with the indirect fluorescent antibody (IFA) test, *Am. J. Trop. Med. Hyg.,* 32, 747, 1983.

44. **Weiss, N., Hussain, R., and Ottesen, E. A.,** IgE antibodies are more species-specific than IgG antibodies in human onchocerciasis and lymphatic filariasis, *Immunology,* 45, 129, 1982.

45. **Voller, A. and De Savigny, D.**, Diagnostic serology of tropical parasitic diseases, *J. Immunol. Methods*, 46, 1, 1981.

46. **Williams, J. F., El Khalifa, M., Mackenzie, C. D., and Sisley, B.**, Antigens of *Onchocerca volvulus*, *Rev. Infect. Dis.*, 7, 831, 1985.

47. **Schulz-Key, H., Albiez, E. J., and Büttner, D. W.**, Isolation of living adult *Onchocerca volvulus* from nodules, *Tropenmed. Parasitol.*, 28, 428, 1977.

48. **Marcoullis, G. and Gräsbeck, P.**, Preliminary identification and characterization of antigen extracts from *Onchocerca volvulus*, *Tropenmed. Parasitol.*, 27, 314, 1976.

49. **Philipp, M., Gómez-Priego, A., Parkhouse, R. M., Davies, M. W., Clark, N. W., Ogilvie, B. M., and Beltrán-Hernández, F.**, Identification of an antigen of *Onchocerca volvulus* of possible diagnostic use, *Parasitology*, 89, 295, 1984.

50. **Laemmli, U. K.**, Cleavage of structural proteins during the assembly of the head of bacteriophage T4, *Nature (London)*, 227, 680, 1970.

51. **Towbin, H., Staehelin, T., and Gordon, J.**, Electrophoretic transfer of proteins from polyacrylamide gels to nitrocellulose sheets: procedure and some applications, *Proc. Nat. Acad. Sci. U.S.A.*, 76, 4350, 1979.

52. **Hussain, R., Kaushal, N. A., and Ottesen, E. A.**, Comparison of immunoblot and immunoprecipitation methods for analyzing cross-reactive antibodies to filarial antigens, *J. Immunol. Methods*, 84, 291, 1985.

53. **Lobos, E. and Weiss, N.**, Identification of non-crossreacting antigens of *Onchocerca volvulus* with lymphatic filariasis serum pools, *Parasitology*, 93, 389, 1986.

54. **Cabrera, Z. and Parkhouse, R. M. E.**, Identification of antigens of *Onchocerca volvulus* and *Onchocerca gibsoni* for diagnostic use, *Mol. Biochem. Parasitol.*, 20, 225, 1986.

55. **Taylor, D. W., Goddard, J. M., and McMahon, J. E.**, Surface components of *Onchocerca volvulus*, *Mol. Biochem. Parasitol.*, 18, 283, 1986.

56. **Lucius, R., Büttner, D. W., Kirsten, C., and Diesfeld, H. J.**, A study on antigen recognition by onchocerciasis patients with different clinical forms of disease, *Parasitology*, 92, 569, 1986.

57. **Lucius, R., Prod'Hon, J. Kern, A., Hébrard, G., and Diesfeld, H. J.**, Antibody responses in forest and savanna onchocerciasis in Ivory Coast, *Tropenmed. Parasitol.*, 38, 194, 1987.

58. **Parkhouse, R. M. E., Cabrera, Z., and Harnett, W.**, *Onchocerca* antigens in protection, diagnosis and pathology, in *Filariasis* (Ciba Foundation Symposium 127), John Wiley & Sons, Chichester, 125, 1987.

59. **Weiss, N. and Karam, M.**, Evaluation of a specific enzyme immunoassay for onchocerciasis using a low molecular weight antigen fraction of *Onchocerca volvulus*, *Am. J. Trop. Med. Hyg.*, 40, 261, 1989.

60. **Weiss, N., Speiser, F., and Hussain, R.**, IgE antibodies in human onchocerciasis. Application of a newly developed radioallergosorbent test (RAST), *Acta Trop.*, 38, 356, 1981.

61. **Weil, G. J., Ogunrinade, A. F., Chandrashekar, R., and Kale, O. O.**, IgG4 subclass antibody serology for onchocerciasis, *J. Infect. Dis.*, 161, 549, 1990.

62. **Ottesen, E. A., Skvaril, F., Tripathy, S. P., Poindexter, R. W., and Hussain, R.**, Prominence of IgG4 in the IgG antibody response to human filariasis, *J. Immunol.*, 134, 2707, 1985.

63. **Hussain, R., Poindexter, R. W., Ottesen, E. A., and Reimer, C. B.**, Use of monoclonal antibodies to quantify subclasses of human IgG. II. Enzyme immunoassay to define antigen specific (anti-filarial) IgG subclass antibodies, *J. Immunol. Methods*, 94, 73, 1986.

64. **Hussain, R., Grögl, M., and Ottesen, E. A.**, IgG antibody subclasses in human filariasis. Different subclass recognition of parasite antigens correlates with different clinical manifestations of infection, *J. Immunol.*, 139, 2794, 1987.

65. **Greene, B. M. and Unnasch, T. R.**, Molecular biologic approaches to research in onchocerciasis, *J. Infect. Dis.*, 154, 1024, 1986.

66. **Perler, F. B. and Karam, M.**, Cloning and characterization of two *Onchocerca volvulus* repeated DNA sequences, *Mol. Biochem. Parasitol.*, 21, 171, 1986.

67. **Shah, J. S., Karam, M., Piessens, W. F., and Wirth, D. F.**, Characterization of an *Onchocerca*-specific DNA clone from *Onchocerca volvulus*, *Am. J. Trop. Med. Hyg.*, 37, 376, 1987.

68. **Erttmann, K. D., Unnasch, T. R., Greene, B. M., Albiez, E. J., Boateng, J., Denke, A. M., Ferraroni, J. J., Karam, M., Schulz-Key, H., and Williams, P. N.**, A DNA sequence specific for forest form *Onchocerca volvulus*, *Nature (London)*, 327, 415, 1987.

69. **Unnasch, T. R., Gallin, M. Y., Soboslay, P. T., Erttmann, K. D., and Greene, B. M.**, Isolation and characterization of expression cDNA clones encoding antigens of *Onchocerca volvulus* infective larvae, *J. Clin. Invest.*, 82, 262, 1988.

70. **Lobos, E., Altmann, M., Mengod, G., Weiss, N., Rudin, W., and Karam, M.**, Identification of an *Onchocerca volvulus* cDNA encoding a low-molecular-weight antigen uniquely recognized by onchocerciasis patient sera, *Mol. Biochem. Parasitol.*, 39, 135, 1990.

71. **Jamnback, H.**, The origins of blackfly control programmes, in *Blackflies: The Future for Biological Methods in Integrated Control*, Laird, M., Ed., Academic Press, London, 1981, 71.

72. **Walsh, J. F., Davies, J. B., and Cliff, B.,** World Health Organization Onchocerciasis Control Programme in the Volta River Basin, in *Blackflies: The Future for Biological Methods in Integrated Control,* Laird, M., Ed., Academic Press, London, 1981, 85.

73. **Lea, A. O. and Dalmat, H. T.,** A pilot study of area larval control of blackflies in Guatemala, *J. Econ. Entomol.,* 48, 378, 1955.

74. **Davies, J. B., LeBerre, R., Walsh, J. F., and Cliff, B.,** Onchocerciasis and *Simulium* control in the Volta River Basin, *J. Am. Mosq. Cont. Assoc.,* 38, 466, 1978.

75. **Philippon, B.,** Problems in epidemiology and control of West African onchocerciasis, in *Black Flies,* Kim, K. C. and Merritt, R. W., Eds., Pennsylvania State University, University Park, 1986, 363.

76. **Guillet, P., Escaffre, H., Ouedraogo, M. M., and Quillevéré, D.,** Mise en évidence d'une résistance au téméphos dans le complexe *Simulium damnosum (S. sanctipauli* et *S. soubrense)* en Côte d'Ivoire, *Cah. ORSTOM Sér Entomol. Med. Parasitol.,* 18, 291, 1980.

77. **Kurtak, D.,** Insecticide resistance in the Onchocerciasis Control Programme, *Parasitology Today,* 2, 19, 1986.

78. **Cheke, R. A. and Garms, R.,** Reinfestations of the southeastern flank of the Onchocerciasis Control Programme by wind borne vectors, *Phil. Trans. R. Soc. London Ser. B,* 302, 471, 1983.

79. **Baker, R. H. A., Baldry, D. A. T., Boakye, D., and Wilson, M.,** Measures aimed at controlling the invasion of *Simulium damnosum* Theobald s.l. (Diptera: Simuliidae) into the Onchocerciasis Control Programme Area. III. Searches in the Upper Niger Basin of Guinea for additional sources of flies invading south eastern Mali, *Trop. Pest Manag.,* 33, 336, 1987.

80. **Kurtak, D., Meyer, R., Ocran, M., Ouedraoogo, M., Renaud, P., Sawadogo, R. O., and Télé, B.,** Management of insecticide resistance in control of the *Simulium damnosum* complex by the Onchocerciasis Control Programme, West Africa: potential use of negative correlation between organophosphate resistance and pyrethroid susceptibility, *Med. Vet. Entomol.,* 1, 137, 1987.

81. **Kurtak, D. C., Grunewald, J., and Baldry, D. A. T.,** Control of blackfly vectors of onchocerciasis in Africa, in *Black Flies,* Kim, K. C. and Merritt, R. W., Eds., Pennsylvania State University, University Park, 1986, 341.

82. **Goldberg, L. J. and Margalit, J.,** A bacterial spore demonstrating rapid larvicidal activity against *Anopheles sergentii, Uranotaenia unguiculata, Culex univittatus, Aedes aegypti,* and *Culex pipiens, Mosq. News,* 37, 355, 1977.

83. **Van Rie, J., McGaughey, W. H., Johnson, D. E., Barnett, B. D., and Van Mellaert, H.,** Mechanisms of insect resistance to the microbial insecticide *Bacillus thuringiensis, Science,* 247, 72, 1990.

84. **Barjac, H. de,** Une nouvelle variété de *Bacillus thuringiensis* très toxique pour les moustiques: *B. thuringiensis* var. *israelensis* serotype 14, *C. R. Acad. Sci. Ser. D.,* 286, 797, 1978.

85. **Lacey, L. A., Escaffre, H., Philippon, B., Seketeli, A., and Guillet, P.,** Large river treatment with *Bacillus thuringiensis* (H-14) for the control of *Simulium damnosum s.l.* in the Onchocerciasis Control Programme, *Z. Tropenmed. Parasitol.,* 33, 97, 1982.

86. **Lacey, L. A. and Undeen, A. H.,** Microbial control of blackflies and mosquitoes, *Annu. Rev. Entomol.,* 31, 265, 1986.

87. **Copplestone, J. F. and Dobrokhotov, B.,** Biological control of vectors, in *Tropical Disease Research: A Global Partnership,* Maurice, J. and Pearce, A. M., Eds., UNDP/World/Bank/WHO Special Programme for Research and Training in Tropical Diseases, 8th Programme Rep., World Health Organization, Geneva, 1987, 125.

88. **Molloy, D., Ed.,** Biological control of blackflies with *Bacillus thuringiensis* var. *israelensis* (serotype 14), *Misc. Publ. Entomol. Soc. Am.,* 12, 1982.

89. **Lacey, L. A. and Undeen, A. H.,** The biological control potential of pathogens and parasites of blackflies, in *Black Flies,* Kim, K. C. and Merritt, R. W., Eds., Pennsylvania State University, University Park, 1986, 327.

90. **Molloy, D. P.,** The ecology of blackfly parasites, in *Black Flies,* Kim, K. C. and Merritt, R. W., Eds., Pennsylvania State University, University Park, 1986, 315.

91. **Pearson, D. and Ward, O. P.,** Effect of culture conditions on growth and sporulation of *Bacillus thuringiensis* subsp. *israelensis* and development of media for production of the protein crystal endotoxin, *Biotechnol. Lett.,* 10, 451, 1988.

92. **Meredith, S. E. O., Kurtak, D., and Adiamah, J. H.,** Following movements of resistant populations of *Simulium soubrense/sanctipauli* (Diptera: Simuliidae) by means of chromosome inversions, *Trop. Med. Parasitol.,* 37, 290, 1986.

93. **Post, R. J. and Kurtak, D.,** Identity of the OP-insecticide resistant species in the *Simulium sanctipauli* subcomplex, *Ann. Soc. Belge Med. Trop.,* 67, 71, 1987.

94. **Magnin, M., Kurtak, D., and Pasteur, N.,** Charactérisation des estérases chez des larves du complexe *Simulium damnosum* résistantes aux insecticides organophosphores, *Cah. ORSTOM Ser. Ent. Med. Parasitol.,* 15, 57, 1988.

95. **Walsh, J. F., Philippon, B., Hendrick, J. E. E., and Kurtak, D. C.,** Entomological aspects and results of the Onchocerciasis Control Programme, *Trop. Med. Parasitol.,* 38, 57, 1987.

96. **Roush, R. T.,** Designing resistance management programs: How can you choose?, *Pestic. Sci.,* 26, 423, 1989.

97. **Taylor, H. R. and Greene, B. M.,** The status of ivermectin in the treatment of human onchocerciasis, *Am. J. Trop. Med. Hyg.,* 41, 460, 1989.

98. **Greene, B. M., Brown, K. R., and Taylor, H. R.,** Use of ivermectin in humans, in *Ivermectin and Abamectin,* Campbell, W. C., Ed., Springer-Verlag, New York, 1989.

99. **Mackenzie, C. D., Boland, M., El Sheikh, H., and Dick, W.,** Adverse reactions to chemotherapy in onchocerciasis, in *Proc. Upjohn WHO/OCP Symp. Onchocerciasis/Filariasis,* Conder, G. A. and Williams, J. F., Eds., Kalamazoo, Mich., 1986, 153.

100. **DeSole, G., Awadzi, K., Remme, J., Dadzie, K. Y., Ba, O., Giese, J., Karam, M., Keita, F. M., and Opoku, N. O.,** A community trial of ivermectin in the onchocerciasis focus of Asubende, Ghana. II. Adverse reactions, *Trop. Med. Parasitol.,* 40, 375, 1989.

101. **Zea-Flores, R. and Richards, F., Jr.,** personal communication, 1990.

102. **Rothova, A., van der Lelij, A., Stilma, S. A., Wilson, W. R. and Barbe, R. F.,** Side effects of ivermectin in treatment of onchocerciasis, *Lancet,* 8650, 1439, 1989 (Vol. I).

103. **Bradshaw, H.,** Onchocerciasis and the Mectizan donation programme, *Parasitology Today,* 5, 63, 1989.

104. **Freedman, D. O., Zierdt, W. S., Luján, A., and Nutman, T. B.,** The efficacy of ivermectin in the chemotherapy of gastrointestinal helminthiasis in humans, *J. Infect. Dis.,* 159, 1151, 1989.

105. **Cupp. E. W., Bernardo, M. J., Kiszewski, A. E., Collins, R. C., Taylor, H. R., Aziz, M. A., and Greene, B. M.,** The effects of ivermectin on transmission of *Onchocerca volvulus, Science,* 231, 740, 1986.

106. **Prod'Hon, J., Lardeux, F., Bain, O., Hébrard, G., and Prud'Hom, J. M.,** Ivermectine et modalité de la réduction de l'infection des simulies dans un foyer forestier d'onchocercose humaine, *Ann. Parasitol. Hum. Comp.,* 62, 590, 1987.

107. **Bissan, Y., Yingtain, P., Doucoure, K., Doumbo, O., Dembele, D., Ginoux, J., Cozettes, P., and Ranque, P.,** L'ivermectine (MK-933) dans le traitement de l'onchocercose, son incidence sur la transmission d'*Onchocerca volvulus* en savane soudanienne au Mali, *Med. Afr. Noire,* 33, 81, 1986.

108. **Cupp, E. W., Ochoa, A. O., Collins, R. C., Ramberg, F. R., and Zea-Flores, G.,** The effects of multiple ivermectin treatment on infection of *Simulium ochraceum* (Diptera: Simuliidae) with *Onchocerca volvulus* (Filarioidea: Onchocercidae), *Am. J. Trop. Med. Hyg.,* 40, 501, 1989.

109. **Remme, J., Baker, R. H. A., DeSole, G., Dadzie, K. Y., Walsh, J. F., Adams, M. A., Alley, E. S., and Avissey, H. S. K.,** A community trial of ivermectin in the onchocerciasis focus of Asubende, Ghana. I. Effect on the microfilarial reservoir and the transmission of *Onchocerca volvulus, Trop. Med. Parasitol.,* 40, 367, 1989.

110. **Trpis, M., Childs, J. E., Fryauff, D. J., Greene, B. M., Williams, P. N., Muñoz, B. E., Pacque, M. C., and Taylor, H. R.,** Effect of mass treatment of a human population with ivermectin on transmission of *Onchocerca volvulus* by *Simulium yahense* in Liberia, West Africa, *Am. J. Trop. Med. Hyg.,* 42, 148, 1990.

111. **Cupp, E. W.,** personal observations, 1990.

112. **Anderson, R. M.,** Epidemiological models and predictions, *Trop. Geo. Med.,* 40, 530, 1988.

113. **Harnett, W., Chambers, A. E., Renz, A., and Parkhouse, R. M. E.,** An oligonucleotide probe specific for *Onchocerca volvulus, Mol. Biochem. Parasitol.,* 35, 119, 1989.

114. **Ogilvie, B. M.,** Vaccines: around which corner?, *Immunol. Lett.,* 19, 245, 1988.

115. **Mitchell, G. F.,** Problems specific to parasite vaccines, *Parasitology,* 98, S19, 1989.

116. **Parkhouse, R. M. E. and Harrison, L. J. S.,** Antigens of parasitic helminths in diagnosis, protection and pathology, *Parasitology,* 99, S5, 1989.

117. **Donelson, J. E., Duke, B. O. L., Moser, D., Zeng, W., Erondu, N. E., Lucius, R., Renz, A., Karam, M., and Flores, G. Z.,** Construction of *Onchocerca volvulus* cDNA libraries and partial characterization of the cDNA for a major antigen, *Mol. Biochem. Parasitol.,* 31, 241, 1988.

118. **Limberger, R. J. and McReynolds, L. A.,** Filarial paramyosin: cDNA sequences from *Dirofilaria immitis* and *Onchocerca volvulus, Mol. Biochem. Parasitol.,* 38, 271, 1990.

119. **Ward, D. J., Nutman, T. B., Zea-Flores, G., Portocarrero, C., Luján, A., and Ottesen, E. A.,** Onchocerciasis and immunity in humans: enhanced T cell responsiveness to parasite antigen in putatively immune individuals, *J. Infect. Dis.,* 157, 536, 1988.

120. **Arasu, P., Nutman, T. B., Steel, C., Mulligan, M. M., Abraham, D., Tuan, R. S., and Perier, F. B.,** Human T-cell stimulation, molecular characterization and *in situ* mRNA localization of a *Brugia malayi* recombinant antigen, *Mol. Biochem. Parasitol.,* 36, 223, 1989.

121. **Cupp, E. W., Bernardo, M. J., Kiszewski, A. E., Trpis, M., and Taylor, H. R.,** Large scale production of the vertebrate infective stage (L$_3$) of *Onchocerca volvulus* (Filarioidea: Onchocercidae), *Am. J. Trop. Med. Hyg.,* 38, 596, 1988.

122. **Cupp, M. S., Cupp, E. W., and Poulopoulou, C.,** The use of cell-conditioned medium for the *in vitro* culture of *Onchocerca* spp. larvae (Nematoda: Filarioidea), *Trop. Med. Parasitol.,* 41, 20, 1990.

123. **Duke, B. O. L.,** Special requirements for use of adult *Onchocerca volvulus* in molecular biology, in *The Onchocerca Nodule and the Adult Filariae: Normal Structure, Changes During Chemotherapy and Optimal Recovery of Worm Material,* Schulz-Key, H., Ginger, C. D., Duke, B. O. L., and Büttner, D. W., Eds., *Trop. Med. Parasitol.,* 39 (Suppl. 4), 463, 1988.

124. **Townson, S.,** The development of a laboratory model for onchocerciasis using *Onchocerca gutturosa: in vitro* culture, collagenase effects, drug studies and cryopreservation, in the *Onchocerca Nodule and the Adult Filarial: Normal Structure, Changes During Chemotherapy and Optimal Recovery of Worm Material,* Schulz-Key, H., Ginger, C. D., Duke, B. O. L., and Bütner, D. W., Eds., *Trop. Med. Parasitol.,* 39 (Suppl. 4), 475, 1988.

125. **Gutteridge, W. E.,** Parasite vaccines versus anti-parasite drugs: rivals or running mates?, *Parasitology,* 98, S87, 1989.

# Mass Production of Microbial and Viral Biocontrol Agents

Chapter 14

# MASS PRODUCTION OF *BACILLUS THURINGIENSIS* AND *B. SPHAERICUS* FOR MICROBIAL CONTROL OF INSECT PESTS*

Clayton C. Beegle, Robert I. Rose, and Yu Ziniu

## TABLE OF CONTENTS

---

* Prepared by U.S. government employees as part of their official duties, and not subject to copyright.

# I. INTRODUCTION

One of the most underreported aspects of *Bacillus thuringiensis* and *B. sphaericus* in the public literature is that of production and formulation. Extensive research has been conducted in these two areas by companies producing and selling products based on *B. thuringiensis,* but due to the proprietary nature of this information, essentially none of it reaches the public literature. Unfortunately, much of the published information on producing *B. thuringiensis* is of limited value because of lack of replication, use of spore counts rather than bioassay as a measure of results, or the use of mortalities at a single dosage rather than potencies or $LC_{50}$s determined by bioassay using a graded series of dosage dilutions. This unfortunate situation exists because few, if any, public researchers have the resources necessary to do fermentation research properly.

There is a need for information in these two areas at present. This is due to renewed interest in *B. thuringiensis* because of end user concerns regarding increasing insect resistance to synthetic chemical insecticides and potential regulatory action to remove these products from the market. There is also considerable potential to construct more effective strains of this organism using genetic engineering. At present there are many small venture capital firms and large established agrochemical companies that are interested in entering, or are entering the *B. thuringiensis* market. These organizations have need of information on producing and formulating *Bacillus*-based products.

# II. ASSEMBLING A CULTURE COLLECTION

Until the early to mid-1980s, the vast majority of *B. thuringiensis* isolates were in collections of public researchers at such places as the Institut Pasteur in Paris, France; the Czechoslovak Academy of Sciences in Prague, Czechoslovakia; Institut für Biologische Schadlingsbekampfung in Darmstadt, Germany; Glasshouse Crops Research Institute in West Sussex, England; Kyushu University in Fukuoka, Japan; Ohio State University in Columbus, U.S.; the U.S. Department of Agriculture laboratories in Brownsville, TX and Peoria, IL; and Huazhong Agricultural University, Wuhan, People's Republic of China. Isolates in these collections were accumulated from isolations from diseased insects (especially stored products from Lepidoptera) silkworm rearing litter, soil, and grain dust. Since many of the public researchers exchanged isolates, there is often a duplication of isolates in such collections. An initial working collection can be created by requesting isolates from these laboratories (the former Brownsville collection is now kept in Peoria).

Since the mid-1980s, there has been an increase in searching for new isolates by commercial companies, either by contracting with public scientists or through their own efforts. Large-scale searching efforts are practical only with soil and grain dust from flour mills and grain storage facilities. Feral *B. thuringiensis* diseased insects are rare and difficult to find. Until very recently, selective isolation of *B. thuringiensis* and *B. sphaericus* from soil was accomplished by the use of selective media,[1,2] A new technique[3] that works well for some[4,5] utilizes sodium acetate-containing medium which reportedly inhibits germination of *B. thuringiensis* and *B. sphaericus*, while allowing germination of other *Bacillus* spp. The medium is then pasteurized to kill nonsporeformers and germinated *Bacillus* spp. The surviving *B. thuringiensis* and *B. sphaericus* spores are then plated out on a rich agar medium. This technique reportedly works especially well with soils containing high background levels of undesired bacteria. However, some workers have had variable results with this technique. Evidently the sodium acetate and sodium acetate solutions must be fresh, and the glassware used must be very clean.[6]

# III. SELECTION OF ISOLATES

Traditionally when searching for an isolate with high activity against a particular pest insect, as many isolates as possible were blindly screened against that insect. For example, in the *Bacillus thuringiensis* International Cooperative Program, 317 isolates were screened against 23 species of insects.[7] From that program some generalities emerged. Isolates most active against *Trichoplusia ni*, *Heliothis virescens*, *Ostrinia nubilalis*, and *Lymantria dispar* are all subsp. *kurstaki*; those most active against *Galleria mellonella* are subsp. *galleriae* and *aizawai*; and those most active against *Spodoptera exigua* and *S. litura* are subsp. *aizawai*. Recently the actual genes which code for the crystal toxins responsible for insecticidal activity towards the different target insect groups have been identified and named.[8]

With *B. sphaericus*, Krych et al.[9] identified five homology groups within the species by DNA homology analysis. All the mosquito active isolates grouped together into one group (group IIA). Also, de Barjac et al.[10] found that entomopathogenic *B. sphaericus* isolates could be clearly distinguished from nonentomopathogenic isolates by H (flagellar) antigen analysis.

In selecting isolates to be tested against target pest insects, two strategies can be used to increase testing efficiency. For insects which are adequately susceptible to *B. thuringiensis* and whose pattern of activity correlates well with one of the above techniques, only isolates belonging to the most active grouping need to be tested. For example, it makes little sense to test any isolates not belonging to subsp. *kurstaki* when searching for superior isolates for use against *T. ni* or *H. virescens*. Conversely, when searching for isolates that would be active against an insect species that is recalcitrant to all currently known isolates, it would be most efficient to test only those newly isolated cultures which are unlike any known isolate. Some companies searching for isolates active against a recalcitrant pest insect test each newly isolated culture with gene probes of all known *B. thuringiensis* crystal toxin genes and bioassay only those that are different.

# IV. MAINTAINING A CULTURE COLLECTION

The most common method of maintaining a *Bacillus* culture collection is to maintain working cultures on agar slants, *B. thuringiensis* on a simple nutrient agar such as Difco Nutrient Agar and *B. sphaericus* on a modified nutrient agar.[11] The slants are subcultured at approximately 6-month intervals when stored at 5°C. Stock cultures are usually lyophilized and kept at 5°C until needed to replenish the working cultures. Other methods include the use of filter paper dots or strips, and soil. Sporulated cultures are applied to sterile filter paper strips or dots, or sterile soil, which are then dried and stored. Heckly[12] and Perlman and Kikuchi[13] reviewed the methods for preservation of microorganisms. Troitskaya[14] compared the effect of ten preservation methods over a period of 7 years on spore survival of *B. thuringiensis* subsp. *galleriae*. Survival was highest for spores stored under petroleum and lowest for spores stored in filter paper strips and in crystalline salts. Jixin and Zaiyon[15] obtained similar results. Toxicity change was least for isolates that had been lyophilized in 10% serine or in sterile insect tissues and stored cold.[14,16]

The recent tremendous advances in elucidating the genetics and protein chemistry of the *B. thuringiensis* crystal toxins, the so-called delta endotoxins, has provided information that points out the dangers of continuous subculturing. The genes that code for the crystal toxins are located on plasmids which can be lost or transferred by a conjugation-like process between bacteria.[17] The crystals of some isolates are made up of a single P-1 component; some have a P-1 and P-2 component,[18] and some have several P-1 components. Some crystal multitoxin isolates are unstable in P-1 composition and in relative amounts of P-1 and P-2.[19] When stability of isolate characteristics is desired, it is obvious that subculturing be kept to an absolute minimum.

**TABLE 1**
**Examples of Media for Shake Flask Production**
**of *Bacillus thuringiensis* (g/l)**

| Component | T-3[6] | Peptonized milk[18] | Soybean flour | B-4[28] |
|---|---|---|---|---|
| Peptonized milk | — | 10.0 | — | — |
| Soybean flour[a] | — | — | 18.0 | — |
| Cottonseed flour | — | — | — | 10.0 |
| Peptone | — | — | — | 2.0 |
| Tryptose | 2.0 | — | — | — |
| Tryptone | 3.0 | — | — | — |
| Glucose | — | 5.0 | 5.0[b] | 15.0 |
| Yeast extract | 1.5 | 2.0 | — | 2.0 |
| $KH_2PO_4$ | — | 1.0 | | |
| $K_2HPO_4$ | — | — | 0.7 | — |
| $FeSO_4·7H_2O$ | — | 0.02 | 0.01 | 0.02 |
| $ZnSO_4·7H_2O$ | — | 0.02 | 0.01 | 0.02 |
| $MnSO_4·H_2O$ | 0.002 | 0.02 | — | — |
| $MgSO_4·7H_2O$ | 0.02 | 0.3 | 0.3 | 0.3 |
| $Na_2HPO_4$ | 1.4 | — | — | — |
| $NaH_2PO_4·H_2O$ | 1.2 | — | — | — |
| $CaCO_3$ | — | — | 0.5 | 1.0 |
| Adjust pH to: | 6.8 | 7.0 | 7.2 | — |

[a]    50% protein, defatted, uncooked.
[b]    Autoclaved separately.

# V. LABORATORY AND PILOT PLANT PRODUCTION AND RECOVERY

## A. EQUIPMENT

The basic equipment needed for laboratory production and recovery are an autoclave, laminar flow hood, temperature-controlled shaker, high-speed batch centrifuge, and/or a tangential flow membrane filtration system, and a freeze-dryer. Additional equipment necessary for pilot plant production and recovery are small stirred fermentors (usually 4 to 20 l working volume), a source of compressed air and chilled water, a small continuous-flow centrifuge or larger tangential flow membrane filtration system, and a small spray dryer.

## B. MEDIA
### 1. *Bacillus thuringiensis*

To grow, *B. thuringiensis* needs a source of carbon, nitrogen, and trace minerals. As mentioned earlier, *B. thuringiensis* can be successfully grown on slants of nutrient agar (3 g beef extract, 5 g peptone, 15 g agar, 1 l water). Tryptose phosphate broth (20 g tryptose, 2 g dextrose, 5 g sodium chloride, 2.5 g disodium phosphate, 1 l water) is satisfactory for growing first- and intermediate-passage seed fermentations,[20] but the final-passage seed should be grown on a medium as close to the production medium as possible to minimize the lag phase time in the final fermentation.[21,22]

Laboratory production media for *B. thuringiensis* vary from synthetic defined,[23,24] synthetic supplemented with yeast extract, casamino acids, tryptone,[25] phytone,[26] or peptonized milk,[18] to those similar to commercial media. Examples of media for shake flask production of *B. thuringiensis* are shown in Table 1. T-3 is a very light soluble medium that has been found useful in growing a wide variety of isolates.[6] Peptonized milk[18] medium has been used in a number of studies of the nature of the crystal toxins of *B. thuringiensis*; it is also a soluble medium. A possible disadvantage of the peptonized milk medium is that there are

**TABLE 2**
**Examples of Media for Stirred Tank Production of *Bacillus***
***thuringiensis* (g/l)**

| Component | Com.A | Com.C | Com.D | B-8b[20] | B-13[20] | Megna[29] |
|---|---|---|---|---|---|---|
| Soybean flour | 25.0 | 22.5 | 30.0 | — | 40.0 | — |
| Cottonseed flour | — | — | — | 30.0 | — | 14.0 |
| Peptone | — | — | — | 2.0 | — | — |
| Corn steep liquor | — | 20.0[a] | 20.0[a] | 30.0[a] | 20.0[a] | 16.7[b] |
| Glucose | 10.0 | — | — | 45.0 | 30.0 | — |
| Corn syrup | — | — | 45.0 | — | — | — |
| Corn starch | 10.0 | 15.0 | — | — | — | — |
| Molasses | — | — | — | — | — | 18.6 |
| Yeast extract | — | — | — | 2.0 | — | — |
| $KH_2PO_4$ | 1.0 | — | — | — | — | — |
| $K_2HPO_4$ | 1.0 | — | — | — | — | — |
| $FeSO_4$ | 0.02 | — | — | — | — | — |
| $FeSO_4 \cdot 7H_2O$ | — | — | — | 0.02 | — | — |
| $MgSO_4 \cdot 7H_2O$ | — | — | — | 0.3 | — | — |
| $MnSO_4 \cdot H_2O$ | — | — | — | 0.02 | — | — |
| $ZnSO_4 \cdot 7H_2O$ | — | — | — | 0.02 | — | — |
| $CaCO_3$ | 1.0 | 1.5 | 1.0 | — | — | 1.0 |
| Hodag FD62 antifoam | 3.0 | — | — | — | — | — |
| SAG 5693 antifoam | — | 1.25 | 0.5 | — | — | — |
| Dow Corning AF antifoam | — | 0.1 | — | — | — | — |

[a] Wet weight.
[b] Dry weight of solids.

indications that higher levels of proteinases are produced by some *B. thuringiensis* isolates when grown in peptonized milk than when grown in other media.[19] The soybean flour and B-4[28] media shown in Table 1 are heavier media that more closely resemble media used in commercial production of *B. thuringiensis*. Their disadvantages are insolubles in the soybean and cottonseed flours that make the media turbid and unutilized insolubles that remain with the spore-crystal complexes when they are recovered. Media heavier than those shown in Table 1 — or example, those shown in Table 2 — are not satisfactory in shake flasks because of the limited aeration that can be achieved in them. Two media that have been successfully used in pilot plant-size stirred vessel fermentors are B-8b and B-13,[20] shown in Table 2, with *B. thuringiensis* isolate 263[31] yielding particularly well in B-13. Dubois[26] used a very light soluble phytone (0.2%), glucose (0.2%), plus minerals medium in a 7.5-l fermenter to produce clean spores and crystals.

## 2. *Bacillus sphaericus*

*B. sphaericus* has different nutritional requirements than does *B. thuringiensis*. *B. sphaericus* does not utilize sugars as a carbon source[33] and does not grow well with starch and several other carbohydrates as carbon sources.[34]

Apparently amino acids are the best source of both nitrogen and carbon, while pyruvate, lactate, acetate, or some Krebs cycle intermediates may serve as carbon source supplements.[33] Industrial proteinaceous digest solutions had to be supplemented with yeast extract (2 g/l) to achieve optimal fermntation; evidently yeast extract provided vitamins and/or other growth factors. Fridlender and co-workers also found that media containing the most digested proteins (average peptide chain length 3.4 to 5.7) resulted in higher toxin production than media containing undigested protein (average peptide chain length 35.14 to 55.6). The addition of urea, uric acid, or citric acid had no effect on either toxin or biomass yield. Acetate and lactate addition resulted in a two- to threefold increase in biomass, but no

**TABLE 3**
**Examples of Media for Stirred Tank Production of *Bacillus***
***sphaericus* (g/l)**

| Component | Suggested commercial | Obeta[30] | Yousten[32] | Pasteur 1[82] | Pasteur 2[82] |
|---|---|---|---|---|---|
| Peptonized milk | — | 10.0 | — | — | — |
| Soybean flour | 25—35 | — | — | — | — |
| Beef extract | — | — | 3.0 | 5.0 | — |
| Peptone | — | — | 5.0 | 5.0 | 10.0 |
| Corn steep liquor | 15—25 | — | — | — | — |
| Yeast extract | — | 2.0 | 0.5 | 10.0 | 3.0 |
| Glycerol | — | — | — | 10.0 | — |
| $CaCO_3$ | 1.5 | — | — | — | — |
| $MgSO_4 \cdot 7H_2O$ | — | 0.3 | — | — | 0.2 |
| $FeSO_4 \cdot 7H_2O$ | — | 0.02 | — | — | 0.1 |
| $MnSO_4 \cdot H_2O$ | — | 0.02 | — | — | — |
| $MnCl_2$ | — | — | — | — | 0.05 |
| $ZnSO_4 \cdot 7H_2O$ | — | 0.02 | — | — | — |
| $MnCl_2 \cdot 4H_2O$ | — | — | 0.01 | — | — |
| $CaCl_2 \cdot 2H_2O$ | — | — | 0.1 | — | — |
| $MgCl_2 \cdot 6H_2O$ | — | — | 0.2 | — | — |
| $KH_2PO_4$ | — | — | — | — | 1.0 |
| $K_2HPO_4$ | — | — | — | — | 1.0 |
| NaCl | — | — | — | 3.0 | 5.0 |

increase in toxicity. With some media, addition of glycerol (5 g/l) improved toxin production tenfold. In general they found that consumption of glycerol correlated well with toxin yield. While amino acid consumption averaged about 60%, individual amino acids were utilized to different degrees. For example, L-proline was depleted from media more than average, while L-arginine was hardly utilized, and L-histidine accumulated in the fermentation beer above its initial concentration. It was noticed that amino acids were depleted more in glycerol-supplemented media. They also found that adding corn steep liquor to media improved toxin yields and that there appeared to be a synergistic relationship between glycerol and corn steep liquor.

Several shake flask media have been reported for laboratory production of *B. sphaericus*.[30,32,35] Five examples of pilot plant stirred tank media[30,32,82] are shown in Table 3.

## C. PROCEDURES

It is recommended to start a fermentation with a double passage inoculum seed. Start by inoculating a fresh slant and incubate the slant at 26 to 30°C for 48 to 72 h (to lysis). Wash the slant in 5 ml sterile water and heat-shock the suspension for 15 min at 75°C. This activates the spores and kills vegetative cells and free bacteriophage. Transfer one loopful of the heat-shocked suspension to a test tube containing 5 ml of appropriate broth — tryptose phosphate broth for *B. thuringiensis*, and nutrient broth supplemented with 0.05% yeast extract[32] for *B. sphaericus* — and incubate at 26 to 30°C for 5 to 6 h. Inoculation volumes of subsequent transfers should be in the 0.5 to 5% range. The final seed medium should be as close as possible to the medium used in the production fermentation to minimize the lag phase. Seed culture incubation times have traditionally been for 14 to 16 h, but this is because that time period is convenient — not because it is best for seed production. It is important that seed cultures be transferred when they are actively growing, i.e., in primary fermentation. If *Bacillus* cells are transferred after they have entered into secondary fermentation, it may be as long as 8 h before they start actively growing again. An easy way to determine if a culture is still in primary fermentation when using a medium containing

**TABLE 4**
**Effect of Flask Loading, pH, and Inoculum Size on**
***Bacillus thuringiensis* Cell Numbers**

| Factor level | Flask loading (ml/500 ml flask) | Medium pH | Inoculum volume (%) | Cell/ml ($\times 10^8$) |
|---|---|---|---|---|
| 3—1 | 25 | 7.5 | 4 | 22.2 |
| 3—2 | 25 | 8.5 | 10 | 18.5 |
| 3—3 | 15 | 7.5 | 10 | 38.4 |
| 3—4 | 15 | 8.5 | 4 | 35.4 |
| r | 33.1 | 6.7 | 0.7 | |

Modified from Ziniu.[37]

glucose is to test the culture for the presence of glucose, using reagent strips such as Dextrostix® and Clinistix®. If the culture tests positive for glucose, it is still in primary fermentation and can be used to inoculate another culture. In the senior author's experience, glucose is nearly always depleted by 14 to 16 h.

It is difficult for *Bacillus* cultures to receive sufficient aeration in shake flasks, even deep triple-baffled flasks. When working with *B. thuringiensis* subsp. *darmstadiensis* isolates,[36] it was found that a maximum of 160 ml medium could be used in 2000 ml deep triple-baffled flasks when optimum growth and lysis was desired. In a factorial investigation of the effect of aeration (flask loading), pH, and inoculation quantity on *B. thuringiensis* cell densities in shake flasks, it was found that aeration was by far the most important (Table 4).[37] This type of factorial experiment is very useful in determining the relative effects of medium ingredients and fermentation conditions. In the experience of the Wuhan Microbial Pesticide Factory, the volume of medium should be 5 to 10% of the flask volume. Foda et al.[83] found that *B. thuringiensis* subsp. *entomocidus* grew much better at 5% flask loading than at 10%. Shaker speed should be 250 to 340 rpm. Aeration rates in laboratory and pilot plant stirred vessel fermentors is usually 1 volume air per volume medium per minute. Stir rates are 150 to 700 rpm, depending upon the agitator type and fermenter size. Foaming in shake flask fermentation, although a problem and undesirable because it reduces the oxygen-transfer rate, is not normally controlled. However, foam control is essential in stirred vessel fermentation, and is normally accomplished by using foam sensors and injection of chemical antifoams when needed. Silicone antifoams, although effective, should be used with caution because *B. thuringiensis* isolates vary in their tolerance to silicone antifoams. Polypropylene glycol antifoams are safer to use with *B. thuringiensis*. It is best to keep the use of antifoams to an absolute minimum because antifoams can reduce the rate of oxygen transfer up to 50%,[48] and can cause problems in formulating the resulting product.[49]

Fermentation temperature is a tradeoff between isolate stability and the time it takes the fermentation to run to completion. It takes less time for cells to divide than it takes plasmids to divide,[22] and at elevated temperatures plasmids can be lost. Since the *B. thuringiensis* crystal toxin genes are carried on plasmids, plasmid loss could result in reduced toxin production. The relationship between incubation temperature, run time, and activity yield in pilot plant-size stirred vessel fermentors is shown in Table 5. It is noteworthy that, for isolate *Btk*-263, the highest activity yields occurred at 26°C.

Initial pH of the medium should be from 6.8 to 7.2. Data in Table 6 indicate that 6.8 is optimum.[37] Under normal fermentation conditions the pH is near 7 when the bacteria start growing, can drop to 5.8 as acetic acid is produced during glucose metabolism, and rises again as the acetic acid is utilized. If the carbon/nitrogen ratio is not proper, the acetic acid which has been produced will not be metabolized, so the pH will not rise and spores and crystals will not be formed.[39] $CaCO_3$, in the range of 0.5 to 1.5 g/l, can be useful as a

## TABLE 5
### Relationship Between Fermentation Temperature, Activity Yield, Time of Peak Activity Yield, and Spore Count Using *Btk*-263 in Pilot Plant Stirred Vessel Fermentors

| Temperature (°C) | Activity yield (IU × $10^3$/ml) | Peak yield (h) | Spore count (× $10^7$) |
|---|---|---|---|
| 37 | 283 | 39—42 | 80 |
| 34 | 1150 | 18—24 | 170 |
| 30 | 1500 | 34—39 | 220 |
| 30 | 1450 | 39 | 240 |
| 26 | 1630 | 39—42 | 200 |

Modified from Dulmage.[38]

## TABLE 6
### Effect of Medium pH on Sporulation, Lysis, and Toxicity of 48 h *Bacillus thuringiensis* Cultures

| pH | Vegetative cells | Free spores and crystals | % insect mortality |
|---|---|---|---|
| 4.5 | + | — | 0 |
| 5.0 | + | — | 10 |
| 5.5 | + | — | 20 |
| 6.0 | — | + | 95 |
| 6.5 | — | + | 95 |
| 6.8 | — | + | 100 |
| 7.0 | — | + | 90 |
| 7.2 | — | + | 95 |
| 7.4 | — | + | 35 |

Modified from Ziniu.[37]

buffering agent in fermentations. Proteins, peptides, and amino acids can also provide some buffering capacity.[42] The final pH at harvest should be between 8 and 9 if the pH is not controlled by the addition of acid or base during fermentation. Arzumanov[40] studied the effect of pH on the growth of *B. thuringiensis* and proposed an ideal pH profile. It was found with both *B. thuringiensis* subsp. *israelensis*[41] and *B. sphaericus*[32] that controlling the pH during fermentation resulted in lower activity yield than when the pH was allowed to fluctuate as a result of metabolism.

Recovery of the spore-crystal complex from shake flask fermentations is nearly always done by centrifuging the final broth and discarding the resulting supernatant. 10,000 × *g* for 20 min is usually adequate, and 500 to 600 ml plastic or metal centrifuge bottles are relatively efficient and convenient. If it is desired to rid the spore-crystal complex of as much of the endogenous proteinases as possible, wash the pellets twice with 0.5 *M* NaCl, then several times with distilled or deionized water. If the isolate being grown produces the heat-tolerant β-exotoxin, and the medium being used contains appreciable amounts of minerals such as calcium and magnesium, significant amounts of bound β-exotoxin can precipitate with the spore-crystal complex and cause considerable mortality to insects tested against the spore-crystal complex. Such inadvertent recovery of bound β-exotoxin has con

## TABLE 7
### Flow Chart for Recovery and Drying of
### *B. thuringiensis* Spore-Crystal Complexes
### from Stirred Vessel Fermentors

Fermentation beer
↓
Recover by tubular bowl continuous flow centrifuge
or
tangential flow cassette filter
↓
Discard filtrate
↓
Weigh retentate
↓
Prepare slurry containing 10% lactose, 1% wetting agent, 1%
dispersant, and 20% solids based on retentate weight
↓
Spray dry
Centrifugal atomization feed
125—150°C inlet temperature
75—100°C outlet temperature
↓
Formulate

Modified from World Health Organization.[20]

fused more than one *B. thuringiensis* screening program. Another method of recovery which is rapidly gaining in popularity is tangential flow filtration, such as the Pharmacia Minisette and the Millipore Pellicon cassette systems. Such systems are fast and convenient, and recovery and washing of the spore-crystal complex can be accomplished in the same operation. The recovered spore-crystal complexes can be dried by either coprecipitation with lactose-acetone[43] or lyophilization. Recovery from pilot plant-size stirred vessel fermentors is usually done with tubular bowl continuous centrifuges or larger tangential flow filtration systems. If it is necessary to hold the fermentation beer or recovered concentrate for any length of time (up to 48 h), the pH should be lowered to 4 to 5 with phosphoric or acetic acid and stored at 4°C. Pilot plant-sized spray driers are very useful for drying the spore-crystal complexes from such fermentors. A flow chart for recovery and spray drying is shown in Table 7.

# VI. LARGE-SCALE PRODUCTION AND RECOVERY

## A. SEMISOLID FERMENTATION

Semisolid fermentation is the most ancient technique used by man to produce microorganisms or their products, possibly 2500 years old.[47] It was used by Nutrilite Products to produce their *B. thuringiensis*-based product Biotrol for a number of years and is currently being used in the People's Republic of China to produce both *B. thuringiensis* and *B. sphaericus*. This technique is normally considered somewhat primitive compared to deep-tank fermentation, but in Japan great quantities of enzymes are produced in large, fully automated commercial semisolid fermentation facilities.[47]

## 1. Equipment

Equipment used in semisolid fermentation can be shallow trays, aerated static bins, or aerated revolving drums. In the People's Republic of China, mesh-bottomed trays (52 × 35 × 5 cm) are used, held five wide in racks with eight angle iron shelves. In static

**TABLE 8**
**Examples of Media for Semisolid Fermentation of *Bacillus***
***thuringiensis* and *B. sphaericus* (%w/w)**

| Component | *B. thuringiensis* subsp. *thuringiensis*[44] | *B. thuringiensis* subsp. *israelensis*[45] | *B. sphaericus*[46] |
|---|---|---|---|
| Wheat bran | 52.9 | 40.0 | 14.3 |
| Expanded perlite | 37.1 | — | — |
| Soybean meal | 6.0 | — | — |
| Peanut flour | — | — | 57.4 |
| Cottonseed flour | — | 32.0 | — |
| Corn flour | — | 12.0 | 14.3 |
| Rice husks | — | 14.0 | 12.0 |
| Glucose | 3.5 | — | — |
| CaO | 0.35 | — | — |
| CaCl$_2$ | 0.03 | — | — |
| CaCO$_3$ | — | — | 2.0 |
| Ca(OH)$_2$ | — | 2.0 | — |
| NaCl | 0.09 | — | — |
| pH | 6.9 | 7.4 | 7.4 |

bins the bottoms are perforated and humidified warm air is forced through the medium. Rotating drum fermentors also are aerated. Additional equipment will be required if aerated static bins or rotating drum fermentors are used. A source of steam, compressed air, air heaters, air humidifiers, air dryers, and a product grinder are necessary.

## 2. Media

Coarse materials such as wheat bran, rice husks, and expanded perlite form the basis of semisolid fermentation. The coarse materials are mixed with nutrients and water until the coarse particles are coated with moist nutrients. As is noted in WHO,[20] each coarse particle functions as a miniature fermenter. Media which have been used to commercially produce *B. thuringiensis* and *B. sphaericus* are shown in Table 8. The medium listed for *B. thuringiensis* subsp. *thuringiensis* was developed and patented by Mechalas[44] and was used by Nutrilite Products to produce the *B. thuringiensis*-based product Biotrol. The media listed for *B. thuringiensis* subsp. *israelensis* and *B. sphaericus* were developed by Jingyuan et al.,[45,46] and are currently used to produce products based on those organisms in the People's Republic of China.

## 3. Procedures

In the Mechalas process, the seed inoculum is prepared in shake flasks and a small stirred vessel fermenter. The dry medium ingredients are mixed with water and treated with steam for 1 h. After the mix is inoculated with the seed inoculum (32% w/w) the pH of the mix should be 6.9 and the moisture content 60% (w/w). After being placed into static bins, the mix is aerated with 95 to 100% RH and 30 to 34°C air at 0.4 to 0.6 v/v/min for 3 h. After 3 h aeration is increased to 1.0 to 1.2 v/v/min. After 36 h, sporulation should have occurred, and the mix is then aerated with dry air at 50 to 55°C for another 36 h. The mix should then have about 4% by weight moisture. It is then ground and sieved to 80 mesh.

In the Jingyuan et al. process, inoculum seeds are produced in stirred vessel fermentors, recovered, and dried. Seed inoculum powders should contain about $1 \times 10^{10}$ live spores per gram. The semisolid medium (50 to 60% moisture) is heated at 100°C for 3 to 4 h, and after it has cooled, it is inoculated at 0.5% (w/w) with the respective inoculum powder. The inoculated medium is placed 2.5 cm deep in the mesh-bottomed shallow trays (750 g per tray) and held at 28 to 30°C for 48 h. The medium containing the sporulated *Bacillus* sp.

is dried at 55 to 60°C to a moisture content of less than 6%. The material is then ground and sieved until it passes through 60 to 80-mesh sieves. The material is then bioassayed to ensure that it is of adequate potency. $LC_{50}$s of both *B. sphaericus* and *B. thuringiensis* $H_{14}$ semisolid fermentation products should be between 0.05 to 0.5 mg/ml when assayed against 3rd instar *Culex pipiens fatigans* larvae. Such products have achieved 88 to 100% control of both *Culex* and *Aedes* spp. in the field when applied at 1 to 2 ppm. A *B. sphaericus* isolate CS-8-based semisolid fermentation product gave effective control from 15 to 32 days.[45,46]

## B. SUBMERGED FERMENTATION

Submerged fermentation in very large stirred-vessel fermentors (up to 100,000 l) is the method of choice used to produce the majority of commercial *B. thuringiensis* products. For this reason it is very impractical to use exotic sugars, peptone, or plant extracts. The majority of media are composed of agricultural products as sources of carbon, nitrogen, and minerals. Sources of carbon are normally glucose, starch, and hydrolyzed corn products. Nitrogen sources are soybean flour, cottonseed flour, fish meal, corn steep liquor, yeast autolysates, and casein.[47] Sufficient minerals are usually contained in the tap water and the agricultural products used in the fermentation.

## 1. Equipment

Manufacturing equipment used to produce *B. thuringiensis* is standard for many types of industrial batch fermentations and consists of main fermentation vessels of total capacities commonly up to 70,000 l or more, made of stainless steel, and fractionally smaller auxiliary seed, pilot, or germinator vessels, and charging tanks. Production vessel contents are mixed with submerged impellers, usually on vertical shafts, and are sterilized in place by internal steam radiator pipes and jackets or by direct steam sparging. Temperature control during fermentation is accomplished by circulating chilled water in a jacket or radiator pipes. Air for vessel aeration is either filter or heat sterilized.

Recovery equipment are either large industrial centrifuges or preferably large microfiltration devices which are plumbed directly to the fermentors. Product drying is accomplished with large industrial spray driers heated with natural or LP gas.

The U.S. Agency for International Development in 1987 commissioned a *B. thuringiensis* and *B. sphaericus* study to extrapolate Western European and U.S. industrial fermentation technology and practices to the available technology, equipment, and renewable agricultural resources of a developing South Asian country. The resulting recommendation was to utilize multiple, easily made, monitored, and controlled fermentation tanks of 200 l operating capacity each, instead of the massive, complex, and very costly single batch vessels used in the developed countries to produce *B. thuringiensis* as well as industrial enzymes and pharmaceuticals. Although operation of comparatively small multiple fermentation tanks is far more labor intensive than large single vessels, they would allow for local control and management in developing countries, as well as the utilization of locally available agricultural products as media components. The proposed system was composed of 20 tanks of 250 l total capacity each, with each tank having a separate gas burner beneath it for sterilization of the vessel and medium before inoculation and boiling of any batches found to be contaminated. Loss from contamination would thus be limited to single small units rather than a large vessel. The tanks were designed to have positive pressure combined aeration and air lift agitation from a multiply perforated copper inlet coil emitting sterile air near the bottom of each tank. Air lines were designed with valves to allow testing for contaminants in the air supply. Fermentation tanks were designed to be made with locally welded stainless steel sheet metal. Locally available fermentation media components were soybeans, broken rice, fish meal, and sugarcane molasses.

## 2. Media

As mentioned above, media components for industrial production of *Bacillus* spp. are, for economic reasons, agricultural products. In Table 2 are some examples of media that have been used commercially to produce *B. thuringiensis*-based products. The Megna medium[29] was used by the Bioferm Corp., the first producers of Thuricide® in the United States. Com. A, C, and D in Table 2 are media that have been used to produce *B. thuringiensis* commercially in the U.S. and Europe. While B-13[20] has only been used in pilot plant-size fermentors, it should also be useful in large industrial fermentors. In Table 3 are media which should be useful in producing *B. sphaericus*. The Obeta,[30] Yousten,[32] and Pasteur[82] media would not be practical in commercial scale fermentations because of the cost of their components, but they have been used in pilot plant-size fermentors.

In the U.S. and Europe, it is most often the relative price of the comparable media components that dictate which are used, rather than availability. In other parts of the world though, availability and price are constraints which influence the practicality of media components. As a result, a large amount of research has been done in developing countries to find reasonably priced, locally available media components to substitute for the ones listed above and in Tables 2 and 3. Chilcott and Pillai[50] investigated the use of coconut wastes for the production of *B. thuringiensis* subsp. *israelensis*. Dharmsthiti et al.[51] were able to reduce substantially the cost of production of both *B. thuringiensis* subsp. *israelensis* and *B. sphaericus* by using a by-product of a monosodium glutamate factory. Ejiofor and Okafor[52] found that 10% cowpea steep liquor was superior to Singer's[53] basal medium for producing *B. sphaericus* isolate 2362. They also evaluated fermented cassava, maize, and cowpea for production of *B. thuringiensis* subsp. *israelensis*, and found that a 1:3 mix of cowpea steep liquor:maize steep liquor was best.[54] Lee et al.[55] were able to grow *B. thuringiensis* on water which had been used to cook cocoons, presumably *Bombyx mori* cocoons. Obeta and Okafor[56] were able to successfully grow *B. thuringiensis* subsp. *israelensis* on several media containing cow blood, mineral salts, and ground legume seeds. In Egypt, Salama and co-workers have extensively evaluated locally available media components for production of *B. thuringiensis*. They found that the type of legume used in media had a significant effect on potency of subsp. *entomocidus*. Kidney beans, chick peas, and peanuts were better than horse beans or soybeans.[57] They also investigated using whey, with and without supplements, as a medium to produce *B. thuringiensis*.[58] They found that the best medium for producing subsp. *entomocidus, kurstaki*, and *galleriae* was 1% horse beans and 1% fodder yeast (dried cells of *Saccharomyces cerevisiae*), plus mineral salts.[59]

Sakharova et al.[60] determined that a shortage of glucose during fermentation had the greatest negative effect on the activity of the resulting spore-crystal complex of *B. thuringiensis* compared to shortages of other nutrients. Arcas et al.[61] found that *B. thuringiensis* spore counts and toxin levels increased when the concentration of glucose was increased from 8 to 56 g/l, with a corresponding increase in the rest of medium components. Holmberg and Sievanen[62] likewise found that β-exotoxin production varied directly with sugar concentration, up to about 3% molasses. Higher sugar levels appeared to inhibit exotoxin production. β-Exotoxin production appeared to be rather insensitive to changes in temperature, pH, aeration, and agitation within the levels studied. Abrosimova et al.[63] examined the influence of minerals, both individually and collectively, on production of β-exotoxin by *B. thuringiensis*.

## 3. Procedures

Care must be taken to insure that the particle size of all ingredients which are to be used to produce by fermentation any microbial insecticide are less than 50-mesh in size, to prevent clogging of ground or aerial spray equipment. This applies particularly to plant seed flours, fish meal, and corn steep liquor. Fermentors are filled to about 75% of their total volume and sterilized by either direct steam sparging or by steam radiator pipes and/or jackets. Bring the temperature up to 120°C and hold for 30 to 35 min. If sterilizing is done by direct steam sparging, allow for about a 15% by volume steam condensate water gain. Starting pH is 6.8 to 7.2, adjusted with NaOH or $H_3PO_4$. The pH is not normally controlled in large fermentors. The medium is cooled to the fermentation temperature by circulating chilled water in a jacket or radiator pipes, and inoculated from a seed fermenter or germinator typically 5% of the size of the final vessel with a pure culture near the end of the log phase of vegetative cell growth and reproduction. Agitate at 2.5 to 3.0 kW/m³, which is about 150/160 rpm, depending on the agitator type. Fermentation temperature is 28 to 32°C, normally 30 to 32°C, because of the energy costs of operating large fermentors. The higher the operating temperature (to a point) the shorter the fermentation. In the People's Republic of China, it has been found to be beneficial to control the fermentation temperature at 30°C before the log phase, 28°C during the log phase, and 32°C after log phase. This temperature regime results in an extended log phase, speeds up sporulation, and reduces the total fermentation time. The heat transfer necessary to maintain fermentation temperature at 32°C is about 120,000 to 150,000 kCal/$M^3$ of fermentation beer. Internal pressure (back pressure) is maintained at about 5 psi to suppress foam, increase oxygen transfer, and to help prevent contamination. The most common antifoam agents used are polypropylene/polyethylene glycols. Aeration rates can be up to 1 v/v/m, but normally are 0.3 to 0.5 v/v/m in large fermentors. In the People's Republic of China the optimum aeration rates have been found to be 1 to 1.7 v/v/m during the lag phase and 0.8 to 0.9 v/v/m during the log phase and stationary phase. It is difficult to supply sufficient air to a *Bacillus* spp. fermentation during the log growth phase. It is more effective to increase the agitation rate, increase the back pressure, or lower the fermentation temperature when it is desired to increase the amount of dissolved oxygen rather than to increase the volume of air delivered.

Typical fermentation times are about 30 to 35 h for *B. thuringiensis* subsp. *israelensis* and about 40 to 44 h for the lepidopterous active *B. thuringiensis* subspecies. At harvest, lower the temperature to 15°C, lower pH to 4 to 5 with $H_3PO_4$, and recover solids with an industrial centrifuge (at about 4,800 rpm or 3,800 × g) or by ultrafiltration. If recovered by centrifuging, wash solids once with water or with a 5% aqueous NaCl solution, lower pH to 5 and recentrifuge. Some active solids may be lost. The centrifuge slurry will be about 10 to 18% solids, depending upon the equipment used. Microfiltration is now considered a superior method for solids concentration because it allows for a convenient water wash (2 to 4 v) by diafiltration which removes soluble fermentation products, including proteinases, which may adversely affect end use product stability. In addition, industrial centrifuges can produce fine aerosols containing spores which can be a major source of bacterial contamination of equipment, the facility, or other products in an industrial fermentation plant or factory.

The recovered acidified (pH 4 to 5) slurry can be stored temporarily at 4°C if necessary. The slurry can be spray-dried if a dry powder is desired, or it can be used directly with microbial stabilizers to produce a liquid flowable concentrate or a few other end use product formulations. Commercial spray driers operate by introducing spray droplets into a heated cyclone of air to evaporate water as they circulate in the vortex. Natural or LP gas is the usual fuel used to produce the large amount of heat required to evaporate the large volumes of water. Stationary or spinning nozzles are used to produce relatively uniform spray droplets. Inlet temperatures should be 180 to 190°C, and outlet temperatures should be 75 to 80°C. The spray drier should be adjusted to produce a product of about 5% moisture. Attempts to get a moisture content less than 5% may result in burned powder. From about 2700 l of slurry, 330 kg of powder may typically be produced. About 20 to 40 g of powder should be produced per liter of fermentation beer, depending upon the isolate and medium used. This powder is the manufacturing use product, technical grade active ingredient, or primary powder from which wettable powder, granular, pellet, briquette, and some flowable concentrate end use products are formulated by blending with diluents, carriers, wetting agents, and preservatives (mainly microbial growth inhibitors). Spray driers are frequently the source of severe bacterial spore contamination of other products and equipment in their vicinity, and at greater distances in the direction of air flow. It is strongly recommended that spray drying be done outside and distant from facilities and establishments where food, pharmaceuticals, and other sensitive products are produced, packaged, or stored — including multipurpose fermentation plants without adequate provisions to avoid cross contamination or lacking effective disinfection procedures.

Another source of severe *B. thuringiensis* production problems are bacterial viruses called bacteriophages. These can be either introduced or carried in the genes of the bacteria and then activated by growth or chemical or physical conditions. Broth cultures in shake flasks, side-arms, carboys, or seed fermenters should be carefully scrutinized for aberrations from the typical brown cloudy growth, including unexplained clearing or other abnormal visual changes. Containers and vessels suspected of being contaminated by bacteriophages should not be opened if possible, but rather sterilized immediately and thoroughly to minimize contamination of other colonies and equipment. Once established, bateriophages are difficult to eliminate except by the use of resistant/tolerant strains which may lack desirable production or activity characteristics of the original strain.

## VII. FORMULATION

Several review articles have been written on formulation of microbial insecticides,[64-67] and the information presented in those articles will not be repeated here. The focus of this section will be on actual commercial formulation of *B. thuringiensis*.

End use products containing *B. thuringiensis* as the active ingredient are formulated from spray-dried powders of fermentor harvests, or from liquid centrifuged or microfiltered fermentor harvests. Equipment used to mix wettable powder and granular products commonly consists of dry material mixers such as ribbon blenders, Nauta® mixers, and even cement mixers. With dry product mixers, the equipment must be covered or sealed to contain generated dust and/or equipped with bags or comparable dust collectors. Blending times for wettable powders after addition of all ingredients vary from 10 to 60 min, depending on equipment and available power. Granular formulations, such as those containing *B. thuringiensis* subsp. *israelensis* for larval mosquito control, usually require less mixing time and produce much less dust. All components used to formulate any wettable powder end use product must be 50-mesh or, preferably, smaller, to prevent clogging nozzles and filter screens of spray equipment. Pellet formulations require pellet presses and possibly water-jacketed extrusion dies where heat generation is a problem. Formation of briquettes requires custom made molds and presses specialized for their shape and purpose. Care must be taken to avoid prolonged exposure to high temperatures which may denature the spore-crystal complex.

Liquid water-based flowable concentrate end-use products can be formulated directly following the recovery/concentration process by addition of stabilizers or microbial inhibitors, spreader-stickers, UV protectants, and rheological agents to control viscosity and prevent settling. Water or mineral oil carrier liquid products are formulated from manufacturing use or technical grade powders in mixing tanks with both high and low speed agitation to insure thorough blending of ingredients. Care should be taken to avoid exposure of *B. thuringiensis* subsp. *israelensis* to high-sheer mixers or homogenizers since activity may be reduced. All ingredients used in liquid product formulation should be 50-mesh or smaller to prevent sprayer screen and nozzle clogging, and the final flowable concentrate product should also be filtered to ensure removal of particulates that could clog spray systems. The final pH of water-based liquid formulations should be 4 to 5 to enhance stability.

## A. *B. THURINGIENSIS* FORMULATION EXAMPLES

| Wettable powders | % by weight |
|---|---|
| *B. thuringiensis* technical grade powder | 20.0—80.0 |
| Calcium/sodium lignosulfonate dispersant | 3.0—10.0 |
| Cornstarch (optional) | 5.0 |
| Nonionic or other wetting agent | 3.0—10.0 |
| Carrier/diluent (usually silicates such as attapulgite; kaolinite, or montmorillonite clays; talc; or pyrophyllite) | to 100.0 |

| Granules (typically *B. thuringiensis* subsp. *israelensis*) | % by weight |
|---|---|
| *B. thuringiensis* subsp. *israelensis* technical grade powder | 1.5—3.4 |
| Dry quartz or other clean sand | 94.5—97.6 |
| *or* | |
| 14- to 20-mesh corn cob grits or granules | 92.5 |
| Vegetable oil (palm, sunflower, or soybean) | 0.9—2.1 |
| *or* | |
| Mineral oil (when used with corn cob granules) | 5.0 |

Granular *B. thuringiensis* subsp. *israelensis* products made with vegetable oil binders are less stable than those made with mineral oil, and should be used within 1 to 2 weeks after formulation. Sand granule recipes are principally for mosquito control agencies that make their own granular formulations in cement mixers. Corn cob granule formulations are commercial and have EPA registrations. Granular formulations are used to penetrate vegetation in order to reach water where mosquito larvae are present.

| Water-based liquid flowable concentrates | % by weight |
|---|---|
| *B. thuringiensis* technical grade powder or slurry (dry weight) | 8.0—15.0 |
| Glycerol/glycerine (optional) | 5.0 |
| Sorbitol, 70% aqueous, for evaporation control (optional) | 35.0—45.0 |
| Propionic acid (some products have a greater amount) | 0.3 |
| Sorbic acid | 0.2 |
| Methylparaben | 0.1 |
| Water | To 100.0 |

*Note:* Xylene at 3.0 to 6.0% and ICI America's PROXEL® up to 0.1% have also been used as bacteriostat stabilizers, but are less environmentally desirable than food grade stabilizers.

## B. *B. SPHAERICUS* FORMULATION

Formulation information and end use product examples given for *B. thuringiensis* subsp. *israelensis* apply also to the formulation of *B. sphaericus* liquid flowable concentrate, wettable powder, granular, pellet, and briquette end use products for mosquito larval control. Since *B. sphaericus* is notably active against *Culex* spp. mosquito larvae, attention should be given to their specific habitats in the design of end use products to achieve targeted delivery of the active ingredient to the larval feeding zone, and also to products which prolong effective toxin presence by providing slow or delayed release. An example of such a formulation is given below:

| Floating slow-release pelleted formulation[68] | % by weight |
|---|---|
| *B. sphaericus* technical grade powder | 30.0 |
| Powdered sugar (release agent) | 30.0 |
| Polypropylene foam (buoyancy agent) | 40.0 |

# VIII. STANDARDIZATION

The rate use recommendations in the labels of *B. thuringiensis* products are based on the field use experience of the manufacturing companies using materials of known potencies. In order to meet the legal requirements of label accuracy (in the U.S. rarely, if ever, enforced), and to ensure that the product has the insecticidal activity per unit weight or volume needed to perform satisfactorily at the recommended rates of application, the primary technical powder or slurry must be bioassayed and the potency determined before the final end use product can be formulated to label potency. There is no reliable relationship between spore count and potency of either lepidopterous active[69] or dipterous active[41] *B. thuringiensis* preparations. Neither can chemical methods of determining the amount of *B. thuringiensis* crystal protein or active fragment protein reliably predict the killing power of a *B. thuringiensis* preparation, since chemical methods measure the quantity, not the quality of protein present. Nor can they discriminate between damaged and undamaged toxin protein.[70]

There are four standard bioassay techniques which have been proposed, two lepidopteran and two dipteran. In addition, two bioassay techniques have been described and compared for bioassaying *B. thuringiensis* subsp. *tenebrionis* against coleopterous larvae.[71] The first lepidopteran bioassay, developed by a joint USDA-industry effort,[72] was based on using *Trichoplusia ni* larvae and a diet incorporation technique using semisynthetic diet. Although this standardized technique served well for nearly a decade, it had two problems. The first was that antibiotic was specified to be used in the semisynthetic diet; and the other was that the ability of a standard to correct for differences in assay methods was overestimated, in that it was agreed by the creators that each could deviate from the standard bioassay as they wished. It was later found that the use of a standard will not in all cases correct for differences in bioassay techniques.[70] The use of antibiotic in bioassay diet can have variable effects, depending on the age of larvae used in the bioassay.[73,74] There is no effect when neonate larvae are used, but when 4-day-old larvae are used, the resulting $LC_{50}$s are raised 3 to 70 times, depending on the insect species, when antibiotic is used in the bioassay diet. It was observed that the shorter the assay period, the greater the effect.

Because of these problems, the USDA and U.S. producers and/or marketers of *B. thuringiensis* products held a meeting in 1982 and decided that a new standardized lepidopterous *B. thuringiensis* bioassay was needed that everybody would follow. The major changes made were that the Brownsville diet would be used without antibiotic or KOH, that sample homogenization would be achieved by sonicating in a bath sonicator for 5 min at 3 ml/w sonicator loading (excessive sonicating increases the activity of *B. thuringiensis* preparations[75,75]), and that test larvae would be multiply or singly held for 4 d at 30°C.

There are two standardized bioassays for mosquito-active *B. thuringiensis* preparations. The first is the WHO method,[77] designed with flexible protocols in view of the differing conditions and availability of materials throughout the world. The second is the U.S. standard bioassay[78] with rigid protocols for U.S. conditions. In the U.S. bioassay, early 4th instar *Aedes aegypti* is specified as the bioassay insect.

Bioassay quality control criteria are an indispensable part of conducting bioassays to reliably determine the insecticidal activity of *B. thuringiensis* preparations. As in other areas of experimental research, invalid conclusions can be drawn from invalid data that results from invalid experiments. The following bioassay quality control criteria have been developed over the nearly 30 years of bioassaying *B. thuringiensis*. The first are the invalidating criteria, which, if not met or if they are violated, result in invalidation of the assay. The results of such assays are only used to estimate the dosages for the repeat bioassay, they are never used in the calculation of a mean $LC_{50}$ or potency of the preparation. The invalidating criteria are

1.  Greater than 10% untreated control mortality in lepidopterous bioassays, and greater than 5% untreated control mortality in mosquito bioassays
2.  Nonsignificant F value (nonsignificant regression) when an $LC_{50}$ is being calculated by probit analysis
3.  Very atypical slope for the insect species being assayed
4.  Over $3 \times$ 95% confidence limits when an $LC_{50}$ is calculated; this is determined by dividing the upper 95% confidence limit by the lower limit (such confidence limits are termed ''very wide confidence limits'' [VWCL]).

Typical slopes are what normally occur for a particular insect in a particular laboratory. In the former Brownsville USDA laboratory, the typical slope when using *T. ni* larvae was about 3 when bioassaying recovered spore-crystal complexes; invalidating slopes were those less than 1 and over 5. In the case of bioassays using *Heliothis virescens* larvae, the typical slopes were about 2, and invalidating slopes were those below 1 and above 4. Bioassays of fermentation beer samples usually have slopes about one unit higher than the bioassays of spore-crystals recovered from those beers. The reason for this slope difference is not known — perhaps there are substances in the fermentation beer which mask the antifeedant and/or avoidance factors associated with *B. thuringiensis* spore-crystal complexes. Other laboratories may have different typical slopes for the same insects.

The second set of bioassay quality control criteria are the discriminatory criteria. Assays with these shortcomings are viewed with suspicion, but the results are utilized if appropriate. The discriminatory criteria are

1.  A significant chi-square value is obtained when the $LC_{50}$ is calculated by probit analysis.
2.  The assay is off curve in that there are not at least two mortality points above and below the calculated $LC_{50}$.
3.  Atypical slope: at Brownsville, atypical slopes for *T. ni* were 1 to 2 and 4 to 5; for *H. virescens* they were 1 to 1.2 and 3 to 4.
4.  Erratic dosage-mortality points.
5.  $2 \times$ to $<3 \times$ 95% confidence limits: such confidence limits are termed ''wide confidence limits'' (WCL).

Potencies are determined by dividing the $LC_{50}$ of a concurrently bioassayed standard by the $LC_{50}$ of the preparation and multiplying that ratio by the potency of the standard. Standards are used to determine relative toxicities of different preparations, correct for day-to-day $LC_{50}$ fluctuations, and to some extent correct for different assay conditions and methodologies between different laboratories. There are five *B. thuringiensis* assay standards being used at the present time:

1.  E-61, *B. thuringiensis* subsp. *thuringiensis*, potency 1000 IU/mg, and available from the Institut Pasteur.
2.  1-S-1980,[79] *B. thuringiensis* subsp. *kurstaki*, potency 16,000 IU/mg, available from the USDA-ARS, Peoria, Illinois. Unfortunately, the 1-S-1980 stored at Peoria has lost about 2000 IU/mg potency, bioassaying now at about 14,000 IU/mg.[80] This loss of potency is probably the result of that lot of 1-S-1980 having been shipped around the U.S. several times from 1986 to 1988. Full potency lots of various quantities of 1-S-1980 still exist at Abbott Laboratories; USDA-ARS at Beltsville, Maryland; USDA-FS at Hamden, Connecticut; and the Environmental Studies Laboratory at Quebec, Canada. Other companies or laboratories that obtained the 1-S-1980 standard from the USDA-ARS at Brownsville, Texas before 1986 also probably have fully active material if it has been stored properly.

3.    IPS-82, *B. thuringiensis* subsp. *israelensis*, potency 15,000 IU/mg, and available from the Institut Pasteur.

4.    635-S-1987,[81] *B. thuringiensis* subsp. *entomocidus*, potency 10,000 IU/mg, and available from the National Research Center, Cairo, Egypt.

5.    CSBt5ab-87, *B. thuringiensis* subsp. *galleriae*, potency 8600 IU/mg, People's Republic of China.

The *B. sphaericus* bioassay standard is RB 80,[82] isolate 1593, potency 1000 IU/mg, and available from the Institut Pasteur.

To determine a mean potency or $LC_{50}$ value for a preparation, it should be bioassayed at least 3 times over at least 2 days (3 preferable). The mean potency or $LC_{50}$ value should consist of at least three values from valid assays whose group coefficient of variation (CV) is under 20%. It is permissible to throw out (discriminate against) valid but suspect bioassays to get the CV below 20% as long as three values remain. This is the value of determining which bioassays are suspect bioassays. In the event there is an outlier value that is from a perfectly good bioassay, then the 4:1 rule is invoked. That is when there are four values out of five for which, when taken together, the calculated CV is less than 20%, but when the fifth value is included the CV is 20% or greater, the outlier value is then discarded. The ratio can be 4:1, 8:2, 12:3, etc. This rule is necessary because mistakes happen, such as an incorrect weighing of the sample, the wrong sample is weighed out, etc. Good assays may be run on such samples, so there are not quality control criteria reasons for throwing out the assay(s).

## IX. REGISTRATION

The U.S. Environmental Protection Agency (EPA) Office of Pesticides and Toxic Substances March 1989 draft of the Pesticide Assessment Guidelines, Subdivision M for Microbial Pest Control Agents presents detailed guidelines for testing that includes *B. thuringiensis* and *B. sphaericus*. Another relevant document available from EPA entitled "Guidance for the Reregistration of Pesticide Products Containing *Bacillus thuringiensis* as the Active Ingredient", dated December 1988, pertains specifically to registration/reregistration requirements for *B. thuringiensis* products.

The following are the current EPA basic and Tier I toxicology and nontarget organism safety testing registration data information or study requirements for microbial pest control agents:

●    Product analysis

    1.  Product identity and disclosure of ingredients
    2.  Manufacturing process
    3.  Discussion of formation of unintentional ingredients
    4.  Analysis of samples
    5.  Certification of ingredient limits
    6.  Physical and chemical properties
    7.  Submittal of sample
    8.  Analytical methods

●    Toxicology

    1.  Acute oral toxicity/pathogenicity study
    2.  Acute dermal toxicity study*

---

\*   Not presently required for *B. thuringiensis* in the Reregistration Guidance/Standard.

3. Acute pulmonary toxicity/pathogenicity study
4. Acute intravenous toxicity/pathogenicity study
5. Primary eye irritation/infection study*
6. Hypersensitivity incidents
7. Cell culture tests with viral pest control agents*

- Nontarget organisms

  1. Avian oral test
  2. Avian respiratory pathogenicity test
  3. Wild mammal toxicity and pathogenicity testing*
  4. Freshwater fish toxicity and pathogenicity testing
  5. Freshwater aquatic invertebrate toxicity and pathogenicity testing
  6. Estuarine and marine animal toxicity and pathogenicity testing
  7. Plant studies
  8. Nontarget insect testing for toxicity/pathogenicity to insect predators and parasites
  9. Honey bee toxicity/pathogenicity testing

- Residue analysis*
- Environmental fate for microbial pesticides*

Further safety data requirements for EPA registration are described in Tiers II, III, and IV of the March 1989 draft Pesticide Assessment Guidelines, Subdivision M for Microbial Pest Control Agents that are found to require testing beyond Tier I. Residue analysis data requirements apply when Tier II or Tier III toxicology data needs are triggered by Tier I acute oral, acute pulmonary, or acute intravenous study data. Environmental fate data requirements apply when Tier II or Tier III testing is triggered by results of nontarget organism Tier I tests.

Registration/reregistration requirements specific for *B. thuringiensis* products described in the December 1988 Guidance for the Reregistration of Pesticide Products Containing *Bacillus thuringiensis* as the Active Ingredient are primary Tier I studies and descriptive data with five Tier III and two Tier IV study requirements reserved pending results of Tier I testing to be done and submitted to EPA. This EPA document also requires end use product analysis for the presence of β-exotoxin and specific end use product label statements.

Microbial pesticide registration requirements vary considerably between countries and the information presented here applies only to those for the U.S. One frequent major difference is that most other countries require field testing of efficacy in addition to various safety studies while in the U.S., efficacy data are usually only required for those products having animal health-related uses. All applications such as agriculture, forestry, stored product, structural, and household, for products not having a direct impact on public health, may have efficacy requirements waived.

# REFERENCES

1. **Saleh, S. M., Harris, R. F., and Allen, O. N.,** Method for determining *Bacillus thuringiensis* var. *thuringiensis* Berliner in soil, *Can. J. Microbiol.,* 15, 1101, 1969.
2. **Yousten, A. A., Jones, M. E., and Benoit, R. E.,** Development of selective/differential bacteriological media for the enumeration of *Bacillus thuringiensis* serovar *israelensis*(H-14) and *Bacillus sphaericus* 1593, *WHO/VBC,* 82, 344, 1982.

---

* Not presently required for *B. thuringiensis* in the Reregistration Guidance/Standard.

3. **Travers, R. S., Martin, P. A. W., and Reichelderfer, C. F.,** Selective process for efficient isolation of soil *Bacillus* spp., *Appl. Environ. Microbiol.*, 53, 1263, 1987.

4. **Couch, T. L.,** personal communication, 1990.

5. **Peferoen, M.,** personal communication, 1990.

6. **Martin, P. A. W.,** personal communication, 1990.

7. **Dulmage, H. T. et al.,** Insecticidal activity of isolates of *Bacillus thuringiensis* and their potential for pest control, in *Microbial Control of Pests and Plant Diseases,* Burges, H. D., Ed., Academic Press, New York, 1981, chap. 11.

8. **Höfte, H. and Whiteley, H. R.,** Insecticidal crystal proteins of *Bacillus thuringiensis, Microbiol. Rev.,* 53, 242, 1989.

9. **Krych, V. K., Johnson, J. L., and Yousten, A. A.,** Deoxyribonucleic acid homologies among strains of *Bacillus sphaericus, Int. J. Syst. Bacteriol.,* 30, 476, 1980.

10. **de Barjac, H., Veron, M., and Dumanoir, V. C.,** Caractérisation biochimique et sérologique de souches de *Bacillus sphaericus* pathogènes ou non pour les moustiques, *Ann. Microbiol. (Inst. Pasteur),* 131 B, 191, 1980.

11. **Myers, P. and Yousten, A. A.,** Localization of a mosquito-larval toxin of *Bacillus sphaericus, Appl. Environ. Microbiol.,* 39, 1205, 1980.

12. **Heckly, R. J.,** Preservation of micro-organisms, *Adv. Appl. Microbiol.,* 24, 1, 1978.

13. **Perlman, D. and Kikuchi, M.,** Culture maintenance, in *Annual Reports on Fermentation Processes,* Vol. 1, Perlman, D., Ed., Academic Press, New York, 1977, 41.

14. **Troitskaya, E. N.,** Comparison of storge methods for the culture of *Bacillus thuringiensis* var. *galleriae, Proc. Biochem. Microbiol.,* 55, 402, 1979.

15. **Jixin, W. and Zaiyon, C.,** Study on storage methods of some strains of *Bacillus thuringiensis, J. Microbiol. (China),* 5, 27, 1985.

16. **Ziniu, Y. and Yong, Y.,** Culture collection, in *Bacillus thuringiensis,* Ziniu, Y., Ed., Science Press, Beijing, 1990, chap. 10.

17. **Gonzalez, J. M., Jr., Brown, B. J., and Carlton, B. C.,** Transfer of *Bacillus thuringiensis* plasmids coding for delta-endotoxin among strains of *B. thuringiensis* and *B. cereus, Proc. Natl. Acad. Sci. U.S.A.,* 79, 6951, 1982.

18. **Yamamoto, T. and Iizuka, T.,** Two types of entomocidal toxins in the parasporal crystals of *Bacillus thuringiensis kurstaki, Arch. Biochem. Biophys.,* 227, 233, 1983.

19. **Smith, R. A.,** personal communication, 1990.

20. **World Health Organization,** Guidelines on local production for operational use of *Bacillus thuringiensis,* especially serotype H-14, in *Guidelines for Production of Bacillus thuringiensis H-14,* Vandekar, M. and Dulmage, H. D., Eds., UNDP/WORLD BANK/WHO, Geneva, 1983, 124.

21. **Underkofler, L. A.,** Microbial enzymes, in *Industrial Mircobiology,* Miller, B. M. and Litsky, W., Eds., McGraw-Hill, New York, 1976, 128.

22. **Yates, R.,** personal communication, 1990.

23. **Nickerson, K. W. and Bulla, L. A., Jr.,** Physiology of sporeforming bacteria associated with insects: minimal nutritional requirements for growth, sporulation and parasporal crystal formation of *Bacillus thuringiensis, Appl. Microbiol.,* 28, 124, 1974.

24. **Kuznetsov, L. E. and Khovrychev, M. P.,** Optimization of a synthetic culture medium for *Bacillus thuringiensis, Mikrobiologiya,* 53, 54, 1984.

25. **Andrews, R. E., Jr., Faust, R. M., Wabiko, H., Raymond, K. C., and Bulla, L. A., Jr.,** The biotechnology of *Bacillus thuringiensis,* in *Critical Reviews in Biotechnology,* Stewart, G. G. and Russell, I., Eds., CRC Press, Boca Raton, FL, 1987, 163.

26. **Dubois, N. R.,** Laboratory batch production of *Bacillus thuringiensis* spores and crystals, *Appl. Microbiol.,* 16, 1098, 1968.

28. **Dulmage, H. T.,** Production of δ-endotoxin by eighteen isolates of *Bacillus thuringiensis,* serotype 3, in 3 fermentation media, *J. Invertebr. Pathol.,* 18, 353, 1971.

29. **Megna, J. C.,** Preparation of microbial insecticide, U.S. Patent 3,073,749, 1963.

30. **Davidson, E. W., Urbina, M., Payne, J., Mulla, M. S., Darwazeh, H., Dulmage, H. D., and Correa, J. A.,** Fate of *Bacillus sphaericus* 1593 and 2362 spores used as larvicides in the aquatic environment, *Appl. Environ. Microbiol.,* 47, 125, 1984.

31. **Beegle, C. C.,** Status of development of superior isolates of *Bacillus thuringiensis* for use against *Heliothis* spp. on cotton, in *Proc. Beltwide Cotton Prod. Res. Conf.,* Brown, J. M., Ed., National Cotton Council of America, Memphis, 1983, 225.

32. **Yousten, A. A. and Wallis, D. A.,** Batch and continuous culture of the mosquito larval toxin of *Bacillus sphaericus* 2362, *J. Ind. Microbiol.,* 277, 1987.

33. **Fridlender, B., Keren-zur, M., Hofstein, R., Bar, E., Sandler, N., Keynan, A., and Braun, S.,** The development of *Bacillus thuringiensis* and *Bacillus sphaericus* as biocontrol agents: from research to industrial production, *Mem. Inst. Oswaldo Cruz (Rio de J.),* 84 (Suppl. 3), 123, 1989.

34. **Lacey, L. A.,** Production and formulation of *Bacillus sphaericus, Mosq. News,* 44, 153, 1984.

35. **Singer, S.,** Potential of *Bacillus sphaericus* and related spore-forming bacteria for pest control, in *Microbial Control of Pests and Plant Diseases 1970-1980,* Burges, H. D., Ed., Academic Press, New York, 1981, chap. 14.

36. **Beegle, C. C.,** unpublished data, 1989.

37. **Ziniu, Y.,** Technology of fermentation, in *Bacillus thuringiensis,* Ziniu, Y., Ed., Science Press, Beijing, 1990, chap. 15.

38. **Dulmage, H. D.,** Productioin of bacteria for biological control of insects, in *Biological Control in Crop Production,* Papavisas, G. C., Ed., Allenheld Osmun, Totowa, N.J., 1981, chap. 9.

39. **Zhizhong, H.,** The effect of carbon, nitrogen and phosphate on *Bacillus thuringiensis* growth, *Microbiology* (China), 7, 7, 1980.

40. **Arzumanov, E. N.,** The effect of pH on the growth of *Bacillus thuringiensis, Mikrobiologiya,* 48, 65, 1979.

41. **Smith, R. A.,** Effect of strain and medium variation on mosquito toxin production by *Bacillus thuringiensis* var. *israelensis, Can. J. Microbiol.,* 28, 1089, 1982.

42. **Stanbury, P. F. and Whitaker, A.,** *Principles of Fermentation Technology,* Pergamon Press, New York, 1984, 83.

43. **Dulmage, H. T., Correa, J. A., and Martinez, A. J.,** Coprecipitation with lactose as a means of recovering the spore-crystal complex of *Bacillus thuringiensis, J. Invertebrate Pathol.,* 15, 115, 1970.

44. **Mechalas, B. J.,** Method for the production of microbial insecticides, U.S. patent 3,086,922, 1963.

45. **Jingyuan, D., Ziniu, Y., Shimei, W., and Jibin, Z.,** Study on semi-solid fermentation of *Bacillus thuringiensis* mosquito larvicide, *J. Huazhong Agric. Univ.,* Suppl. 6, 121, 1989.

46. **Jingyuan, D., Jibin, Z., and Ziniu, Y.,** Study on semi-solid fermentation of *Bacillus sphaericus* mosquito larvicide, *Insect. Microbiol.,* Academic Periodical Press, Beijing, in press.

47. **Kenney, D. S. and Couch, T. L.,** Mass production of biological agents for plant disease, weed and insect control, in *Biological Control in Crop Production,* Papavisas, G. C., Ed., Allenheld Osmun, Totowa, N.J., 1981, chap. 10.

48. **Solomons, G. L. and Perkins, M.,** The measurement and mechanism of oxygen transfer in submerged culture, *J. Appl. Chem.,* 8, 251, 1958.

49. **Spear, B. B.,** personal communication, 1989.

50. **Chilcott, C. N. and Pillai, J. S.,** The use of coconut wastes for the production of *Bacillus thuringiensis* var. *israelensis, Mircen J.,* 1, 327, 1985.

51. **Dharmsthiti, S. C., Pantuwatana, S., and Bhumiratana, A.,** Production of *Bacillus thuringiensis* subsp. *israelensis* and *Bacillus sphaericus* strain 1593 on media using a byproduct from a monosodium glutamate factory, *J. Invertebr. Pathol.,* 46, 231, 1985.

52. **Ejiofor, A. O. and Okafor, N.,** The production of *Bacillus sphaericus* 2362 using fermented cowpea *(Vigna unguiculata)* medium containing mineral substitutes from Nigeria, *Mircen J.,* 4, 455, 1988.

53. **Singer, S.,** *Bacillus sphaericus* for the control of mosquitoes, *Biotechnol. Bioeng.,* 22, 1335, 1980.

54. **Ejiofor, A. O. and Okafor, N.,** Production of mosquito larvicidal *Bacillus thuringiensis* serotype H-14 on raw material media from Nigeria, *J. Appl. Bact.,* 67, 5, 1989.

55. **Lee, J. K., Kim, K. C., and Kim, D. Y.,** A study on the development of a microbial insecticide, *J. Korean Agric. Chem. Soc.,* 22, 123, 1979.

56. **Obeta, J. A. N. and Okafor, N.,** Medium for the production of primary powder of *Bacillus thuringiensis* subsp. *israelensis, Appl. Environ. Microbiol.,* 47, 863, 1984.

57. **Salama, H. S., Foda, M. S., Dulmage, H. T., and El-Sharaby, A.,** Novel fermentation media for production of δ-endotoxins from *Bacillus thuringiensis, J. Invertebr. Pathol.,* 41, 8, 1983.

58. **Salama, H. S., Foda, M. S., El-Sharaby, A., and Selim, M. H.,** A novel approach for whey recycling in production of bacterial insecticides, *Entomophaga,* 28, 151, 1983.

59. **Salama, H. S., Foda, M. S., Selim, M. H., and El-Sharaby, A.,** Utilization of fodder yeast and agro-industrial by-products in production of spores and biologically active endotoxins from *Bacillus thuringiensis, Zbl. Mikrobiol.,* 138, 553, 1983.

60. **Sakharova, Z. V., Ignatenko, Y. N., Khovrychev, M. P., Lykov, V. P., Rabotnova, I. L., and Shevtsov, V. V.,** Sporulation and crystal formation in *Bacillus thuringiensis* with growth limitation via the nutrient sources, *Microbiology,* 53, 221, 1984.

61. **Arcas, J., Yantorno, O., and Ertola, R.,** Effect of high concentration of nutrients on *Bacillus thuringiensis* cultures, *Biotech. Lett.,* 9, 105, 1987.

62. **Holmberg, A. and Sievanen, R.,** Fermentation of *Bacillus thuringiensis* for exotoxin production: process analysis study, *Biotechnol. Bioeng.,* 22, 1707, 1980.

63. **Abrosimova, L. I., Babaeva, P. V., Zubareva, G. M., and Shevtsov, V. V.,** Influence of mineral salts on the level of exotoxin production and productivity of a culture of *Bacillus thuringiensis, Microbiology,* 55, 337, 1986.

64. **Angus, T. A. and Luthy, P.,** Formulation of microbial insecticides, in *Microbial Control of Insects and Mites,* Burges, H. D. and Hussey, N. W., Eds., Academic Press, New York, 1971, chap. 28.

65. **Couch. T. L. and Ignoffo. C. M.,** Formulation of insect pathogens, in *Microbial Control of Pests and Plant Diseases 1970-1980,* Burges, H. D., Ed., Academic Press, New York, 1981, chap. 34.

66. **Most, B. H. and Quinlan, R. J.,** Formulation of biological pesticides, in *Fundamental and Applied Aspects of Invertebrate Pathology,* Samson, R. A., Vlak, J. M., and Peters, D., Eds., Foundation of the 4th Int. Colloq. Invertebr. Pathol., Wageningen, Netherlands, 1986, 624.

67. **Lacey, L. A.,** Production and formulation of *Bacillus sphaericus, Mosq. News,* 44, 153, 1984.

68. **Lacey, L. A., Urbina, M. J., and Heitzman, C. M.,** Sustained release formulations of *Bacillus sphaericus* and *Bacillus thuringiensis* (H-14) for the control of container-breeding *Culex quinquefasciatus, Mosq. News,* 44, 26, 1984.

69. **Dulmage, H. D. and Rhodes, R. A.,** Production of pathogens in artificial media, in *Microbial Control of Insects and Mites,* Burges, H. D. and Hussey, N. W., Eds., Academic Press, New York, 1971, chap. 24.

70. **Beegle, C. C.,** Bioassay methods for quantification of *Bacillus thuringiensis* delta-endotoxin, in *Analytical Chemistry of Bacillus thuringiensis,* Fitch, W. and Hickle, L., Eds., American Chemical Society, Washington, D.C., 1990, chap. 3.

71. **Riethmuller, U. and Langenbruch, G. A.,** Two bioassay methods to test the efficacy of *Bacillus thuringiensis* subspec. *tenebrionis* against larvae of the Colorado Potato Beetle *(Leptinotarsa decemlineata), Entomophaga,* 34, 237, 1989.

72. **Dulmage, H. D., Boening, O. P., Rehnborg, C. S., and Hansen, G. D.,** A proposed standardized bioassay for formulations of *Bacillus thuringiensis* based on the international unit, *J. Invertebr. Pathol.,* 18, 240, 1971.

73. **Ignoffo, C. M., Garcia, C., and Couch, T. L.,** Effect of antibiotics on the insecticidal activity of *Bacillus thuringiensis, J. Invertebr. Pathol.,* 30, 277, 1977.

74. **Beegle, C. C., Lewis, L. C., Lynch, R. E., and Martinez, A. J.,** Interaction of larval age and antibiotic on the susceptibility of three insect species to *Bacillus thuringiensis, J. Invertebr. Pathol.,* 37, 143, 1981.

75. **Bai, C. and Degheele, D.,** The influence of ultrasonication on the morphological structure and bioactivity of the parasporal body of *Bacillus thuringiensis* Berliner, *Med. Fac. Landbouww. Rijksuniv. Gent,* 49, 1307, 1984.

76. **Lee, B. L.,** unpublished data, 1983.

77. **World Health Organization (WHO),** Mosquito Bioassay Method for *Bacillus thuringiensis* subsp. *israelensis* (H-14), Annex 5 in WHO Report TDR.VEC-SWG(5)/81.3.

78. **McLaughlin, R. E., Dulmage, H. T., Alls, R., Couch, T. L., Dame, D. A., Hall, I. M., Rose, R. I., and Versoi, P. L.,** U.S. standard bioassay for the potency assessment of *Bacillus thuringiensis* serotype H-14 against mosquito larvae, *Bull. Entomol. Soc. Am.,* 30, 26, 1984.

79. **Beegle, C. C., Couch, T. L., Alls, R. T., Versoi, P. L., and Lee, B. L.,** Standardization of HD-1-S-1980: U.S. standard for assay of lepidopterous-active *Bacillus thuringiensis, Bull. Entomol. Soc. Am.,* 32, 44, 1986.

80. **Martinat, P. J.,** personal communication, 1989.

81. **Salama, H. S., Foda, M. S., and El-Sharaby, A.,** A proposed new biological standard for bioassay of bacterial insecticides vs. *Spodoptera* spp., *Trop. Pest Manage.,* 35, 326, 1989.

82. **Bourgouin, C., Larget-Thiery, I., and de Barjac, H.,** Efficacy of dry powders from *Bacillus sphaericus*: RB 80, a potent reference preparation for biological titration, *J. Invertebr. Pathol.,* 44, 146, 1984.

83. **Foda, M. S., Salama, H. S., and Selim, M.,** Factors affecting growth physiology of *Bacillus thuringiensis, Appl. Microbiol. Biotechnol.,* 22, 50, 1985.

Chapter 15

# MASS PRODUCTION OF VIRAL INSECTICIDES

**Spiros N. Agathos**

## TABLE OF CONTENTS

# I. INTRODUCTION

Hundreds of insect species are susceptible to viral infections resulting in one or more of over two dozen diseases that are lethal to the insect or its progeny. Insect species affected by viral entomopathogens include pests in agriculture and forestry, as well as vectors of human and veterinary diseases. Entomopathogenic viruses form the basis for the development, over the last three decades, of bioinsecticides which have emerged as important new weapons in biological pest control. Viral insecticides, primarily baculoviruses, are an environmentally attractive alternative to synthetic chemical insecticides, because they do not pollute the environment, do not persist in the food chain, are specific only to their pest insect targets, and do not damge beneficial insects or the ecosystem.

In order to be economically viable, especially in competition with the newer chemical insecticides (e.g., pyrethroids), viral insecticides must be produced on an industrial scale cheaply and efficiently, under strictly controlled and reproducible conditions. These fundamental requirements are not yet met fully in practice, since the viral agents are still being produced mostly *in vivo* (in whole insects). Production on this basis is expensive, labor-intensive, and not easy to automate and scale up. Therefore, *in vitro* processes using insect cell cultures are currently recognized as being more suitable for large-scale production of viral insecticides. The early development of techniques for the growth of insect cells in tissue culture, particularly since the 1970s, plus subsequent work aimed at obtaining significant amounts of virus free from microbial contamination, have provided a strong motivation towards research and development efforts in the area of mass virus production in bioreactor systems. The clear possibility of routinely producing high quantities of viral insecticides through insect cell cultivation *in vitro* by the application of state-of-the-art bioprocess technology and engineering will be the main subject of this chapter. Emphasis will be placed on the conceptual and technological problems presented by the large-scale propagation of insect cells and the solutions proposed and tested in process development.

# II. VIRAL PATHOGENS AND CURRENT PRACTICES IN THEIR MASS PRODUCTION

The development of viral insecticides and their use for biological control in agriculture and forestry have been documented in numerous scientific journal articles, technical bulletins, and books, and have been assembled in single-volume treatises.[1-3] The latter, as well as other chapters in the present volume, address the myriad of factors and considerations that are involved in the successful development and application of these agents. In this section, I will attempt to summarize briefly the types of viruses used as biological insecticides and the main drawbacks of their *in vivo* production processes.

Although the viral agents known to infect insects belong to several categories, including cytoplasmic polyhedrosis viruses (CPVs), entomopoxviruses (EPVs) and baculoviruses,[4,5] the latter constitute the group of choice for formulating bioinsecticides, due to their specificity and widespread occurrence among economically important insect pests, their virulence, and their safety for nontarget species.[6] Moreover, numerous cell lines, primarily from lepidopteran insects, have been shown to be susceptible to baculovirus infection.[5,7] The group of baculoviruses (family: Baculoviridae) consists of rod-shaped, enveloped viruses, whose nucleic acid portion is double-stranded DNA, and are often occluded in proteinaceous crystal bodies. The baculoviruses are classified in subgroups:

1.    Subgroup A includes viruses which are occluded in polyhedral-shaped protein crystals within the nucleus of infected insect cells. These viruses are known as nuclear polyhedrosis viruses (NPVs) and represent the majority of agents used as bioinsecticides.[6,8,9]

The most widely studied NPV is the *Autographa californica* (alfalfa looper) nuclear polyhedrosis virus (AcNPV), which is not only the basis for pest control but, over the last 5 years, has been used as a novel vector for expressing heterologous genes at high levels in biotechnological (primarily therapeutic) applications.[10,11]

2.  Subgroup B includes viruses which tend to become occluded in granular-shaped protein crystals, often within the nucleus of target insect cells. Viruses of this type are referred to as granulosis viruses (GVs). An extensively studied granulosis virus with considerable bioinsecticide potential is the *Cydia pomonella* granulosis virus (CpGV), which infects the codling moth.[9]

3.  Subgroup C includes viruses which are not or are only slightly occluded, and are simply recognized as nonoccluded baculoviruses. An example of virus of this type is the nonoccluded baculovirus of the coconut palm rhinoceros beetle, *Oryctes rhinoceros*, used for biological control in tropical countries.[12]

Table 1 summarizes the best-known viral insecticides that have been developed and produced commercially or semi-commercially to date, including the products used in experimental field test applications. Only four viral formulations have been registered with the U.S. Environmental Protection Agency (EPA) for use in agriculture and forestry.[6,8,9] Among them, only one preparation has been produced industrially, i.e., the *Heliothis zea* NPV (commercial name: Elcar®), by Sandoz, Inc., for the control of pests devastating annual crops including cotton bollworm, corn earworm, and tobacco budworm infestations. The other three registered viral insecticides are the NPVs of the Douglas fir tussock moth *(Orgyia pseudotsugata)*, the gypsy moth *(Lymantria dispar)*, and the European pine sawfly *(Neodiprion sertifer)*, which are produced by the U.S. Forest Service. One more preparation whose registration is expected in the near future is the GV of the codling moth *(Cydia pomonella)*, which has shown promising results in orchards of apple, pear, and walnut trees.

Despite the convincing record of Elcar® as an effective and competitively priced agricultural insecticide, its commercial production has ceased since the mid-1980s. Similarly, the three NPVs registered for biological control of forest insect pests are not for sale to the public. Today no major industrial firm produces viral insecticides, and the products available (either permanently registered or temporarily approved for experimental use in field tests) are obtained by subcontracting. This is a result of many complex and interacting socioeconomic reasons, including costs for development, registration, and production; perceived small size of the market and low profit margin; competition from synthetic chemical products (e.g., pyrethroids) and from other microbial formulations (e.g., *Bacillus thuringiensis* parasporal toxins, entomocidal fungal products); perceived complexities in application; persistent misevaluations of their effectiveness at controlling large insect populations, etc.[8,9] Fortunately, the ecological soundness of viral insecticides and their capacity to control major pest species over long periods of time, coupled with accelerated advances in the basic science and technology of their mass production (see Section III, below), underscore their increased potential as innovative replacements of chemical pesticides at the present time.[6,8,9]

Despite the slow pace in commercialization and registration of viral insecticides in the U.S. and most other industrialized countries, it is anticipated that these agents will find progressively broader acceptance and applications, either singly or, more likely, in combination with chemical and other microbial pesticides, as constituents of biorational programs of integrated pest management.[9] This prospect is more realistic today, because of the slowly developing resistance to synthetic chemical insecticides among many insect pest populations, and also because today's consumer is willing to pay higher prices for agricultural products, such as fruits, that have been grown "organically". In addition, baculoviruses have perhaps even greater potential for pest control applications in developing nations, especially in tropical climates, where agricultural and forest ecosystems are particularly sensitive to disturbances from broad-spectrum chemical pesticides.[9]

**TABLE 1**
**Viral Insecticides in Commercial or Field Test Applications[6,8,9]**

| Application | Target insect | Virus | Trade name | Registration status in U.S. by EPA |
|---|---|---|---|---|
| **Annual crops** | | | | |
| Cotton, corn, sorghum, soybeans, tobacco, tomato | Cotton bollworm Corn earworm Tobacco budworm | *Heliothis zea* NPV | Elcar® (Sandoz) | First registered 1975 |
| Various annual crops | Alfalfa looper | *Autographa californica* NPV | SAN 404 (Sandoz) | — |
| Various annual crops | Fall armyworm | *Spodoptera frugiperda* NPV | Several | — |
| Various annual crops | Cabbage looper | *Trichoplusia ni* NPV | SAN 405 (Sandoz) | — |
| **Orchards** | | | | |
| Apple, pear, walnut | Codling moth | *Cydia pomonella* GV | SAN 406 (Sandoz) | Registration expected in near future |
| **Plantations** | | | | |
| Coconut palm | Palm rhinoceros beetle | *Oryctes rhinoceros* nonoccluded virus | — | — |
| **Forest** | | | | |
| Douglas fir | Douglas fir tussock moth | *Orgyia pseudotsugata* NPV | TM-Biocontrol 1 (U.S. Forest Service) | First registered 1976 |
| Many species of forest, shade, ornamental and orchard trees | Gypsy moth | *Lymantria dispar* NPV | Gypchek (U.S. Forest Service) | First registered 1978 |
| Pine | Pine sawfly | *Neodiprion sertifer* NPV | Neochek-S (U.S. Forest Service) and others | First registered 1983 |
| Spruce | Spruce budworm | *Christoneura fumiferana* NPV | — | — |

Irrespective of the general methodology used for mass production of viral insecticides, a central goal of any large scale process is to furnish the greatest amount of biologically active virus per unit time, at the lowest cost under strict quality and safety standards. The twin requirements of low cost and large amount for field applications imply that the efficient production of the baculovirus of interest must be demonstrated at the pilot plant scale, whereas biological activity (virulence) and safety must be demonstrated in the field. The biological activity is influenced by several in-plant and on-field factors, but generally recognized prerequisites of the preparation are its half-life of infectivity in the field and its shelf stability, both of which must be maximized.

The predominant methodology in practice today for large-scale baculovirus production is the *in vivo* process.[13-16] Active material (primarily polyhedrin inclusion bodies, PIB) for insecticide formulations is obtained by infecting living host insects with the baculovirus strain of interest. This methodology has been developed over many years into a full-fledged industrial technology, and Shapiro[16] has provided an excellent detailed review on the intricacies of this production process and its optimization. It is possible to produce large quantities of baculovirus by introducing an appropriate seed inoculum of the virus into a large target population and then harvest the infected individuals, or simply obtain the virus by collecting large numbers of insects from a naturally infected population in the wild. This approach is often fraught with drawbacks, such as the possibility of contamination with undesirable adventitious agents, and the occurrence of weaker insect hosts compared to laboratory-grown colonies, but it may be useful in some cases as a source of new inoculum. Therefore, laboratory colonization of host insects and controlled introduction of the desired virus is the preferred method for large-scale production of viral insecticides. In order to achieve the goals of efficiency and qualty control expected in this process, many biological and environmental parameters must, in turn, be optimized.[16] Briefly, the host insect stock, the viral inoculum, the environmental conditions of mass rearing (temperature, humidity, air flow, photoperiod, diet), the individual operations of production (inoculation, virus incubation, harvesting, extraction, purification, storage, and quality assurance) and, finally, the physical layout of the production facility, are all key factors influencing the overall quality of virus production. For a detailed discussion of these factors, the reader is referred to Shapiro.[16]

Despite this existing and reasonably streamlined technology of mass rearing of colonized insects for *in vivo* production of viral agents, the sheer number and frequent interdependence of the factors enumerated above present many problems. The obvious alternative is the development of insect tissue culture systems for *in vitro* production of viral agents, in a manner analogous to that used predominantly today for commercial vaccine production (mammalian cell culture). Growth and large-scale production of virus in insect cell culture has many attractive features that constitute clear advantages over *in vivo* production methodologies:

- Reduced capital costs for cell propagation equipment (bioreactors) compared to the elaborate facilities required for growing and maintaining whole insect colonies
- Reduced labor costs
- Reduced costs of various consumable supplies, especially with the advent of low-serum and serum-free media for cell cultivation
- Reduced costs for harvesting and downstream processing operations
- Production of clean viral products, substantially free from microbial contamination (often occurring in insect larvae) and from unwanted impurities (larval proteins, cuticles, etc.), that may be a safety hazard
- Closer monitoring and control of nutritional and physiological environment leading to reliable quality control

- Potential for plant automation and for scale-up based on the principles of chemical reactor engineering
- Increased productivity of a cell production facility by applying emerging breakthroughs in genetic engineering and in bioreactor design (e.g., perfusion, hollow fiber reactors, microencapsulation, etc.)

## III. PROGRESS IN INSECT CELL CULTURE SYSTEMS FOR LARGE-SCALE VIRUS PRODUCTION

### A. GENERAL CONSIDERATIONS FOR BIOPROCESS DEVELOPMENT

It is clear from the foregoing that the commercial exploitation of insect cell systems for the production of viral agents requires efficient cultivation methods *in vitro*, and that efficient scale-up of cell production implies the growth of the insect cells in one or a few large containers (bioreactors). The propagation of insect cells in monolayers or stationary culture (glass or plastic tissue culture flasks, roller bottles) appears unsuitable for large-scale production schemes, due to the decrease in the surface-to-volume ratio upon increase in scale of operation, with a concomitant decrease in the profitability of the process. Even if a process consisting of large numbers of T-flasks or roller bottles could be envisioned and justified by the possibility of containing an accidental contamination to one or a few individual vessels, such a process could not be managed as effectively as the large vessel(s) of equivalent volume and surface area. Therefore, the most promising approaches in process design and development have been oriented predominantly towards cultivation of insect cells in suspension, using spinner vessels or a variety of fermentation systems. Efforts are focused on the development of efficient insect cell culturing processes by adopting bioreactor technology and engineering principles that have been successfully demonstrated in microbial and mammalian cell product manufacturing.

Crucial early steps in this direction have been the pioneering of insect tissue culture media[17-19] and the *in vitro* establishment of continuous cell lines or strains,[18,20,21] which catalyzed the development of standardized techniques for the cultivation of insect cells in the laboratory.[22,23] These lessons were applied later to projects of more massive insect cell cultivation, making use of vessels in the range of 100 ml to 10 l, but typically below 3 l. The primary goal of such projects was to obtain large amounts of baculoviruses free from microbial contamination.[24-28] These developments could not have been attained without the substantial improvements in insect cell culture media and in aseptic techniques for the establishment and propagation of insect cell lines, that have occurred over the last four decades. This progress in culturing techniques allows today's biotechnologist to produce almost routinely substantial amounts of insect cells in short periods of time through bioprocess technology and engineering, whether the cells are destined for viral insecticide production or for manufacturing of various recombinant proteins expressed by modified baculovirus vectors.[10,11]

Conceptually, large-scale insect cell culture is very similar to the mass cultivation of vertebrate animal cells (e.g., mammalian fibroblasts, hybridomas, etc.) on an industrial scale. In terms of equipment, instrumentation, kinetic descriptions and control strategies, large-scale insect cell technology is but a variation of fermentation technology. However, there is still little understanding and a dearth of published information on the behavior of insect cells growing in bioreactors. In principle, analogies between mammalian cells and insect cells in culture guide the general design of large-scale insect cell cultivation processes. However the detailed design and development of an efficient and scalable insect cell cultivation system must be based on several theoretical and practical considerations: adequate understanding of the nutritional and physical requirements of the candidate cell line for growth and virus proliferation, coupled with information on the behavior of the cells in different culture configurations (free suspension or attached growth or matrix-confined growth), in long-term cultivation (sustained viability, "passage" effects on virus attenuation), under environmental conditions inherent in large-scale vessels (mechanical agitation, liquid medium circulation, aeration by sparging or over free liquid surface or through indirect means, etc.), plus a plethora of other factors influencing both upstream (i.e., bioreactor level) and downstream processing (i.e., separation and purification operations). Although a considerable body of such information is already available on industrial mammalian cell culture,[29-32] there is a serious lack of comparable data for insect cell culture.

A cursory comparison of cultured insect and mammalian cells in terms of properties that affect the technology of their propagation reveals a number of similarities and differences. Because insect cell lines originate from undifferentiated embryonic or specific organ tissues, the same cell line can be grown (often after some adaptation) as either surface attached (monolayer) or suspended.[33] This versatility in mode of growth which is in contrast with the majority of mammalian cell lines, increases the choices of bioreactor type and culturing strategy for industrially attractive insect cell lines. The contact inhibition manifested by many mammalian cell lines (except for a number of transformed and lymphoid cell lines) is considerably milder or absent in insect cells, which, instead, have a natural tendency to form aggregates (clumps) both in attached and in suspension culture.[34] The attachment forces between insect cells and solid surface for growth are also weaker than for mammalian cells — thus gentle mechanical force can be used for their detachment, instead of trypsin, the standard enzymatic approach in mammalian cell culture. Both mammalian and insect cells require strict aseptic operation of their cultivation vessel and both categories of cells require a minimum seeding density (inoculum size) for successful initiation of proliferative growth in most suspension and surface attached cultures. In practice, insect cell cultures are initiated at inoculum levels of 1 to 2 $\times$ $10^5$ cells per ml of medium.

The design and scale-up of bioreactors for mass production of insect cells and their viral products is based on the aerobic nature of these cells' metabolism and on the physical, mechanical and geometric constraints imposed upon the devices used to supply the oxygen required for cell growth and product formation. In discussions of scale-up, "scale" does not always mean bioreactor size, but may refer to increased productivity, provided that the concentration of viable and/or product-forming cells is increased with increased scale of operation. Today, there is a recognized trend towards increased cell concentrations by one or two orders of magnitude for industrial mammalian cell bioreactors over and above what is typically achieved in simple batch suspension cultures.[31,35] As a rule of thumb, a final insect cell density of 2 to 5 $\times$ $10^6$ cells per ml of medium (equivalent to about $10^8$ PIB/ml) is routinely achieved in laboratory spinner flasks. It has been calculated that the treatment of a million acres of an annual crop, such as cotton, would require an amount of viral PIB obtainable from approximately 60,000 l of bioreactor capacity, assuming that the final virus titer is equivalent to that obtained in the small scale spinner ($10^8$ PIB/ml). Therefore, it is obvious that an increase by ten- to one-hundredfold in the NPV titer would bring about a proportional decrease in the required bioreactor capacity, making the process much more attractive to industry. This also means that the oxygen demand of denser cell cultures will be correspondingly higher.

In order to satisfy the oxygen demand of production systems employing insect or mammalian cells, the standard approaches of established microbial fermentation technology, i.e., mechanical agitation and aeration by introduction of a swarm of air bubbles (sparging) are considered too aggressive, in view of the larger size of animal cells (8 to 10 $\mu$m) compared with microbial cells (1 to 10 $\mu$m) and their lack of cell wall. The rates of oxygen mass transfer required for large-scale production of cultured insect cells are generally thought to result in cell damage and death due to hydrodynamic shear stress. The hydrodynamic forces frequently referred to as "shear" stem from mass transfer-assisting operations, like agitation and air sparging in freely suspended culture and also medium (re-)circulation for surface-grown cells (e.g., on microcarriers), and are generally considered as the most significant bottleneck in insect cell culture scale-up. Interestingly, these assessments are made in the still-limited research literature, without the benefit of extensive data from appropriate experimental studies. Reports on unusually high shear sensitivity of insect cells,[28,34] on the susceptibility of these cells to damage and death by direct interaction with air bubbles[27,36] and on the unusually high oxygen demand of insect cells compared to mammalian cells[37,38] are often cited, to substantiate the perceived need for special bioreactor designs or for modifications of existing vessels. However, there are also emerging reports pointing to disagreements on the high degree of shear sensitivity and on the numerical values of oxygen demand of cultured insect cells.[39-42] Nonetheless, because these concerns have been guiding most documented efforts to develop production-level insect cell culture systems to date, I will address these two inextricably connected factors, shear stress and oxygen supply, further below in Section C. In the next section I will review briefly the considerations surrounding insect cell lines and media for large scale cultivation, since, as pointed out before, the development of cell types with appropriate biotechnological characteristics and the elaboration of inexpensive and convenient media are as important as bioprocess engineering improvements for the efficient use of insect cell culture in industry.

## B. CELL LINES AND MEDIA DEVELOPMENT

Ever since Grace[18] established the first insect cell lines in 1962 using the medium named after him (a modification of Wyatt's[17] semisynthetic medium) which is still in widespread use today, great strides have been made in the establishment and maintenance of continuous insect cell lines. Today more than 200 insect cell lines are available (more than 70 among them from lepidopteran species) and regular compilations of newly established cell lines are given by Hink.[43,44] For the establishment and serial propagation of these strains, over 60 culture media have been formulated to date, many of them specifically designed for insect cell culture, some originally used for vertebrate cell cultivation, supplemented or not with serum. Established insect cell lines are continuous in the sense that they are "immortal", although, unlike continuous mammalian cell lines, they do not appear to be transformed ( = propagating in a tumor-like manner). Established insect cell lines originate from a variety of tissues, but mostly from undifferentiated ovarian or embryonic tissue. As a result, they may consist of a mixture of cell types often differing in morphology and physiology (e.g., lepidopteran cell susceptibility to baculovirus infection),[45,46] but efforts are underway to develop clonally pure cell lines that would be useful for fundamental research and for the production of viral insecticides and other biologicals in insect cell culture.

The establishment of new cell lines and of strains with biotechnically desirable phenotypes (e.g., higher baculovirus infectivity and stronger attachment dependence) has advanced markedly in the last decade. Granados and Hashimoto[7] have reviewed the numerous lepidopteran cell lines that are susceptible to baculovirus infection and have listed over two dozen of baculovirus species grown *in vitro*, including several that are the basis for viral insecticide preparation. Notable in this respect are the new cell lines of *Trichoplusia ni* that are susceptible to replication of NPV and of GV,[45] *Cydia pomonella* susceptible to replication of GV,[47] as well as a *Cydia pomonella* line designated IZD-Cp 4/13, which represents the first reported stable hybrid cell line originating from somatic cell fusion of lepidopteran cells, and which is significantly less susceptible to infection by *Choristoneura muriana* NPV than its parent.[48] By far the most widely studied cell line for industrial scale propagation is *Spodoptera frugiperda* line Sf9, a clonal isolate derived from the pupal ovary of the fall armyworm, which is used for propagation of wild type as well as genetically engineered baculovirus AcNPV,[10,11,39,42] but *Trichoplusia ni* cell lines are also shown recently to be suitable for large-scale production of the same baculovirus.[49] There is ample room for further developments in selection and construction of insect cell lines from a broader variety of host species and even from distinct organ tissues, that will ensure optimal formation of desirable viral products in large-scale cell culture. A number of genetic traits that are currently considered for incorporation into baculovirus vectors for a new generation of viral insecticides will be addressed in the concluding portion of this chapter (Section IV).

The dramatic increase in the number of continuous cell lines established in the last decade is largely due to the improvement of culture media commercially available.[50] Chemically defined media are desirable for exact metabolic studies, and simpler, low-cost media are continually sought for use in practical applications of insect cell culture, including large-scale production of viral pesticides.[51] Despite the progress made to date, the nutritional requirements of insect cells in culture have not been studied as extensively as those of cultured mammalian cells. Hink[52] and Mitsuhashi[50,53] have provided excellent summaries of current knowledge on nutritional requirements of cultured insect cells.

Briefly, most insect cell lines utilize sugars, proteins, and peptides, amino acids, organic acids, vitamins, lipids, and inorganic salts. A number of vitamins, especially those of the water-soluble B vitamin group, tend to be contained in most media. Among lipids, exogenous sources of polyunsaturated fatty acids and sterols (e.g., cholesterol) are included. Inorganic salts are important ingredients of insect cell culture media, because they contribute to maintaining the correct range of ion balance and osmotic pressure. Insect cells in culture are generally flexible to the ionic conditions of the medium, and usually grow at somewhat higher osmotic pressures than mammalian cells. Proteins and peptides are incorporated in insect cell culture media predominantly by supplementing the defined components with complex chemically undefined substances. Among the latter, serum (primarily fetal bovine serum, FBS) is the single most widely used complex supplement. It is a source not only of proteins and smaller peptides, but also of several free amino acids as well as vitamins, lipids, and metal ions. Serum is the most expensive component of the majority of insect cell culture media. In addition, it has variable composition from lot to lot, may contain cytotoxic factors, is susceptible to mycoplasma and virus contamination, and contributes to large-scale bioprocessing difficulties both upstream (e.g., foaming in bioreactor) and downstream (e.g., product separation and purification). Most significantly, serum is *not* necessary for insect cell growth, as demonstrated by the development of many serum substitutes, and there are continuing and successful efforts to replace serum with cheaper defined and undefined compounds. Defined proteins proposed for serum replacement include globulins, bovine serum albumin, transferrin, etc., whereas undefined substances include peptone, yeast extract, tryptose phosphate broth, and protein hydrolysates like lactalbumin hydrolysate (LH) and yeastolate (YL). Finally, a number of miscellaneous media components with no direct nutritional role are also included in media formulations, because they are thought to contribute to the survival and protection of the cells both during static maintenance and subculturing and, especially, under the intense conditions of cultivation in agitated and aerated vessels (see below): methylcellulose, PVP-40, Darvan, Pluronic polyols (thickening agents presumed to protect cells from hydrodynamic shear-induced cell damage and death, and to reduce cell clumping), silicones and Tweens (antifoam and surface-active agents), α-tocopherol acetate (antioxidant), etc. Occasionally, antibiotics (penicillin, streptomycin) are added to the media to protect against microbial contamination.

Hink and Hall[44] have summarized the most commonly used media for insect cell cultivation *in vitro*. Most cell lines from lepidopteran insects are cultured in Grace's medium[18] and its modifications (TNM-FH,[20] IPL-41,[28] TC-100,[54] BML/TC-10,[54] etc.) as well as on MM medium.[19]

The quest for media simplification and cost reduction has led to a number of low-serum, low-serum low-protein, and serum-free media formulations. For instance, Hink and co-workers[55] developed a serum-free and protein-free medium for culture of *Trichoplusia ni* (cell line TN-368) cells, by supplementing Grace's basal medium with LH, YL, tryptose, peptone, glucose, glutathione, oleic acid, cholesterol (plus PVP-40 and methylcellulose as thickening agents), i.e., FBS was totally replaced by chemically undefined and defined ingredients. The same cell line was cultivated in serum-free medium by Vail et al.[56] These workers succeeded in adapting the cells to Hink's TNM-FH medium[20] from which whole egg ultrafiltrate, bovine serum albumin and FBS were successively deleted. Modifications of media may contribute to simplification of preparation and savings in production cost, themselves significant considerations in large-scale applications, but these same modifications may also deprive the cells of their ability to maintain differentiated functions and to produce heterologous proteins, since often these functions do not exhibit the same nutritional requirements as cell proliferation.[57] In the example of Vail's medium development,[56] this serum-free medium supported lower cell growth but good virus production. Roder[58] cultivated three lepidopteran cell lines in BML/TC-10 medium[54] from which the concentration of FBS was gradually decreased and replaced by egg yolk emulsion, until the cells became fully adapted to the same medium now containing 1% egg yolk and no FBS. This medium's utility was demonstrated in bioreactor cultivation (up to 10 l agitated vessels) of the above cell lines in terms of both good growth and NPV baculovirus formation. Production costs of viral insecticide output were reduced by 95% thanks to this serum-free medium. In the cases of serum-free media cited here, the serum is replaced by more or less undefined components. Such "semidefined" media are generally the least expensive kinds of serum-free media, because crude natural substances are used to replace costlier purified chemicals. Serum-free MM medium (MM-SF)[59] serves as a good prototype of semidefined media with the lowest cost of ingredients and the simplest preparation. Recent work by Mitsuhashi[51] has brought about further simplification and cost reduction of MM-SF medium: the organic salt mix has been replaced with diluted sea water, and glucose has been replaced by table sugar. The new medium, designated MTCM-1601, consists of 25 ml sea water, 0.65 g LH, 0.5 g YL, 0.8 g table sugar of supermarket grade, and distilled water added to a total volume of 100 ml. The resulting solution has a pH of 6.5; thus no pH adjustment is necessary. The medium is sterilized by autoclaving for 15 min. Mitsuhashi[51] showed that this medium MTCM-1601 was a step towards the ideal of a "totipotent" medium, in that it could support the growth of at least 15 lepidopteran cell lines. In a similar manner, Koike and Sato[60] have developed a series of autoclavable insect culture media by further simplifying the salt components of serum-free MM medium (MM-SF). These media, designated No. 8, No. 10, and No. 15, supported good growth of 9 lepidopteran cell lines not only in stationary cultures but also in spinner flask (500 ml) and jar fermentor (40 l) suspension cultures.[60] Also noteworthy for its simplicity and demonstrated capacity to support growth of *Spodoptera frugiperda* cells and production of recombinant protein is the serum-free protein-free semidefined medium recently developed by Maiorella et al.[39] In this work, IPL-41[28] basal medium (i.e., containing only inorganic salts, sugars, amino acids, organic acids, trace elements, and vitamins, but no serum or other supplements) was supplemented with ultra-filtered yeast extract and a specially prepared stable Pluronic polyol-lipid microemulsion containing Pluronic F-68 and Tween 80 as a dual detergent system along with cod liver oil and cholesterol (lipids) plus α-tocopherol acetate (antioxidant). In this medium, Maiorella et al.[39] were able to reach the same cell growth rate and extent and the same titer of recombinant colony stimulating factor as in IPL-41 medium supplemented with 10% FBS. Moreover, this medium was successfully used in a 21-liter airlift bioreactor, due to the inherent suitability of its formulation for sparged cultivation (see below). An example of a new commercial serum-free medium with low protein content ($<$10 µg/ml) optimized for airlift and stirred bioreactor cultivation of insect cells is EX-CELL 400™, which was recently shown to support good growth of Sf9 cells whether uninfected or infected with a recombinant AcNPV vector.[42,61] Finally, an important landmark in serum-free media formulation has been the development of a few completely chemically defined media for insect cell culture. Among them, the often cited CDC medium developed by Wilkie et al.[62] consists of almost

70 chemical compounds and supports the growth of cell lines from the lepidopteran *Spodoptera frugiperda* and the dipteran *Aedes aegypti* and *Aedes stephensi* (mosquitoes). Remarkably, this serum-free protein-free medium allowed the propagation of these three cell lines without any previous adaptation step. Although this and other chemically defined media have extremely complex compositions, they are very useful for exact biochemical studies of cultured cells.

Even though media formulations play an important role in the efficient and economical cultivation of insect cells on a large scale, the growth physiology of these cells is also affected by physicochemical requirements such as temperature, pH, gas exchange, and osmolarity. The optimal temperature range for the growth of most insect cell lines in culture is between 25 and 30°C, i.e., considerably below that for mammalian cells. For most applications of insect cell cultures, including cultivation of insect cells in bioreactors, the control of the temperature at the optimum set point (e.g., 28°C) does not present excessive cooling requirements for metabolic heat removal. The pH required for optimal *in vitro* growth of lepidopteran cells falls in the range between 6.0 and 6.25.[34] The pH in culture tends to change, usually monotonically, to more basic or more acidic values, as various medium components are consumed and metabolic products excreted. Thus, it is important to maintain the pH within the optimal range. This is done usually by incorporating buffers in the media, typically phosphate or bicarbonate. In the latter case the dissociation equilibrium is stabilized under an atmosphere of 5% $CO_2$, especially in stationary cultures, a practice widely employed in vertebrate tissue culture. However, as with ionic balance in general insect cells are known to tolerate pH changes easier than mammalian cells. Nonetheless, an optimized insect cell cultivation in large scale should include effective pH control, which can adopt the addition of acid or alkali on demand, as in well-mixed microbial fermentations, or could be exerted through continuous adjustment of $CO_2$-containing gaseous environment of the bioreactor. The osmolarity of insect cell culture media is generally higher than that of mammalian cell media, as insect cells have been shown to grow in higher osmotic pressure, and also to tolerate a wider range of osmotic pressure.[52] Thus the osmolarity of insect cell media can be typically about 340 mOsm/l but also sometimes it can exceed 400 mOsm/l.[52]

## C. BIOREACTOR GROWTH OF CULTURED INSECT CELLS IN SUSPENSION
### 1. Control of Agitation and Aeration under Low Mechanical Shear (Stirred Tank Bioreactors)

The use of bioreactors of standard design, e.g., stirred tank (jar) fermentors, for insect cell growth, is an economically desirable goal. Hink's group pioneered efforts in this direction since the mid-1970s, working with *Trichoplusia ni* (TN-368) as a useful system for the production of viral insecticides. Hink and Strauss[25] compared insect cell growth in jar bioreactors of 400 ml and 2 l with well-established growth kinetics in 100 ml laboratory spinner flasks. The scaled-up systems equipped with agitators allowed the same growth extent ( $= 3 \times 10^6$ cells per ml) as the 100-ml spinner, but at a considerably slower growth rate (6 to 7 days vs. 3 to 4 days in spinner). A fourth scale-up system, a 400 ml vibromixer, gave even slower growth and the cells exhibited unhealthy morphology. These phenomena were attributed to mixing-associated shear in the larger bioreactors. The same workers addressed successfully the problem of cell clumping by supplementing the medium with 0.1% methylcellulose to increase viscosity. Hink and Strauss[25] also compared oxygen supply between the surface-aerated spinner flask and the larger vessels and found that by combining air sparging with surface aeration, the 400 ml and 2-l stirred reactors allowed for higher cell growth than the spinner.

The oxygen requirements of *Trichoplusia ni* cells were investigated by Streett and Hink[63] who found that the oxygen uptake rate (OUR) of these cells after being infected with AcNPV baculovirus increased to twice the OUR value of exponentially growing uninfected cells, reaching a maximum of 0.26 $\mu l$ $O_2$ per $10^6$ cells per min. In contrast to these measurements (and to the high OUR reported by other workers[37,38]) Maiorella et al.[39] reported that there was no difference between the oxygen demand of *Spodoptera frugiperda* cells before and after infection with AcNPV baculovirus ($4.3 \times 10^{-17}$ mol $O_2$ per cell per second and $4.6 \times 10^{-17}$ mol $O_2$ per cell per second, respectively), although these values, as expected, were about three times as high as for cells in stationary phase ($1.4 \times 10^{-17}$ mol $O_2$ per cell per second).

Further work by Hink's group[26,34] revealed that *Trichoplusia ni* cells could grow well in 2 to 3 l jar bioreactors of standard stirred tank design, equipped with marine impellers, under sparged aeration, provided 0.1 to 0.3% methylcellulose (or other viscosity-enhancing agents like Darvan™) was included in the medium. Under these conditions they reported routine attainment of final cell densities of $5 \times 10^6$ cells per ml in 5 d, with initial densities of $2 \times 10^5$ cells per ml, and a corresponding doubling time of only 14 h during exponential growth. Maintaining growth comparable with 100 ml spinners (surface aeration) in the scaled-up 2 to 3 l stirred tanks could be achieved by increasing the aeration rate from 5 $cm^3$/min (for the spinners) to 750 $cm^3$/min (after an initial operation at 125 $cm^3$/min in the first 2 days) and the impeller speed from 100 rpm (for the spinner) to 220 rpm. Operating under these apparently aggressive conditions caused cessation of growth and physical damage to the cells, possibly due to combined shear effects from both agitation and aeration, while excessive foaming was also reported. Cell damage was prevented under these conditions by increasing the methylcellulose content to 0.3% for further enhancement of medium viscosity, and growth rate was further enhanced by adding a silicone-based antifoam emulsion, possibly due to better gas mass transfer.

Dissolved oxygen (D.O.) level in the culture medium is an important control variable in many aerobic fermentations. Hink and his team[26,34] found that D.O. also affected the growth of insect cells in the stirred fermentors. When D.O. was allowed to drop freely from an initial 100% to a 2% level after 24 h, the cells stopped growing, while continuous maintenance of D.O. at 100% caused vacuolation and precipitate formation after the cell concentration reached a plateau. Also a 15% D.O. constantly maintained caused cell vacuolation at 120 h, and cell concentration fell rapidly beyond this point. A 50% D.O. level controlled throughout the batch allowed the same growth as 100% D.O., and it was routinely adopted for cell growth and virus production. In recent work with Sf9 cells in laboratory stirred and sparged fermentors, Jain et al.[42] found also that cell growth, baculovirus infection and recombinant protein (antistasin) production were adversely affected by very low (10%) or very high (110%) D.O. levels.

Upon increases in scale of operation, the oxygen requirements of the insect cells may not be satisfied by further increases in agitation and aeration rates without at the same time compromising the cells' growth and functional integrity in the enhanced shear field thus generated. Workers in Germany [27,58] succeeded in culturing lepidopteran cell lines at even higher scales, by using 10-l jar fermentors. Miltenburger and David[27] were able to meet the oxygen demand of the insect cells by introducing a semipermeable silicone rubber tube system for aeration by diffusion in the 10-l vessel. In this way, it was reasoned, the fragile insect cells were not exposed to the shear forces generated by direct interaction with swarms of air bubbles. An added advantage of this system was that foaming was avoided, and thus also the usual problems of cell entrainment and flotation, which are lethal to the cells. This system was developed further by Eberhardt and Schugerl,[64] in anticipation of the limitation of useful tube length for medium oxygenation upon further scale increase. An increase of inlet air pressure from 1.0 to 1.5 bar resulted in a 50% increase in cell yield of *Spodoptera frugiperda* and in considerable enhancement of growth rate. At a low agitation speed (35 rpm), mixing was inefficient and a portion of the cells settled at the bottom of the bioreactor (mass transfer limitation), whereas at high agitation (75 and 100 rpm) a limit in cell yield was reached, due to the adverse effects of higher shear forces. For intermediate agitation (50 rpm) and upon increasing the oxygen transfer rate (OTR) by either introducing pure oxygen in the immersed tubing or using air both through the tubing and over the free fluid surface, final cell density was enhanced to 4 to 5 $\times$ $10^6$ cells per ml, i.e., to values routinely obtained in 50 ml spinner flasks. Tramper and co-workers[36] assessed systematically the shear stress experienced by *Spodoptera frugiperda* cultivated in a 1-l round bottom bioreactor equipped with marine impeller (medium containing 0.1% methylcellulose), by varying agitation and aeration rates and measuring the kinetics of cell growth and cell death. In two different runs, viable cell decrease started at 220 rpm (corresponding to a shear stress of 1.5 N $\cdot$ m$^{-2}$) and at 510 rpm (corresponding to a shear stress of 3 N $\cdot$ m$^{-2}$), respectively. In the same work,[36] Tramper et al. confirmed that the critical shear stress at which cell viability declined was between 1 to 4 N $\cdot$ m$^{-2}$, by employing a rotary viscometer. Our own work with a cell line from *Aedes albopictus* (mosquito) in spinner flasks and a 1-l stirred fermentor indicated that the cells were sensitive to agitation rates corresponding to shear stresses above 1 N $\cdot$ m$^{-2}$. In sum, there is an urgent need for more systematic studies that would address the useful range of operating parameters, such as agitator speeds, in response to accurate evaluations of the effects of hydrodynamic shear on insect cells in culture. Moreover, a better theoretical understanding of the basis for the remedies (e.g., thickening supplements) proposed to counteract the effects of shear will increase our capacity to utilize conventional equipment with minimal modifications for large scale insect cell cultivation.

## 2. Effects of Air Sparging (Bubble Columns and Airlift Bioreactors)

As is apparent from the foregoing discussion, it is difficult to separate the adverse effects on the cells due to agitation-induced shear and due to air sparging-induced shear in bioreactor vessels that are both stirred and sparged. Nonetheless, the effects of air sparging can be observed and quantified in experiments utilizing bubble column or airlift reactors with no moving internals. Tramper et al.[36] used a bubble column reactor (18 cm height and 3.5 cm diameter) to establish quantitatively the effect of air sparging on cell viability. From these experiments they found that a first-order death rate constant for these insect cells was proportional to the air flow rate. On the basis of these findings, and given their repeatedly unsuccessful efforts to grow the cells in an airlift reactor, (an experience shared by our group for mosquito cell growth[65]) Tramper et al.[36] recommended alternative bioreactor designs for scalable oxygen supply, such as membrane bioreactors (i.e., a separate vessel for oxygenation of medium that is to be supplied to the cell growth chamber) and oxygen supply through semipermeable tubing, as described by Eberhardt and Schugerl.[64] Subsequent work by Tramper and co-workers[66,67] extended the findings of their bubble column experiments[36] to a rational bubble column design approach for growth of fragile insect cells. Since measurable loss of viability was seen to occur in the course of bubble rise through the columnar reactor and in the region of bubble bursting at the liquid surface, a "killing volume" hypothesis was advanced, to formulate an explicit equation for the first-order death rate constant and for the minimum specific surface area of the air bubbles to supply sufficient oxygen.[66,67] The "killing volume" represents the hypothetical volume associated with each rising air bubble during its lifetime, in which (volume) all viable cells are killed. The experimental findings of adverse effects on insect cells in directly sparged vessels have been seen also with mammalian cells in sparged systems[68,69] and have been explained in terms of the damage generated by the shear forces acting on the cells in the region of bubble disengagement from the free liquid surface.[68,69] For example, Handa-Corrigan et al.[69] found that higher bubble column bioreactors ensure better cell growth. The same explanations underly also the reported failure of Koike and Sato[60] to grow *Mamestra brassicae* cell lines on their serum-free medium No. 8 in an airlift reactor (1l), in contrast to their successful cultivation of the same cell lines in the same medium in a 40-l agitated jar fermentor (marine impeller, 150 rpm). Wudtke and Schugerl[40] confirmed the importance of bubble disengagement for the loss of insect cell viability in sparged cultivation systems by showing that a bubble column reactor in which the free suspension surface was covered with paraffin oil to prevent bubble bursting generated significantly less cell debris and maintained higher cell viability over a longer time, compared to a control system.

In a significant development in commercial insect cell culture scale-up, Maiorella et al.[39] were able to circumvent the bad prognosis that had been established for airlift reactors with direct air sparging. These workers[39] exploited the inherent scalability of the airlift bioreactor (OTR increase with increased vessel volume)[70] by (1) controlling the air flow stream and its composition, (2) controlling air bubble diameter at 0.5 to 1 cm by judicious choice of sparger orifice diameter (at this size range bubbles do not coalesce and, unlike smaller bubbles, have lower surface tension), and (3) including in the medium Pluronic F-68 polyol, which is known to protect mammalian[68,71] and insect[72] cultured cells from damage due to mechanical agitation as well as to air sparging. In this way, Maiorella et al.[39] were able to scale up the growth of *Spodoptera frugiperda* and the production of recombinant macrophage colony stimulating factor following infection with a modified AcNPV viral vector from a 3-l spinner flask to a 21-l airlift reactor. The protective effects of Pluronic F-68 for insect cells against shear-induced damage due to agitation and/or sparging have been recently demonstrated by Murhammer and Goochee,[72] who reported the polymer's usefulness in both a 3-l agitated and sparged tank bioreactor and in a 670-ml airlift bioreactor for the cultivation of *Spodoptera frugiperda* and the production of recombinant β-galactosidase using a modified AcNPV baculovirus vector. No satisfactory mechanism of the protective effect of Pluronic polyols has been demonstrated. Moreover, there are recent reports[42,61] of experimental insect cell cultivation on a pilot scale (up to 40-l airlift reactors and sparged agitated fermentors of several liters), where air sparging is not deemed problematic. Again, as stated previously, a larger body of information must be accumulated before the relevant mechanisms, as well as the reasons for apparently conflicting reports, are elucidated.

## D. BIOREACTOR GROWTH OF INSECT CELLS IN ATTACHED CULTURE

The quest for high bioreactor productivity can be approached by achieving high cell densities with surface growth dependent cells. Weiss and Vaughn[73] provide an extensive review of attached cell culture systems, that have been proposed and used for large-scale production of baculoviruses. Most notable among them are banks of roller bottles,[24,28] with growth surfaces ranging from 490 cm$^2$ to 1750 cm$^2$ (corresponding to media volumes from 100 ml to 250 ml, respectively) rotating at 1 revolution every 8.5 min. Despite the initial labor required for this setup, remarkably reproducible growth was achieved, and some of the highest specific growth rates (doubling times 8.35 to 10.22 h) were seen for the *Spodoptera frugiperda* cell line cultures. Weiss and Vaughn[73] have also evaluated a surface growth-dependent perfusion system, the Dyna Cell Propagator or bulk culture vessel, for continuous growth of *Spodoptera frugiperda* cells. This system consists of a 1.7 l polystyrene bottle containing a spiral core film that provides 9500 cm$^2$ of internal growth surface for continuously perfused media. The scale-up roller bottles, bulk culture vessels, and similar surface growth-dependent systems is limited by inherent design difficulties in maintaining consistently high cell concentrations, since increases in reactor volume are not followed by commensurate increases in available growth surface (surface-to-volume ratio diminished with scale). This substantial bottleneck has been addressed in mammalian cell culture technology by the development of microcarriers for the propagation of anchorage-dependent cells.[74] A promising recent development of attached growth insect cell culture was reported by Shuler and co-workers,[49] who could culture an attachment-dependent strain of *Trichoplusia ni* in a packed bed configuration using 3-mm glass beads as the packing material of their columnar bioreactor and allowing separately oxygenated medium to circulate through the biocatalyst bed. In this design, aeration is decoupled from the main bioreactor and, hence, both oxygen supply and hydrodynamic shear are not strongly dependent upon bioreactor scale. This provides for easy scale up. The same group showed that in this attached growth bioreactor system, the *Trichoplusia ni* cells could reach high levels of cell density and viability that allowed the production of up to 33% of their total protein content in the form of heterologous β-galactosidase, after being infected with a modified AcNPV vector.[49] Another modern approach to achieving high cell densities and, hence, high bioreactor productivities, while at the same time protecting shear-sensitive animal cells from detrimental mechanical stress, is cell immobilization and especially, entrapment.[75,76] Recent results by King et al.[77] indicate the attainment of $8 \times 10^7$ cells per ml matrix densities when they entrapped in semipermeable microcapsules *Spodoptera frugiperda* cells infected with a temperature sensitive baculovirus. However, the biocompatibility between the cells and the matrix was compromised when sodium alginate, a matrix component, exceeded a concentration of 0.8%.[78] The advent of new cell lines with higher infectivity coupled with stronger attachment dependence bodes well for the industrial application of attached insect cell culture.

## IV. CONCLUSIONS AND FUTURE PROSPECTS

In the previously reviewed developments of large-scale insect cell culture, valuable lessons from experience with mammalian cell technology are adapted to the peculiarities of insect cells in culture. The twin and interconnected factors of hydrodynamic stress and oxygen supply are likely to dominate among the bioprocess engineering considerations for production-level systems. Thus, the design of insect cell bioreactor systems should be based as much as possible on (a) increased basic understanding of insect cell nutrition and physiology, and (b) quantitative kinetic and mass transfer descriptions of insect cells in suspension and in attached or immobilized culture. Current efforts underway to address the latter point are illustrated by work that attempts to elucidate the mechanism(s) of physiological effects, such as cell damage and death, due to the hydrodynamics of turbulent flow.[66,69,79,80] Notable in this respect is the suggestion that cell damage depends functionally on the bioreactor power input per unit fluid mass and on the viscosity of the medium, as components of the characteristic length of turbulent eddies (Kolmogorov scale shear stress).[79,80] According to this theory, cell damage starts to occur when the characteristic eddy length reaches a value smaller than the diameter of the cells (or that of cell aggregates or of cell supports, such as microcarriers, depending on the culture configuration). It can be shown that increased medium viscosity increases the characteristic eddy length, and, therefore, the turbulent shear stress is less damaging to the cells.[79,80] Fundamental studies of this type should not only bring about satisfactory explanations for empirically found ways that circumvent the undesirable physiological effects in animal cell reactors (e.g., use of thickening agents, modifications of oxygenation systems, etc.), but also result in fresh approaches to optimal design, operation and performance of many bioprocessing systems.

Beyond these efforts in state-of-the-art biochemical engineering, parallel progress in the development of genetically engineered baculoviruses is anticipated to play a leading role in the expanded use of these insecticides in the foreseeable future. Miller[11,81] has proposed to expand the effective host range of the AcNPV virus as a pesticide by inserting an insect-specific neurotoxin (or behavior-modifying) gene into the baculovirus genome so that the gene is under the control of a promoter that is expressed early on virus entry in the cell and to retain polyhedrin expression so that occluded viruses (PIB) are produced. Very recent work by O'Reilly and Miller[82] has identified a viral gene, *egt*, of AcNPV (encoding an ecdysteroid UDP-glucosyl transferase). This is the first identification of an insect virus gene which can interfere with normal host insect development by blocking molting in infected larvae of *Spodoptera frugiperda*, and opens up a valuable avenue for tailor-made genetically engineered bioinsecticides with higher effectiveness than wild type NPVs. With a similar rationale, Bishop's group[83,84] has launched a cautious yet innovative program of research with AcNPV that has been genetically modified to alter its host range, persistence in soil, and susceptibility to UV radiation, in anticipation of the safeguards required for release of recombinant organisms in the environment. After the 1986 release of a genetically marked AcNPV in a field test in the United Kingdom to monitor the persistence and spread of the virus in the environment,[83] an AcNPV-based insecticide lacking the polyhedron structure was designed and shown that it could self-destruct after release in the field,[84] thus reducing the likelihood of unwanted spread. Finally, the different genes expressing β-galactosidase, *Bacillus thuringiensis* entomocidal delta-endotoxin and insect juvenile hormone esterase have been introduced into the genome of the AcNPV and the constructs analyzed as potential new-generation insecticides.[84]

# ACKNOWLEDGMENTS

I thank K. Venkat for his interest and contributions during the ongoing studies of bioreactor insect cell cultivation in my laboratory. These studies have been initiated through a Rutgers University Research Council grant and through an institutional Biomedical Research Support Grant (BRSG) from the U.S. Public Health Service. I extend my warm appreciation to Karl Maramorosch for his encouragement and support, and to Helene Agathos for her excellent typing of this manuscript.

# REFERENCES

1. **Burges, H. D.,** *Microbial Control of Pests and Plant Diseases 1970-1980,* Academic Press, New York, 1981.
2. **Kurstak, E.,** *Microbial and Viral Pesticides,* Marcel Dekker, New York, 1982.
3. **Maramorosch, K. and Sherman, K. E.,** *Viral Insecticides for Biological Control,* Academic Press, New York, 1985.
4. **Tinsley, T. W. and Kelly, D. C.,** Taxonomy and nomenclature of insect pathogenic viruses, in *Viral Insecticides for Biological Control,* Maramorosch, K. and Sherman, K. E., Eds., Academic Press, New York, 1985, 3.
5. **Granados, R. R., Dwyer, K. G., and Derksen, A. C. G.,** Production of viral agents in invertebrate cell cultures, in *Biotechnology in Invertebrate Pathology and Cell Culture,* Maramorosch, K., Ed., Academic Press, New York, 1987, 167.
6. **Podgwaite, J. D.,** Strategies for field use of baculoviruses, in *Viral Insecticides for Biological Control,* Maramorosch, K. and Sherman, K. E., Eds., Academic Press, New York, 1985, 775.
7. **Granados, R. R. and Hashimoto, Y.,** Infectivity of baculoviruses to cultured cells, in *Invertebrate Cell System Applications,* Vol. 2, Mitsuhashi, J., Ed., CRC Press, Boca Raton, FL, 1989, 3.
8. **Bohmfalk, G. T.,** Practical factors influencing the utilization of baculoviruses as pesticides, in *The Biology of Baculoviruses,* Vol. 2, Granados, R. R. and Federici, B. A., Eds., CRC Press, Boca Raton, FL, 1986, 223.
9. **Huber, J.,** Use of baculoviruses in pest management programs, in *The Biology of Baculoviruses,* Vol. 2, Granados, R. R. and Federici, B. A., Eds., CRC Press, Boca Raton, FL, 1986, 181.
10. **Luckow, V. E. and Summers, M. D.,** Trends in the development of baculovirus expression vectors, *Biol Technology,* 6, 48, 1988.
11. **Miller, L. K.,** Baculoviruses as gene expression vectors, *Annu. Rev. Microbiol.,* 42, 177, 1988.

12. **Caltagirone, L. E.,** Landmark examples in classical biological control, *Annu. Rev. Entomol.,* 26, 213, 1981.

13. **Shieh, T. R. and Bohmfalk, G. T.,** Production and efficacy of baculoviruses, *Biotechnol. Bioeng.,* 22, 1357, 1980.

14. **Shapiro, M.,** *In vivo* mass production of insect viruses for use as pesticides, in *Microbial and Viral Pesticides,* Kurstak, E., Ed., Marcel Dekker, New York, 1982, 463.

15. **Sherman, K. E.,** Considerations in the large-scale and commercial production of viral insecticides, in *Viral Insecticides for Biological Control,* Maramorosch, K. and Sherman, K. E., Eds., Academic Press, New York, 1985, 757.

16. **Shapiro, M.,** *In vivo* production of baculoviruses, in *The Biology of Baculoviruses,* Vol. 2, Granados, R. R. and Federici, B. A., Eds., CRC Press, Boca Raton, FL, 1986, 31.

17. **Wyatt, S. S.,** Culture *in vitro* of tissue from the silkworm, *Bombyx mori* L., *J. Gen. Physiol.,* 39, 841, 1956.

18. **Grace, T. D. C.,** Establishment of four strains of cells from insect tissues grown *in vitro, Nature (London),* 195, 788, 1962.

19. **Mitsuhashi, J. and Maramorosch, K.,** Leafhopper tissue culture: embryonic, nymphal, and imaginal tissues from aseptic insects, *Contrib. Boyce Thompson Inst.,* 22, 435, 1964.

20. **Hink, W. F.,** Established insect cell line from the cabbage looper, *Trichoplusia ni, Nature,* 226, 466, 1970.

21. **Vaughn, J. L., Goodwin, R. H., Tompkins, G. J., and McCauley, P.,** The establishment of two cell lines from the insect *Spodoptera frugiperda* (Lepidoptera; Noctuidae), *In vitro,* 13, 213, 1977.

22. **Granados, R. R.,** Infection and replication of insect pathogenic viruses in tissue culture, *Adv. Virus Res.,* 20, 189, 1976.

23. **Marks, E. P.,** Insect tissue culture: An overview, 1971-1978, *Annu. Rev. Entomol.,* 25, 73, 1980.

24. **Vaughn, J. L.,** The production of nuclear polyhedrosis viruses in large-volume cell cultures, *J. Invertebr. Pathol.,* 28, 233, 1976.

25. **Hink, W. F. and Strauss, E. M.,** Growth of the *Trichoplusia ni* (TN-368) cell line in suspension culture, in *Invertebrate Tissue Culture: Applications in Medicine, Biology and Agriculture,* Kurstak, E. and Maramorosch, K., Eds., Academic Press, New York, 1976, 297.

26. **Hink, W. F. and Strauss, E. M.,** Semi-continuous culture of the TN-368 cell line in fermentors with virus production in harvested cells, in *Invertebrate Systems in Vitro,* Kurstak, E., Maramorosch, K., and Dubendorfer, A., Eds., Elsevier/North-Holland, Amsterdam, 1980, 27.

27. **Miltenburger, H. G. and David, P.,** Mass production of insect cells in suspension, *Dev. Biol. Stand.,* 46, 183, 1980.

28. **Weiss, S. A., Smith, G. C., Kalter, S. S., and Vaughn, J. L.,** Improved method for the production of insect cell cultures in large volumes, *In vitro,* 17, 495, 1981.

29. **Tolbert, W. R. and Feder, J.,** *Large-Scale Mammalian Cell Culture,* Academic Press, New York, 1985.

30. **Arathoon, W. R. and Birch, J. R.,** Large-scale culture in biotechnology, *Science,* 232, 1390, 1986.

31. **Lydersen, B. K.,** *Large Scale Cell Culture Technology,* Hanser Publishers, Munich/New York, 1987.

32. **Seaver, S. S.,** *Commercial Production of Monoclonal Antibodies,* Marcel Dekker, New York, 1987.

33. **Goodwin, R. H.,** Insect cell culture: improved media and methods for initiating attached cell lines from the Lepidoptera, *In vitro,* 11, 369, 1975.

34. **Hink, W. F.,** Production of *Autographa californica* nuclear polyhedrosis virus in cells from large-scale suspension cultures, in *Microbial and Viral Pesticides,* Kurstak, E., Ed., Marcel Dekker, New York, 1982, 493.

35. **Tyo, M. A. and Spier, R. E.,** Dense cultures of animal cells at the industrial scale, *Enzyme Microb. Technol.,* 9, 514, 1987.

36. **Tramper, J., Williams, J. B., Joustra, D., and Vlak, J. M.,** Shear sensitivity of insect cells in suspension, *Enzyme Microb. Technol.,* 8, 33, 1986.

37. **Stockdale, H. and Gardiner, G. R.,** Utilization of some sugars by a line of *Trichoplusia ni* cells, in *Invertebrate Tissue Culture: Applications in Medicine, Biology and Agriculture,* Kurstak, E. and Maramorosch, K., Eds., Academic Press, New York, 1976, 267.

38. **Weiss, S. A., Ort, T., Smith, G. C., Kalter, S. S., Vaughn, J. L., and Dougherty, E. M.,** Quantitative measurement of oxygen consumption in insect cell culture infected with polyhedrosis virus, *Biotechnol. Bioeng.,* 24, 1145, 1982.

39. **Maiorella, B.,** Large-scale insect cell-culture for recombinant protein production, *Bio/Technology,* 6, 1406, 1988.

40. **Wudtke, M. and Schugerl, K.,** Investigations of the influence of physical environment on the cultivation of animal cells, in *Modern Approaches to Animal Cell Technology,* Spier, R. E. and Griffiths, J. B., Eds., Butterworths, Kent, U.K., 1987, 297.

41. **Chalmers, J., Bae, Y., Brodkey, R., and Hink, W. F.,** Microscopic observations of cells subjected to various levels of shear stress, presented at the American Institute of Chemical Engineers (AIChE) Annu. Meet., Washington, D.C., Nov. 27 to Dec. 2, 1988.

42. **Jain, D., Gould, S., Seamans, C., Wang, S., Lenny, A., and Silberklang, M.,** Production of antistasin using the baculovirus expression vector system, presented at the 198th Natl. Meet. American Chemical Society (ACS), Miami Beach, FL, Sept. 10 to 15, 1989.

43. **Hink, W. F.,** The 1979 compilation of invertebrate cell lines and culture media, in *Invertebrate Systems in Vitro,* Kurstak, E., Maramorosch, K., and Dubendorfer, A., Eds., Elsevier/North-Holland, Amsterdam, 1980, 553.

44. **Hink, W. F. and Hall, R. L.,** Recently established invertebrate cell lines, in *Invertebrate Cell System Applications,* Vol. 2, Mitsuhashi, J., Ed., CRC Press, Boca Raton, FL, 1989, 269.

45. **Granados, R. R., Derksen, A. C. G., and Dwyer, K. G.,** Replication of the *Trichoplusia ni* granulosis and nuclear polyhedrosis viruses in cell cultures, *Virology,* 152, 472, 1986.

46. **Corsaro, B. G. and Fraser, M. J.,** Characterization of clonal populations of the *Heliothis zea* cell line IPLB-HZ 1075, *In vitro,* 23, 855, 1987.

47. **Miltenburger, H. G., Naser, W. L., and Harvey, J. P.,** The cellular substrate: A very important requirement for baculovirus *in vitro* replication, *Z. Naturforsch., C: Biosci.,* 39C, 993, 1984.

48. **Miltenburger, H. G., Naser, W. L., and Schliermann, M. G.,** Establishment of a lepidopteran hybrid cell line by use of a biochemical blocking method *in vitro, Cell Dev. Biol.,* 21, 433, 1985.

49. **Shuler, M. L., Cho, T., Wickham, T., Ogonah, O., Kool, M., Hammer, D. A., Granados, R. R., and Wood, H. A.,** Bioreactor development for production of viral pesticides or heterologous proteins in insect cell cultures, *Ann. N.Y. Acad. Sci.,* 589, 399, 1990.

50. **Mitsuhashi, J.,** Media for insect cell cultures, in *Advances in Cell Culture,* Vol. 2, Maramorosch, K., Ed., Academic Press, New York, 1982, 133.

51. **Mitsuhashi, J.,** Simplification of media and utilization of sugars by insect cells in cultures, in *Invertebrate and Fish Tissue Culture,* Kuroda, Y., Kurstak, E., and Maramorosch, K., Eds., Springer-Verlag, New York, 1988, 15.

52. **Hink, W. F., Ralph, D. A., and Joplin, K. H.,** Metabolism and characterization of insect cell cultures, in *Comprehensive Insect Physiology, Biochemistry, and Pharmacology,* Vol. 10, Biochemistry, Kerkut, G. A. and Gilbert, L. L., Eds., Pergamon Press, New York, 1985, 547.

53. **Mitsuhashi, J.,** Nutritional requirements of insect cells *in vitro,* in *Invertebrate Cell System Applications,* Vol. 1, Mitsuhashi, J., Ed., CRC Press, Boca Raton, FL, 1989, 3.

54. **Gardiner, G. R. and Stockdale, H.,** Two tissue culture media for production of lepidopteran cells and nuclear polyhedrosis viruses, *J. Invertebr. Pathol.,* 25, 363, 1975.

55. **Hink, W. F., Strauss, E. M., and Lynn, D. E.,** Growth of TN-368 insect cells in serum-free media, *In vitro,* 13, 177, 1977.

56. **Vail, P. V., Jay, D. L., and Romine, C. L.,** Replication of the *Autographa californica* nuclear polyhedrosis virus in insect cell lines grown in modified media, *J. Invertebr. Pathol.,* 28, 263, 1976.

57. **Goodwin, R. H. and Adams, J. R.,** Nutrient factors influencing viral replication in serum-free insect cell line culture, in *Invertebrate Systems in Vitro,* Kurstak, E., Maramorosch, K., and Dubendorfer, A., Eds., Elsevier/North-Holland, Amsterdam, 1980, 493.

58. **Roder, A.,** Development of a serum-free medium for cultivation of insect cells, *Naturwissenschaften,* 69, 92, 1982.

59. **Mitsuhashi, J.,** Continuous cultures of insect cell lines in media free of sera, *Appl. Entomol. Zool.,* 17, 575, 1982.

60. **Koike, M. and Sato, K.,** Culture of insect cell lines originated from *Mamestra brassicae* with autoclaved serum-free medium, in *Invertebrate and Fish Tissue Culture,* Kuroda, Y., Kurstak, E., and Maramorosch, K., Eds., Springer-Verlag, New York, 1988, 7.

61. **Weiss, S. A., Belisle, B. W., Chiarello, R. H., DeGiovanni, A. M., Godwin, G. P., and Kohler, J. P.,** Large scale cultivation of insect cells, presented at the Conf. Biotechnology, Biological Pesticides and Novel Plant-Pest Resistance for Insect Pest Management, Boyce Thompson Institute, Cornell University, Ithaca, NY, July 18 to 20, 1988.

62. **Wilkie, G. E. I., Stockdale, H., and Pirt, S. V.,** Chemically-defined media for production of insect cells and virus *in vitro, Dev. Biol. Stand.,* 46, 29, 1980.

63. **Streett, D. A. and Hink, W. F.,** Oxygen consumption of *Trichoplusia ni* (TN-368) insect cell line infected with *Autographa californica* nuclear polyhedrosis virus, *J. Invertebr. Pathol.,* 32, 112, 1978.

64. **Eberhardt, U. and Schugerl, K.,** Investigation of reactors for insect cell culture, *Dev. Biol. Stand.,* 66, 325, 1987.

65. **Agathos, S. N., Jeong, Y.-H., Fallon, A. M., and Venkatasubramanian, K.,** Design considerations of insect cell bioreactor systems, presented at the American Institute of Chemical Engineers (AIChE) Annu. Meet., New York, November 15 to 20, 1987.

66. **Tramper, J., Smit, D., Straatman, J., and Vlak, J. M.,** Bubble-column design for growth of fragile insect cells, *Bioprocess Eng.,* 3, 37, 1988.
67. **Tramper, J., van der End, E. J., de Gooijer, C. D., Kompier, R., van Lier, F. L. J., Usmany, M., and Vlak, J. M.,** Production of baculovirus in a continuous insect-cell culture: bioreactor design, operation, and modeling, *Ann. N.Y. Acad. Sci.,* 589, 423, 1990.
68. **Handa, A., Emery, A. N., and Spier, R. E.,** On the evaluation of gas-liquid interfacial effects on hybridoma viability in bubble column bioreactors, *Dev. Biol. Stand.,* 66, 241, 1987.
69. **Handa-Corrigan, A., Emery, A. N., and Spier, R. E.,** Effect of gas-liquid interfaces on the growth of suspended mammalian cells: mechanisms of cell damage by bubbles, *Enzyme Microb. Technol.,* 11, 230, 1989.
70. **Lambert, K. J., Boraston, R., Thompson, P. W., and Birch, J. R.,** Production of monoclonal antibodies using large scale cell culture, *Dev. Ind. Microbiol.,* 27, 101, 1987.
71. **Mizrahi, A.,** Pluronic polyols in human lymphocyte cell line cultures, *J. Clin. Microbiol.,* 2, 11, 1975.
72. **Murhammer, D. W. and Goochee, C. F.,** Scaleup of insect cell cultures: protective effects of Pluronic F-68, *Bio/Technology,* 6, 1411, 1988.
73. **Weiss, S. A. and Vaughn, J. L.,** Cell culture methods for large-scale propagation of baculoviruses, in *The Biology of Baculoviruses,* Vol. 2, Granados, R. R. and Federici, B. A., Eds., CRC Press, Boca Raton, FL, 1986, 63.
74. **Fleischaker, R.,** Microcarrier cell culture, in *Large Scale Cell Culture Technology,* Lydersen, B. K., Ed., Hanser Publishers, Munich, 1987, 59.
75. **Nilsson, K.,** Methods for immobilizing animal cells, *Trends Biotechnol.,* 5, 73, 1987.
76. **Dean, R. C., Karkare, S., Ray, N. G., Runstadler, P. W., and Venkatasubramanian, K.,** Large scale culture of hybridoma and mammalian cells in fluidized bed bioreactors, *Ann. N.Y. Acad. Sci.,* 506, 129, 1987.
77. **King, G. A., Daugulis, A. J., Faulkner, P., and Goosen, M. F. A.,** Growth of baculovirus-infected cells in microcapsules to a high cell and virus density, *Biotechnol. Lett.,* 10, 683, 1988.
78. **King, G. A., Daugulis, A. J., Goosen, M. F. A., Faulkner, P., and Bayly, D.,** Alginate concentration: A key factor in growth of temperature-sensitive baculovirus-infected insect cells in microcapsules, *Biotechnol. Bioeng.,* 34, 1085, 1989.
79. **Cherry, R. S. and Papoutsakis, E. T.,** Hydrodynamic effects on cells in agitated tissue culture reactors, *Bioprocess Eng.,* 1, 29, 1986.
80. **Croughan, M. S., Hamel, J. F., and Wang, D. I. C.,** Hydrodynamic effects on animal cells grown in microcarrier cultures, *Biotechnol. Bioeng.,* 29, 130, 1987.
81. **Miller, L. K.,** Expression of foreign genes in insect cells, in *Biotechnology in Invertebrate Pathology and Cell Culture,* Maramorosch, K., Ed., Academic Press, New York, 1987, 295.
82. **O'Reilly, D. R. and Miller, L. K.,** A baculovirus blocks insect molting by producing ecdysteroid UDP-glucosyl transferase, *Science,* 245, 1110, 1989.
83. **Bishop, D. H. L.,** UK release of genetically marked virus, *Nature (London),* 323, 496, 1986.
84. **Possee, R. D. and Bishop, D. H. L.,** Development and release of genetically engineered viral insecticides: A progress report 1986-1989 (Abstr.), presented at Int. Symp. Molecular Insect Science, Tucson, AZ, October 1989, 84.

*Regulatory and Environmental Aspects*

Chapter 16

# FEDERAL REGULATION OF BIOTECHNOLOGY: JURISDICTION OF THE U.S. DEPARTMENT OF AGRICULTURE

**Sivramiah Shantharam and Arnold S. Foudlin***

## TABLE OF CONTENTS

*  Disclaimer: The views and opinions expressed in this article are those of the authors alone and do not necessarily represent the regulatory and administrative policies of USDA-APHIS.

*Science is an international enterprise; its practitioners and those who benefit from the knowledge it creates are located throughout the world. The knowledge, which translates into standards, can be used for good or evil, can be distributed fairly or unfairly. The challenge to society is to use those discoveries for the betterment of all.*

Koshland, 1989[1]

# I. INTRODUCTION

The second green revolution is said to have been heralded by the advent of biotechnology. In the modern sense, "biotechnology" is a collection of laboratory techniques such as tissue culture, cell culture, organ culture, somaclonal variation, somatic hybridization, monoclonal antibodies, and techniques of molecular cloning, popularly known as recombinant DNA technology. The application of these techniques has caused a revolution in altering the functioning of biological organisms. Biotechnology has found ready application in the genetic engineering of plants/crops, microbes, and animals. The products of biotechnology in the areas of microbiology and therapeutic agents for animal and human health care have already entered the market place. However, genetically engineered plants are still in the field-testing stage.

Regulations governing a new technology such as biotechnology are imperative in an environmentally conscientious society. The introduction of new technology and its product(s) are bound to be viewed with a great deal of skepticism. The public wants to know every aspect of the impact of products of biotechnology on the environment before they are mass produced and made available to the public. Historically, the development of prudent policies and regulations governing new technologies has been a difficult and daunting process. Particularly, the federal government must present a balanced view by accommodating different opinions. Regulations, when properly drafted and administered, can be a potent catalyst for technology transfer. This is evidenced by the number of documented field trials with genetically engineered organisms approved by the U.S. Department of Agriculture (USDA), Animal and Plant Health Inspection Service (APHIS) (Appendix 1). The number greatly exceeds the approvals granted in any other country in the world. It is a matter of pride that the federal government has facilitated the U.S. gain of the position of world leadership in the development of safe biotechnology by complementing the ingenuity and creativity of corporate and academic research establishments.

This article discusses USDA regulations governing biotechnology of plants, microbes and invertebrate animals (i.e., insects), and the scientific and legal basis for ascertaining the regulatory control.

# II. U.S. DEPARTMENT OF AGRICULTURE REGULATIONS

Since the implementation of a "Coordinated Framework for Regulation of Biotechnology" (51 Federal Register 23302-2350 (1986)) and of specific regulations adapting existing laws to cover potential risks posed by genetically engineered organisms (7 CFR Part 340 (1987)), the approval time for field test applications has been significantly reduced. This is contrary to the view held in some quarters that regulating biotechnology would be a time-consuming and laborious process that impedes research and development.

APHIS has broad regulatory authority under the Plant Quarantine Act (PQA, 1912)[2] and the Federal Plant Pest Act (FPPA, 1957)[2] to protect American agriculture from pests and predators. APHIS regulations enable the agency to make informed decisions in which data necessary for risk identification, evaluation, and management are reviewed. They also provide for state and federal coordination so that duplication can be avoided. The regulatory goals to which APHIS is committed are balanced, scientifically sound, risk-, but not process-based assessment of planned releases of genetically engineered organisms into the environment and one that is protective of American agriculture. They allow the U.S. to maintain leadership and competitiveness in the world and facilitate technology transfer. APHIS is also committed to maintaining flexibility to accommodate the changes in the rapidly developing field of biotechnology.

APHIS regulations can be found in 7 CFR Part 340 (1987), entitled "Introduction of Organisms and Products Altered or Produced through Genetic Engineering Which Are Plant Pests or Which There Is a Reason to Believe Are Plant Pests." These regulations have been promulgated by USDA under the authority of the FPPA[2] and PQA.[2] The responsibility of administering the regulations governing biotechnology rests with Biotechnology, Biologics, and Environmental Protection (BBEP), within APHIS.

A permit is required prior to introducing a regulated article into the environment, for importing, and for interstate movement. A genetically engineered organism is deemed a regulated article if the donor organism, recipient organism, vector, or vector agent used in modifying the organism comes from one of the genera listed in the regulations and is a plant pest. The administration of biotechnology regulations ensures a risk-based process of assessment for proposed field test by critically examining the nature of the genetically engineered organism and the biotic and abiotic elements of the immediate environment into which it is being released. Limited field releases are necessary to gather information for scientific evaluation of the efficacy of the genetic changes. The plants have been tested in the greenhouse to obtain initial data relating to the genetic stability of the plants and preliminary data on efficacy. It is normal for planned field tests to be performed after greenhouse testing to confirm the efficacy data, which can only be validated in the environment using standard agricultural practices. Such limited field testing is normally required to develop a potential agricultural product.

Under FPPA and PQA, *plant pest* is defined as any living state of insects, mites, nematodes, slugs, snails, protozoa, or other invertebrate animals; bacteria, fungi, other parasitic plants or reproductive parts thereof; viruses; or any organisms similar to or allied with any of the foregoing; or any infectious agents or substances, which can directly or indirectly injure or cause disease or damage in or to any processed, manufactured, or other products of plants (7 USD Part 150 aa (jj). Based on this definition and authority, a permit is required prior to introduction of a regulated article (importation, interstate movement, or environmental release) into the environment. *Environment* includes all bodies of water, land, and air (7 CFR Part 340.1, 1987). *Regulated article* is any organism which has been altered or produced through genetic engineering, if the donor organism, recipient organism, vector, or vector agent belongs to any generus or taxon designated in the list of regulated organisms in 7 CFR Part 340.2 (1987), and meets the definition of plant pest, or is an unclassified organism, and/or an organism whose classification is unknown, or any product which contains such an organism, or any other organism or product altered or produced through genetic engineering which is a plant pest or there is a reason to believe is a plant pest. APHIS does not simply consider that all organisms listed under 7 CFR 340.2 are plant pests. In specific instances, for example, members of the genera *Rhizobium, Bradyrhizobium,* and certain nonpathogenic strains of pseudomonads are not considered plant pests. APHIS, after careful study of the engineered organism(s) in question, has opined that they are not plant pests, and as such a permit for their release has not been required.

There is a provision to petition or to amend the list of regulated organisms (7 CFR 340.4, 1987). Exempted from the regulations are nonplant pest recombinant microorganisms containing well-characterized noncoding regulatory sequences from either prokaryotic or eukaryotic plant pests; naked DNA molecules; pest sequences of DNA carried in a sterile *Escherichia coli* K-12 background; or killed organisms. Examples of well-characterized noncoding regulatory regions are operators, promoters, origins of replication, terminators, and ribosome binding sites or regions.

*Genetic engineering* is the modification of organisms by recombinant DNA techniques (7 CFR 340.1, 1987). *Release into the environment* constitutes the use of regulated articles outside the constraints of physical confinement such as a laboratory, greenhouse, or a fermenter, or other contained structure (7 CFR 340.1, 1987). APHIS realizes that it is very difficult to define contained greenhouses. Therefore, decisions regarding greenhouses or other safety containments are made on a case-by-case review basis. *Vectors* or *vector agents* are organisms, molecules, or objects that are used to transfer genetic material from the donor organism to the recipient organism. APHIS, specifically, does not regulate laboratory-contained experiments involving recombinant DNA organisms.

## III. THE PERMIT PROCESS

Basically, the process involves preparation of an environmental assessment (EA), as prescribed by the National Environmental Policy Act (NEPA, 1986),[3] which results in either a "Finding of No Significant Impact" (FONSI) or a determination that an Environmental Impact Statement (EIS)[4] must be prepared.

Under NEPA, an EA or an EIS embodies risk evaluation and is considered a public document. The contents of the application materials deemed to be Confidential Business Information (CBI) will be protected to the extent provided by the Freedom of Information Act (FOIA, 5 U.S.C. 552[b][4]). CBI is generally defined as information for which disclosure would cause substantial competitive harm to a commercial entity from which the information was obtained. In the academic sphere, CBI may be successfully invoked to protect information which the institution is going to patent or to cover research competitiveness. CBI is described in the policy statement on the Protection of Privileged or Confidential Business Information (50 Fed. Reg. 38561 (1985)).

Because the genetically engineered organisms are released under physically and biologically well-contained conditions for experimental purposes only, usually EAs are sufficient. The EA document describes the conditions for the field testing and critically delineates the applicable rules and regulations and the different options available to the agency (i.e., APHIS). It examines the biology of the donor and the recipient organisms, the molecular biology of the genes, gene structure, gene expression, the stability of gene insertion, the vector DNA, the vector agent, the phytogeography, the zoogeography, the size of the plot, and the number of articles to be introduced. It also analyzes potential significant impacts, if any, on biotic and abiotic components of the environment. The geographical resolution to which the EA applies is the county level. If an official Finding of No Significant Impact (FONSI) is forthcoming, a permit will be issued for the proposed field test. However, the permit will contain a set of standard conditions and may contain a set of supplementary conditions which the applicant must follow.

An APHIS representative(s) visits the test site during the course of the experiment to ascertain adherence to protocols and conditions with respect to physical and biological safety. Applicants are usually asked to closely monitor the test plants for development of any unusual phenotypic expression and the field test site for germination of volunteer plants, usually for a year following the termination of the experiment.

This whole process of subjecting an application to the NEPA process is essentially a certification process to assure that there is no significant risk of introduction or dissemination of plant pests into the environment.[5] Federal agencies, including the U.S. Department of Agriculture, subscribe to the philosophy of the National Academy of Sciences (NAS) that neither the use of recombinant DNA (rDNA) techniques nor the transfer of genes between unrelated organisms is inherently hazardous. This philosophy was first presented in 1987 in the National Academy of Sciences publication, Introduction of Recombinant DNA Engineered Organisms into the Environment: Key Issues.[6] It states that the risks associated with the introduction of such organisms are the same as those associated with the introduction into the environment of unmodified organisms or organisms modified by other genetic techniques. This principle is further emphasized in the NAS recent study, Field Testing Genetically Modified Organisms: Framework for Decisions.

APHIS review of permit applications has basically focused on the risks to the environment presented by the rDNA-derived product to be introduced. In fact, APHIS allows field testing with organisms that are plant pests or contain genetic material from plant pests when the agency determines after making a finding of no significant impact, there are no chances for the introduction or dissemination of a plant pest.

## IV. NATIONAL COORDINATION OF BIOTECHNOLOGY REGULATIONS

APHIS strives to maintain an excellent working relationship with state authorities by continuously keeping them informed of its decisions and actions. The key element of this open communication is seeking active input from the state agencies on matters concerning the issuance of permit within their state. In fact, without a formal written concurrence from the State, APHIS does not issue a permit for field-testing of genetically engineered products in that state.

The "Coordinated Framework for Regulation of Biotechnology" (51 Fed. Reg. 23302-23350, 1986) contains policies of the federal agencies involved with the review of biotechnology products passing through various stages of research, development, and marketing. An index of laws applicable to biotechnology was published in the Federal Register on November 14, 1985. The issues that were examined before its publication ranged from long-term ecological impacts to the environment by the release of genetically engineered organisms, to warnings that unwarranted restraints on the biotechnology industry would impede technological progress and competitiveness. This document enumerated several policy imperatives of the federal agencies that share major responsibility for regulating the products of biotechnology, including those agencies that share responsibility for regulating at different stages of development and commercialization of genetically engineered plants. The agencies are the U.S. Department of Agriculture (USDA), the Food and Drug Administration (FDA), and the Environmental Protection Agency (EPA).

Inter- and intra-agency coordination has been the cornerstone for successful implementation of federal biotechnology regulations. The interagency coordination of policy issues was assigned to the Biotechnology Science Coordinating Committee (BSCC), now extinct, comprised of key policymakers from the USDA, FDA, and EPA. The coordination has been achieved at the administrative level in the U.S. Department of Agriculture through the activities of the Committee on Biotechnology in Agriculture (CBA). The Office of Agricultural Biotechnology (OAB) has been established as a focal point for the USDA in the development of policies and procedures for agricultural biotechnology research. OAB provides support for CBA and the Agricultural Biotechnology Research Advisory Committee (ABRAC). ABRAC was established by the U.S. Department of Agriculture to provide scientific advice on biotechnology matters, as required, to research and regulatory agencies, and to review research proposals and guidelines.

In 1984, an interagency working group was established by the White House to coordinate regulatory policies between the federal agencies. The working group was charged with reviewing laws and regulations applicable to biotechnology and determining their adequacy for regulating the products of new technologies. President Bush has an intrinsic interest in biotechnology, as mentioned in his fiscal 1991 budget message: "Biotechnology is a classic case of investing in the future. . . . In drugs, food, agriculture, waste management, and energy, biotechnology advances offer the possibility of improvements that will make a real difference in people's lives . . . Applying biotechnology may allow farmers to deal with pests and diseases in environmentally safe ways, thereby reducing dependence on chemical controls and enhancing the overall safety and quality of the food supply.''[8]

Within APHIS, BBEP as the biotechnology regulatory unit encourages biotechnology and facilitates technology transfer, while at the same time ensuring that new products do not threaten American agriculture, public health, or the environment. BBEP is dedicated to reviewing field tests and product applications for genetically engineered organisms on a case-by-case basis and to providing a thorough analysis of potential effects of these organisms on the environment.

APHIS has sponsored two national conferences with the States for better understanding of both federal and state regulations affecting biotechnology. North Carolina and Minnesota have promulgated their own biotechnology regulations including registration, with which the potential applicant will have to contend. Hawaii, Illinois, Oklahoma, and Wisconsin have state notification statutes which require that applicants notify them at about the same time the federal government is notified.[9]

## V. INTERNATIONAL HARMONIZATION

At the international level, APHIS' position is that efficient technology transfer and international commerce are facilitated via international harmonization of regulations. To that end, APHIS has participated in a number of international activities and has established networking with Canada, Mexico, the United Nations Educational, Scientific, and Cultural Oranization (UNESCO), the Organization for Economic and Cooperative Development (OECD), the European Economic Commission (EEC), the Food and Agricultural Organization (FAO), the World Health Organization (WHO), the United Nations Development Program (UNDP), the U.S. Agency for International Development (U.S. AID), the Third World Academy of Sciences (TWAS), and many other Third World organizations having similar concerns and interests. Collaboration with the OECD has fostered a true international cooperation and communication. The most significant achievement along this line is the development of a document on biotechnology safety entitled ''Good Development Practices for Small Scale Field Research'' (GDP), for OECD member countries.[10] GDP outlines basic scientific principles and conditions for determining the safety of small scale field testing with low or negligible risks. These principles have been drawn from a wealth of knowledge and experience gained from countless field trials in plant breeding, some involving genetically engineered crops plants, and hundreds of field experiments on microorganisms and biological control agents. It is expected that GDP will serve as the foundation document for the scientific basis on which to work toward international harmonization. It will also provide a basis for evaluating proposals from the international community.

## VI. INTRODUCTION OF BIOCONTROL AGENTS

The importation of an insect parasite (*Cotesia glomerata* L.) of the imported cabbage-worm (*Piperis rapae* L.) by the U.S. Department of Agriculture in 1883 heralded the introduction of exotic biocontrol agents.[2] There was an increase in importation of biocontrol agents between 1888 and 1889 which resulted in the highly successful control of cotton cushion scale (*Icerya puchasi* Maskell) on citrus in California, by the vedalia beetle (*Rordolia cardinalis* Mulsant) from Australia.[2]

In the early days of importation, there were no established regulations for the safe introduction of biocontrol agents. In recent years, procedures have evolved to ensure the safety of importations of biocontrol agents. There are provisions for initial receipt and study of exotic organisms under quarantine conditions, subsequent movement under a permit from

USDA, APHIS, and review of research data by interagency groups prior to release.[2] Currently, experimenters must comply with a complex array of federal and state laws such as the California Biological Control Programs of 1931; a three-way Memorandum of Understanding between USDA, California Department of Food and Agriculture, and the University of California; PQA FPPA; Public Health Service Act (PHSA) of 1944; FIFRA of EPA passed in 1942; and NEPA and the Endangered Species Act (ESA) of 1973.[2,3] In addition, there are other regulations promulgated by the Convention on International Trade in Endangered Species (CITES), as amended by Executive Order #11987, May 1977, on "exotic organisms," and under the Federal Food, Drug, and Cosmetic Act (FDCA) of FDA. States like California, Florida, North Carolina, Michigan, Wisconsin, Kansas, New York, and Utah have their own laws to regulate the entry of living organisms. Thus, there is no simple comprehensive piece of legislation that exclusively addresses importation and release of biological control agents. Therefore, the discussion below is offered to explain the principal authorities and procedures.

## VII. CURRENT PROCEDURES FOR BIOCONTROL AGENTS

APHIS has identified biological control as a high priority program area. One measure of its importance to the Agency is the implementation of the Biological Control Institute, an organization within APHIS which develops policy and carries out methods development to implement the increased use of biocontrol agents mandated in APHIS pest eradication and control programs. These programs at APHIS require the use of biocontrol organisms as alternatives to chemical pesticides.

Biocontrol organisms are now regulated by APHIS, as plant pests under the authority of the PPA in regulations codified in 7 CFR 330.200. These regulations have been in place since 1957.

The EPA also regulates many of the uses of biocontrol organisms. The Federal Insecticide, Fungicide, and Rodenticide Act (FIFRA) (7. U.S.C. § 136 *et seq.*) defines a pesticide as "any substance or mixture of substances intended for preventing, destroying, repelling, or mitigating any pest . . . ". The EPA has interpreted "substance or mixture of substances" to include organisms; for instance, the EPA requires microorganisms used in ways that are analogous to chemical pesticides (e.g., repeated applications) to be registered as pesticides under FIFRA. FIFRA Section 25(b) authorizes the EPA to exempt any pesticide which is adequately regulated by another federal agency from the requirements of FIFRA. Under this provision the EPA has specifically exempted "pest control organisms such as insect predators, nematodes, and macroscopic parasites" from registration under FIFRA (40 CFR 158.65 (b)(3)) because they are "adequately regulated" by APHIS.

The EPA has not required the registration review of microorganisms that are used as "inoculative agents," i.e., that are released in small populations and allowed to increase naturally in the environment. This traditional approach is under review because of EPA's policy statement on biopesticides. Currently, under a strict reading of the FIFRA regulations and the policy statement of EPA inoculative agents would need an Experimental Use Permit (EUP) before release. In their new draft regulations under discussion (yet to be formally proposed), the EPA is considering exempting nonindigenous biocontrol agents from review when they are applied to small areas of land. The rationale is to avoid duplication of reviews conducted by APHIS in accordance with FIFRA section 25(b).

## VIII. SAFETY PRECAUTIONS, GUIDELINES, AND PROTOCOLS

A Technical Advisory Group (TAG) on Biological Control of Weeds in APHIS, Plant Protection and Quarantine (PPQ), was established in January 1987, to replace the former interagency Working Group on Biological Control of Weeds. The TAG now includes a member from the National Plant Board.

The introduction of pathogens of weeds presents, at times, a "gray area" between the USDA and the EPA, under FIFRA. For example, a ruling by the Agricultural Research Service (ARS) in 1986, requested an EA for field testing of biocontrol agents (7 CFR Part 520, 521, Reg. 34190, 1986). Certain ARS guidelines require that biological control agents to be introduced or released should be authentically identified; imported into quarantine facilities under permit from APHIS, PPQ; available research data be reviewed by TAG, Canadian, and Mexican officials; released from quarantine by strictly following PPQ permit procedures; and all documentation maintained in proper order.

In 1982, the ARS established the Biological Control Documentation Center to record biocontrol agents that had been in use by the ARS in Beltsville, Maryland. A "Release of Beneficial Organisms in the United States and Territories" (ROBO)[2] database was developed to serve as a centralized repository of information for scientists, agricultural administrators, and regulatory officials; to provide information on collection, importation, or other movement of exotic polinators and biological control organisms, and to facilitate the exchange of taxonomic and nomenclatorial information. ROBO also helps in formal surveys, catalogs, and taxonomic revisions and reviews.

# IX. FUTURE TASKS

Several factors are driving the demand for the introduction into the U.S. of an ever-increasing number of additional biological control organisms. These factors include research toward developing systems for low input sustainable agriculture (LISA), integrated pest management, and organic farming. Biological control research traditionally has been carried out by the academic and governmental research communities, but increasingly it is being developed by entrepreneurs for profit.

Mounting criticisms of the overuse of chemical pesticides have compelled vigorous renewal of efforts to develop biological control agents. Developing biological control agents and studying their environmental impacts is timely, as the rapid development of biotechnology might create the possibility of potentially new risks from their introduction. The introduction of biological control agents has been touted as safe and risk free, but there may not be adequate data based on systematic study to sustain this argument. If the consequences of introduction of both classical and genetically engineered biocontrol organisms are predictable, then the time is ripe to ask the right questions concerning the biota and the environment. This knowledge should help to mitigate potential disasters. Survivability, persistence, host range, habitat range, and genetic plasticity are some of the key factors that will influence the effects of introduced biocontrol agents on nontarget organisms.[11] Environmental risks associated with the introduction of exotic biological control organisms are sufficient to justify the creation of new legal safeguards.[12]

Biotechnology provides some hitherto unforeseen possibilities for novel biocontrol applications by suitably modifying native species of pests. It may be preferable to introduce modified native species rather than modified alien species. It is believed that the risks to the environment from the introduction of genetically engineered biocontrol organisms are no different from the introduction of nonengineered ones. Assuming that there will be negative effects from the release of some novel genetically engineered organisms, appropriate questions must be developed to identify and document the negative effects that might occur. The development and proper enforcement of safety precautions will be necessary. Self-dispersing organisms need to be studied more carefully, as they persist longer and have potential long-term effects on the environment. Nontarget organisms should be identified, as they may be vulnerable to extinction.

With the expected increase in the number of regulatory actions that will be required, it is essential that the existing procedures for biocontrol agents be examined and specific new procedures be instituted, if necessary. However, any procedures specific to biological control organisms should principally facilitate the testing and use of biological control organisms, while protecting agriculture and the environment.

## ACKNOWLEDGMENTS

The authors would like to thank Dr. John Payne for useful discussions on biocontrol agents and critical review thereof; Jane Montgomery for the excellent editorial help, and Anita Walker for her excellent assistance in typing.

### APPENDIX I
### Permits Issued for Release into the Environment under 7 CFR 340 Animal and Plant Health Inspection Service, U.S. Department of Agriculture

| Number | Date issued | Company | Organism | Location |
|--------|-------------|---------|----------|----------|
| 87-229-01 | 11-25-87 | Calgene, Inc. | Bromoxynil-tolerant tobacco | AZ |
| 87-229-02 | 12-11-87 | Calgene, Inc. | Glyphosate-tolerant tobacco | AZ |
| 87-208-01 | 12-21-87 | Calgene, Inc. | Bromoxynil-tolerant tomato | CA |
| 000074 | 12-23-87 | Calgene, Inc. | Glyphosate-tolerant tomato | CA |
| 87-226-01 | 12-28-87 | Du Pont Co. | Sulfonylurea-tolerant tomato | DE |
| 87-331-01 | 3-22-88 | Du Pont Co. | Sulfonylurea-tolerant tomato | FL |
| 87-329-02 | 3-23-88 | Monsanto Co. | Lepidopteran insect-resistant tomato | FL |
| 87-329-01 | 3-23-88 | Monsanto Co. | *TMV-resistant tomato | FL |
| 88-011-01 | 4-25-88 | Monsanto Co. | Glyphosate-tolerant tomato | CA |
| 88-036-01 | 4-27-88 | Sandoz Crop Protection Corporation | Lepidopteran insect-resistant tobacco | NC |
| 88-054-01 | 4-28-88 | Sandoz Crop Protection Corporation | Sulfonylurea-tolerant tobacco | NC |
| 88-041-01 | 5-5-88 | Monsanto Co. | *TMV-resistant tomato | IL |
| 88-041-04 | 5-23-88 | Monsanto Co. | Lepidopteran insect-resistant tomato | IL |
| 88-041-07 | 5-23-88 | Monsanto Co. | Glyphosate-tolerant tomato | IL |
| 88-029-02 | 5-23-88 | Agrigenetics Advanced Science Co. | Lepidopteran insect-resistant tomato | WI |
| 88-028-01 | 5-24-88 | Agrigenetics Advanced Science Co. | **AMV-resistant tomato | WI |
| 87-355-01 | 5-25-88 | Crop Genetics International | *Clavibacter xyli* subsp. *cynodontis* expressing delta-endotoxin gene of *Bacillus thuringiensis* subsp. *kurstaki* in corn | MD |
| 88-027-03 | 6-6-88 | Iowa State University | Tobacco expressing chimeric proteinase inhibitor II promoter-chloramphenicol acetyl transferase gene | IA |
| 88-092-01 | 6-22-88 | Du Pont Co. | Sulfonylurea-tolerant tomato | DE |
| 88-091-01 | 7-28-88 | Du Pont Co. | Tobacco expressing gene-producing chitinase | DE |
| 88-236-01 | 12-14-88 | Calgene | Tomato containing antisense endopolygalacturonase enzyme gene | HI |
| 88-314-05 | 2-22-89 | Monsanto | Tomato expressing delta-endotoxin gene of *Bacillus thuringiensis* subsp. *kurstaki* | FL |
| 88-333-02 | 3-13-89 | Rohm & Haas | Tobacco expressing delta-endotoxin gene of *Bacillus thuringiensis* var. *berliner* | NC |
| 88-351-12 | 3-30-89 | Calgene | Glyphosate-tolerant tomato | CA |

## APPENDIX I (continued)
## Permits Issued for Release into the Environment under 7 CFR 340 Animal and Plant Health Inspection Service, U.S. Department of Agriculture

| Number | Date issued | Company | Organism | Location |
|--------|-------------|---------|----------|----------|
| 88-344-07 | 4-6-89 | Calgene | Tomato containing antisense endopolygalacturonase enzyme gene | CA |
| 88-351-13 | 4-13-89 | Agracetus | Cotton expressing delta-endotoxin gene of *Bacillus thuringiensis* sub. *kurstaki* | MS |
| 89-030-04 | 4-26-89 | Monsanto | Potato/Virus/X/Y and Potato leaf roll | IL |
| 89-047-04 | 4-26-89 | Monsanto | *Bt*-tolerant tomato | IL |
| 88-355-01 | 4-27-89 | CGI | *Cxc*/*Bt* in corn | IL/MD/MN/NE |
| 89-030-02 | 4-28-89 | Monsanto | *Bt*-tolerant tomato | CA |
| 89-030-03 | 5-3-89 | Monsanto | Virus X/Y-tolerant potato | ID |
| 89-034-10 | 5-3-89 | Monsanto | Glyphosate-tolerant cotton | AL |
| 89-034-11 | 5-4-89 | Monsanto | Glyphosate-tolerant soybeans | IL |
| 89-034-12 | 5-8-89 | Monsanto | Glyphosate-tolerant soybeans | AR |
| 89-034-15 | 5-8-89 | Monsanto | Glyphosate-tolerant soybeans | TN |
| 89-065-01 | 5-19-89 | University of Kentucky | Tobacco/mouse metallothionein | KY |
| 89-047-07 | 5-24-89 | Calgene | Bromoxynil-tolerant cotton | MS |
| 89-038-03 | 6-6-89 | Northrup King | Glufosinate-tolerant alfalfa | CA |
| 89-053-01 | 6-22-89 | CGI | *Cxc*/*Bt*/Rice | MD |
| 89-038-01 | 6-30-89 | Northrup King | Alfalfa/glufosinate | MN |
| 89-073-01 | 6-30-89 | Monsanto | Tomato/TmV/ToMV | IL |
| 89-097-01 | 6-30-89 | Iowa State | Tobacco/chloramphenicol acetyl transferase (CAT) gene | IA |
| 89-116-20 | 7-6-89 | Biotechnica | Tobacco/DHDPAS gene | WI |
| 89-074-01 | 7-13-89 | Calgene | Tobacco/BT/CpTi | CA |
| 89-109-03 | 7-28-89 | Iowa State | Poplar/CAT gene | IA |
| 89-136-01 | 8-11-89 | Pioneer | Alfalfa/AMV-resistant | IA |
| 89-136-04 | 8-14-89 | Calgene | Tobacco/glyphosate | CA |
| 89-172-01 | 8-14-89 | New York State Agricultural Experiment Station, Geneva | Cucumber/CMV | NY |
| 89-192-01 | 10-10-89 | Calgene | Cotton/bromoxynil | HI |
| 89-150-01 | 10-11-89 | Monsanto | Cotton/*Bt*/glyphosate | HI |
| 89-208-01 | 11-21-89 | Monsanto | Soybean/glyphosate | PR |
| 89-278-01 | 1-23-90 | Monsanto | Tomato/*Bt* | FL |
| 89-278-02 | 2-2-90 | Monsanto | Tomato/*Bt* | FL |
| 89-320-01 | 2-12-90 | Calgene | Tomato/antisense polygalacturonase gene | FL |
| 89-293-01 | 2-14-90 | Monsanto | Tomato/tobacco MV/tomato MV | FL |
| 89-220-01 | 2-15-90 | UC Davis | Walnut/marker genes | CA |
| 89-290-01 | 2-16-90 | Auburn Univ. | *Xanthomonas*/marker gene | AL |
| 89-257-04 | 2-21-90 | USDA/ARS | Potato/marker gene | ID |
| 89-300-01 | 2-21-90 | UpJohn | Cantaloupe and squash/CMV and/or papaya ringspot virus | MI |
| 89-305-01 | 3-1-90 | UpJohn | Cantaloupe and squash/CMV and/or papaya ringspot virus | CA |
| 89-305-03 | 3-1-90 | UpJohn | Cantaloupe and squash/CMV and/or papaya ringspot virus | CA |
| 89-311-01 | 3-1-90 | UpJohn | Cantaloupe and squash/CMV and/or papaya ringspot virus | GA |
| 90-019-01 | 3-19-90 | Calgene | Tomato/antisense PG gene | CA |
| 89-326-03 | 3-21-90 | Ciba-Geigy | Tobacco/*Bt* | NC |
| 89-339-01 | 4-5-90 | Northrup King | Cotton/*Bt* or glyphosate | MS |
| 90-016-04 | 4-11-90 | Calgene | Cotton/bromoxynil | MS/AZ |
| 90-025-01 | 4-16-90 | Monsanto | Cotton/*Bt* | AZ/IL |

## APPENDIX I (continued)
### Permits Issued for Release into the Environment under 7 CFR 340 Animal and Plant Health Inspection Service, U.S. Department of Agriculture

| Number | Date issued | Company | Organism | Location |
|---|---|---|---|---|
| 89-362-01 | 4-19-90 | Rohm & Haas | Tobacco/*Bt* | NC |
| 90-032-03 | 4-19-90 | Monsanto | Potato/*Bt* or Virus X/Y/and Potato leaf roll | IL |
| 90-043-02 | 4-19-90 | UpJohn | Tomato/TMV | MI |
| 90-032-02 | 4-27-90 | Monsanto | Cotton/*Bt* | AL, AZ, CA, LA, MS, TX |
| 90-038-02 (Renewal of 89-30-02) | 5-7-90 | Monsanto | Tomato/*Bt* | CA |
| 90-032-01 | 5-8-90 | Monsanto | Potato/virus X, Y potato leaf roll | WA |
| 90-016-01 | 5-9-90 | Crop Genetics | *Cxc*/*Bt* in corn | MD |
| 90-038-05 | 5-9-90 | Monsanto | Soybean/glyphosate | AR, IL, MD |
| 90-066-01 | 5-9-90 | Calgene | Tomato/antisense to pectalytic enzyme; or cytokinin pathway | CA |
| 90-025-05 | 5-10-90 | Monsanto | Cotton/glyphosate | AZ, IL |
| 90-044-05 | 5-11-90 | Du Pont Co. | Cotton/sulfonylurea | MS |
| 90-023-01 | 5-15-90 | Monsanto | Cotton/glyphosate | AL |
| 90-038-03 | 5-15-90 | Monsanto | Soybean/glyphosate | IL |
| 90-065-06 | 5-15-90 | University of Kentucky | Tobacco/tobacco vein mottling and tobacco etch viruses | KY |
| 90-031-02 | 5-23-90 | ARS/USDA | Potato/*Bt* | WA |
| 90-038-04 | 5-23-90 | Monsanto | Soybean/glyphosate | AR, GA, IL, IN, IA, KY, MO, NE, OH, TN, VA |
| 90-029-01 | 5-31-90 | Louisiana State University | Rice/marker gene | LA |
| 90-033-01 | 5-31-90 | BioTechnica | Corn/marker genes | IA |
| 90-059-01 | 5-31-90 | NYS Ag. Exp. Station | Cucumber/CMV | NY |
| 90-114-01 (Renewal of 89-136-01) | 6-5-90 | Pioneer | Alfalfa/AMV | IA |
| 90-071-02 (Renewal of 89-065-01) | 6-21-90 | University of Kentucky | Tobacco/mouse metallothionein | KY |

*Note:* TMV = tobacco mosaic virus. AMV = alfalfa mosaic virus. CMV = cucumber mosaic virus. PRV = papaya ringspot virus. WMV-2 = watermelon mosaic virus. ZYMV = zucchini yellow mosaic virus.

\* = Tomato mosaic virus.
\*\* = Alfalfa mosaic virus.

## APPENDIX II
### Pending Applications for Environmental Release as of 06/07/90

| | | | |
|---|---|---|---|
| 90-065-01 | Canners Seed | Tomato/glufosinate | CA |
| 90-088-01 | UpJohn | Cantaloupe and squash/CMV, PRV, watermelon mosaic virus-2 and zucchini yellow mosaic virus | CA, GA, MI |
| 90-088-02 | UpJohn | Cantaloupe and squash/CMV, PRV, WMV-2, ZYMV | GA |
| 90-088-03 (Renewal of 89-300-01, 89-305-01, 89-305-03, 89-311-01) | UpJohn | Cantaloupe and squash/CMV, PRV | CA, GA, MI |
| 90-108-03 | Calgene | Cotton/bromoxynil/*Bt* | HI |
| 90-121-01 | Penn State | Rice/marker gene | AZ |
| 90-135-01 | University of Wisconsin | *Pseudomonas syringae* pv. *syringae*/Tn5 | WI |
| 90-135-02 | Amoco Technology Co. | Tobacco/eukaryotic gene/marker gene | KY |

*Note:* AMV = alfalfa mosaic virus; PRV = papaya ringspot virus; WMV-2 = watermelon mosaic virus; and ZYMV = zucchini yellow mosaic virus.

## REFERENCES

1. **Koshland, D. E., Jr.,** The molecule of the year, *Science,* 4937, 1541, 1989.
2. **Coulson, J. R. and Soper, R. S.,** Protocols for the introduction of biological control agents in the U.S., in *Plant Protection and Quarantine,* Kahn, R. P., Ed., CRC Press, Boca Raton, FL, 1989.
3. Regulations for implementing the procedural provisions of the National Environmental Policy Act, 40 CFR parts 1500-1508, 1986.
4. **Anon.,** *NEPA Deskbook,* Environmental Law Institute, Washington, D.C., 1989.
5. **McCammon, S. L. and Medley. T. L.,** Certification for the planned introduction of transgenic plants into the environment, in *Molecular and Cellular Biology of the Potato,* Vayda, M. E. and Park, W., Eds., CAB International, United Kingdom, 1990, 233.
6. **National Academy of Sciences,** *Introduction of Recombinant DNA- Engineered Organisms into the Environment: Key Issues,* National Academy Press, Washington, D.C., 1987.
7. **National Academy of Sciences,** *Field Testing Genetically Modified Organisms: Framework for Decisions,* National Academy Press, Washington, D.C., 1989.
8. Budget of the United States Government, Fiscal Year 1991, Executive Office of the President, Office of Management and Budget 1990, Washington, D.C., 1583 p; ill.
9. **Anon.,** *Survey of State Government Legislation on Biotechnology, Year-End Review,* Industrial Biotechnology Association, Washington, D.C., 1990.
10. Good Developmental Practices for Small Scale Field Research with Genetically Modified Plants and Microorganisms, Organization for Economic Cooperation and Development, Paris, France, 1990.
11. **Tiedje, J. M., Colwell, R. K., Grossman, Y. L., Hodson, R., Lenski, R. E., Mack, N., and Regal, P. J.,** The planned introduction of genetically engineered organisms: ecological considerations and recommendations, *Ecology,* 70, 298, 1989.
12. **Howarth, F. G.,** Environmental impacts of classical biological control, *Annu. Rev. Entomol.,* 36, 485, 1991.

Chapter 17

# ENVIRONMENTAL IMPACTS OF GENETICALLY ENGINEERED MICROBIAL AND VIRAL BIOCONTROL AGENTS

**Anne K. Hollander**

## TABLE OF CONTENTS

# I. INTRODUCTION

## A. SUMMARY AND DEFINITIONS

Genetically engineered biological control agents can have positive, negative, or essentially no environmental consequences, depending on their construction and use and the environments in which they are used. This chapter identifies both the positive and negative consequences that such agents are capable of having in the environment, and describes how to develop safe biocontrol agents to the extent that is currently known.

For purposes of this chapter, "genetically engineered microbial and viral biocontrol agents" means bacteria, fungi, or viruses that have been genetically modified using molecular techniques,[1,2] and that are released to the environment to control or suppress, but not eradicate, insect or plant pests.[3]

Classical biocontrol methods have successfully controlled approximately 140 insect pests and 40 weed pests.[4] Now that genetic engineering techniques are available, scientists hope to expand both the number of pests controlled and the species affected. For example, future engineered biocontrol agents may be able to help control plant diseases, a goal which largely has eluded biocontrol scientists in the past.[4]

## B. HISTORY

Most biocontrol programs in the past have employed naturally occurring insects and mites rather than bacterial or viral pathogens. There are only a few important exceptions, such as *Bacillus thuringiensis*, which is used widely to control lepidopteran pests of many important agricultural and forest plants. Some work also has occurred with pathogens, such as nucleopolyhedrosis virus to control the European spruce sawfly.[3] Still, microbial examples are rare compared to the hundreds of examples of vertebrate and insect biocontrol. Estimates of success with nonmicrobial biocontrol agents range from approximately 100 to over 300 cases of biological control of insects throughout the world.[3]

The tools of genetic engineering may expand the types of biological control that are possible, the success rate in using biocontrol,[5] and the safe use of such agents, e.g., by removing undesirable characteristics that might otherwise limit their appropriate use.

# II. POTENTIAL ENVIRONMENTAL CONSEQUENCES

Environmental consequences of engineered control agents can be placed into two categories. Some effects are primary and specific, that is, they result directly from the introduction of specific agents into the environment. These effects either can constitute perturbations to the environment such as displacement or extinction of beneficial insects, or avoidance of existing degradation to the environment through the replacement of a more dangerous method of pest control.

In addition to agent-specific effects, cumulative effects will result from releasing genetically engineered biocontrol agents into the environment over time. Some scientists believe that cumulative effects may change the complexion of the environment in much more significant ways than the specific effects of any one control agent.

## A. LIKELIHOOD OF ENVIRONMENTAL EFFECTS

The safety or risk of genetically engineered biological control agents cannot be stated in absolute terms. Effects will depend on many individual decisions made by researchers, biotechnology companies, and society — decisions about which host and donor organisms to use, how much and what kinds of testing to conduct prior to environmental release, and the extent and location of permissible large-scale uses.

Considerable theory has developed about the safety and risks of both genetic engineering and classic biocontrol, but unfortunately, the theories in both areas are insufficiently confirmed with data and often point to conflicting conclusions. Many scientists realize that "one can use the novelty of recombinant DNA organisms to argue as effectively for their being hazardous as their being safe."[6] Similarly with biological control, some state that the record of biocontrol is without risks, while others believe just the opposite.[7]

## 1. Theories of Risk for Genetically Engineered Organisms

Apprehension about genetically engineered organisms originally arose because of concerns that their novelty may make them less predictable than existing organisms. During the last 20 years of experimentation with engineered organisms, a competing and more reassuring perspective has arisen, i.e., that because genetically engineered organisms often differ from their parental strains by only one or a few genes and are introduced into well-characterized, agricultural environments,[8] we have more familiarity and consequent ability to control these organisms than with many exotic species which have been introduced to the environment in the past. Over the next few years, data from environmental tests will be gathered and will help to confirm or rebut this reassuring perspective.

Further theoretical conflicts arise from trying to predict the competitiveness of engineered organisms. Research has shown that modified organisms generally are less fit to compete successfully in the environment due to their "excess baggage," i.e., the metabolic burden of synthesizing additional nucleic acids and proteins.[2,9] On the other hand, important exceptions have been found, and scientists speculate that fitness-enhancing genes may at times inadvertently be transmitted during genetic alteration processes.[9] Also, organisms which have reduced long-term survival capabilities may still cause ecosystem effects during the interim before they disappear.[2] Therefore, it is not possible to make general statements regarding the competitiveness of engineered organisms relative to existing organisms. As detailed later, competitiveness must be assessed on a case-by-case basis.

Even if scientists could agree that genetically engineered organisms are not very different from existing organisms, this either can be reassuring or cause for concern. It can be comforting because it allows us to predict effects better.[1] It may be disturbing because the more closely related an introduced organism is to those with which it must compete, the more likely it may be to survive and thrive, and therefore the more difficult it may be to eradicate if necessary.[10]

Despite the conflicting theories, experts appear to agree that the majority of engineered organisms probably will pose minimal ecological risk, particularly those which are very similar to organisms used extensively in the past.[2] One author has estimated the risk of ecological damage from engineered organisms at roughly 1 in 100 to 1 in 10,000 releases,[11] but the extent of damage from those risk levels, even if the estimate is accurate, is not predictable. As experience accumulates, familiarity with a broader range of organisms will be obtained, thereby increasing the number of organisms that can be used safely in the future.

## 2. Theories of Risk for Biocontrol Agents

Principles about the safety of biocontrol agents are not much more established than for genetically engineered organisms, although more data and experience exist with the use of biocontrol agents. Many scientists believe biocontrol agents are usually safe, or ineffective.[12] Others point to problems that have sometimes occurred, warning that we should not be sanguine about the ability of biological control agents to have untoward effects.[7]

In addition to different interpretations of the historical record of biocontrol efforts, people disagree about whether the records are relevant to bacterial and viral biocontrol agents. Most biological control experience involves vertebrates and insects;[8] we have little data from the use of microbes for biocontrol. Deriving conclusions from animals about the safety of biocontrol using microorganisms therefore may be invalid because they have distinct differences in modes of action, growth, reproduction, dispersal, and ecosystem roles.[13] For pathogen species in particular, we often lack data on the genetic factors affecting virulence, pathogenicity, host range, and other characteristics relevant to population dynamics.[14]

Another theoretical question is whether small-scale tests with biocontrol agents will cause long-term effects in the environment. Some theorists argue that microbes are too limited in their dispersal and survival capabilities for small-scale tests to have effects, even using agents for which larger-scale uses might be ill-advised. In fact, experience with agricultural pathogens suggests this is usually true.[15] On the other hand, historical evidence proves long-term effects from small-scale introductions are possible, if not common. For example, the Japanese fungus *Entomophaga maimaga* was introduced in Massachusetts in 1910–1911 to see if it would control gypsy moths. Although researchers at the time could not reisolate the fungus after its release, subsequent evidence strongly suggests that it did become established over a large area of the northeastern U.S. Further, scientists now suspect that the fungus evolved during the intervening 80 years to attack the U.S. variety of gypsy moth (slightly different from the original Japanese moth hosts).[16] This suggests that long-term effects (in this case positive) are indeed possible from small-scale releases, and that mutations or organisms over time may result in ecological effects that are not discernible at the initial time of release.[11] Experiences with "unofficial" small-scale tests, such as the introduction of Dutch elm disease and chestnut blight, also indicate that small-scale releases can on occasion lead to large-scale effects.

In conclusion, ecological effects of engineered biocontrol agents should be evaluated on a case-specific basis, considering the phenotype of the modified organism and the environment in which it will be introduced.[2] These factors form the basis for the discussion in Section III.

## B. TYPES OF NEGATIVE EFFECTS

### 1. Effects on Nontarget Organisms

Introduced organisms have occasionally disrupted or killed nontarget organisms in addition to their intended hosts. They may do this by directly infecting or attacking species other than those for which they were intended, or they may indirectly kill nontarget organisms. For example, if an introduced organism successfully depletes its target population and the target population served as a major food source for other organisms, these other organisms may in turn suffer population declines or extinction as a result of competition from the introduced organism.

One author has estimated that about 8% of biological control introductions have caused extinctions of resident species.[17] It is unclear whether this rate is greater or smaller than species extinctions caused by chemical control; some scientists estimate that it is smaller.[7]

Genetic engineering of a biocontrol agent can either increase or decrease the chances of nontarget effects. Although single gene changes have been shown at times to increase the virulence of organisms, the effect is not believed to be common.[8] Presumably, gene changes also can be used deliberately to reduce the host range of a biocontrol agent, thus limiting the chances that it will unintentionally damage species closely related to its host.

Although knowledge of pest-pathogen interactions is invariably helpful in avoiding unintended effects to nontarget organisms, there is always some finite chance of longer term effects. All organisms mutate and evolve in adaptation to their environment. Over time, pathogens' host ranges can change to permit exploitation of new food resources. For example, about 30% of the pathogens that attack U.S. crops fed originally on native vegetation but evolved to feed on crop plants.[18]

### 2. Unintended Effects on Beneficial Organisms

A particularly undesirable type of nontarget effect occurs when biological control agents damage or cause the extinction of populations or organisms that control agricultural or human pests ("beneficials"). For example, parasites imported to Hawaii for moth control have destroyed many nonpest, nontarget lepidopteran species. The reduced numbers of native lepidoptera resulted in reduced food supplies and extinction of several important predators, including *Odynerus* wasps and a variety of insect-eating birds.[7]

Pimentel (1985) estimates that because about 99% of all insects are beneficial, the likelihood that beneficial soil insects will be damaged by some biological control agents is quite high.[18] Though some scientists believe that exotic organisms are even more dangerous to beneficial insects than pesticides,[7] this view is not commonly held.

The U.S. Environmental Protection Agency (EPA), as part of its research program on engineered organisms, is sponsoring research to develop and validate procedures for testing the effects of microbial biocontrol agents on nontarget, beneficial arthropods.[19] These tests may aid in predicting and avoiding adverse effects to beneficials in the future.

## 3. Evolution of Resistance

Resistant pest populations may develop with the use of genetically engineered biocontrol agents.[2,20] Over 500 insect species around the world have developed resistance to chemical insecticides,[21] particularly the carbamates, organophosphates, and pyrethroids.[22]

It often is asserted that insect resistance to microbial agents "is less common or develops more slowly than for chemical pesticides."[14] This phenomenon may be explained by concomitant evolution, as each agent evolves in response to the protective measures of the other until some balance is achieved.[18] On the other hand, microbial agents acting continuously in the environment might place greater selection pressure on insect pests than chemicals.

Unfortunately, data to confirm either theory are sparse. Pests are known to develop resistance to biocontrol agents through various mechanisms such as encapsulating parasite eggs, developing thicker cuticles, or behavioral changes.[12] While a few documented cases of host resistance to biocontrol agents exist (e.g., larch sawfly resistance to *Mesoleius tenthredinis*; European corn borer resistance to *Lydella stabulans grisescens*[18]), insufficient information exists to generalize about the overall extent or likelihood of this problem. Even *Bacillus thuringiensis (Bt)* has not been extensively studied for resistance problems, in spite of its widespread use as a biocontrol agent — and the data that do exist on *Bt* are inconsistent. In some studies insects developed rapid resistance to *Bt* (e.g., Indian meal moth, tobacco budworm) while others showed virtually no resistance, even under high levels of selection pressure (e.g., diamondback moth, Egyptian cotton leafworm).[21,23]

It still is an open question whether engineered microbial biocontrol agents will reduce or exacerbate the insect resistance problem. Despite the uncertainty, the biotechnology industry is taking the resistance problem seriously, as evidenced by the formation in the U.S. of an "Industry Working Group on *Bt*" which has as one major goal to "formulate strategies to maintain effectiveness of *Bt*."[24]

## 4. Ecosystem Process Effects

The cycling of nutrients and energy in the environment is intimately connected with many microbial life processes. Some scientists postulate that these processes could be affected by environmental introductions of genetically engineered organisms.[8]

If such ecosystem effects were to occur, most likely they would be caused by organisms which play fundamental roles in ecosystem processes — such as organisms which fix nitrogen or degrade carbon plant materials.[8] To the extent that scientists contemplate using such organisms as biological control agents, their use will require close examination, but most biological control agents contemplated for use at this time do not appear to fall in this category.

EPA is conducting some research on the effects of genetically engineered microbes on ecological processes in soil,[19] but many years of research will be necessary before generalizations about this issue are possible.

## 5. Unintentional Development of Pests

In the past, concerns have been expressed that genetic modifications might inadvertently create a pest or pathogen from a nonpathogen. However, the current scientific consensus is that this is extremely unlikely.[1] Pathogenicity generally is controlled by such a large array of genes that the statistical chance of unintentionally conveying all of those genetic factors to a nonpathogen host is virtually zero.

On the other hand, minor genetic modifications can increase the virulence of organisms that already are pathogens or that are very closely related to pathogens, as are most microbial biocontrol agents. For this reason, genetic engineering with parents that are closely related to pathogens raises more concerns in regards to pest or pathogen "creation" than other types.

## 6. Biological Pollution/Cumulative Effects

Biological pollution, or "the establishment in the wild of foreign or nonnative organisms,"[7] is by definition associated with introducing any modified organism in an environment, if the organism remains in the environment. Varying views exist as to the importance of biological pollution; for many, some biological pollution may be acceptable but its cumulative impacts may not. There are those who point out that changes in nature, including the extinction of organisms, are normal and even beneficial — up to a point. "On the larger scale of things, change within ecosystems should . . . be seen as natural and inevitable, even if it sometimes leads to local extinctions. . . . Conservationists should be interested in persistence, not constancy — that is, the number of species within an ecosystem, not the number of individuals within any particular species."[25]

On the other hand, there are those whose concern about biological pollution is driven by the ethical and aesthetic belief that nature should not be altered just for the benefit of humans. The degree of alteration may also be a consideration. For example, some philosophers argue that we ought to "value and protect nature for its own sake, rather than to manage or alter it for our benefit."[26] According to this view, any introduction of a modified organism may contribute to biological pollution in unacceptable ways, and many introductions are even more likely to be unacceptable. The introduction of large numbers of modified organisms could, over time, alter the makeup of some biotic communities.

Because most biocontrol organisms will be introduced in agricultural environments which already are highly affected by human activity, this concern may be minimized if one also assumes the organisms will stay in those environments rather than moving into surrounding areas. However, more research will be needed to determine whether this is a reasonable assumption. As the National Academy of Science has said, " . . . large-scale or sustained applications might have consequences different from small-scale or single applications . . . the cumulative probability of undesirable effects resulting from repeated applications or frequent introductions must be considered."[1]

## C. TYPES OF POSITIVE EFFECTS

Engineered biological control agents provide some opportunities for important ecological benefits at the same time that they may present some ecological risks. As mentioned above, whether the ecological benefits will outweigh the ecological risks in any individual case depends on the decisions of researchers, as well as on evolving policies and economic factors.

## 1. Replacement of More Dangerous Control Agents

An important benefit of biological control agents is that they can be used to replace more dangerous chemical pest control agents. At the present, chemical controls are far more commonly used in the U.S. than biological controls. Approximately 500,000 kg of chemicals, worth about $3 billion, are applied each year to control crop pests.[18] In some major crops such as cotton, insecticides are applied over ten times per season in some areas.[27] In spite of this, nearly 37% of all crops are lost to pests each year.[18]

Of course, it is unclear whether all chemical pesticides are environmentally harmful, so replacing all of them with biocontrol agents would not guarantee fewer environmental risks. Some authors even have argued that biocontrol may cause more species extinctions than chemical control.[7] Nonetheless, for the numerous chemical pesticides known to have toxic effects beyond their target pests — including toxic effects to animals and humans[28] — the opportunity to substitute safer, more selective, and biodegradable biocontrol agents can provide important ecological benefits.

One of the ecological advantages of biological control agents is that they tend to be highly selective, infecting or killing a very narrow range of target pests.[29] For example, *Bacillus thuringiensis*, which is widely used as a biocontrol agent against leaf-eating lepidopteran pests, is nontoxic to all but a few nonlepidopteran insects, and causes no effects in vertebrates.[30]

Genetic engineering can be used to produce microbial biocontrol agents which are even safer than their natural counterparts. For example, cell culture techniques can be used to produce viruses that have no contaminating agents such as microorganisms, insect proteins and cuticles, which often cause major allergic reactions in humans.[31] Genetic engineering can also be used to enhance the safety of biocontrol agents by reducing their ability to transfer genetic material to nontarget species. For example, a particular strain of *Agrobacterium radiobacter* was recently modified to delete the genes that permitted its plasmid to be transferred to other organisms, making it safer for long-term environmental uses.[1]

### 2. Enhancement of Integrated Pest Management

Integrated Pest Management (IPM) tries to maximize pest mortality using a variety of approaches for controlling pests, including biological, chemical, physical, genetic, and cultural controls. The goal of IPM is to keep pest population densities low enough so they will not cause economic damage, while preserving natural predators and beneficial organisms.

IPM is more environmentally and agriculturally sound than efforts to completely eradicate pests using chemical controls. Because biological controls tend to be host-specific, they are particularly useful in IPM systems.[30] However, the shortage of effective, easy-to-use biological controls has been one of the limiting factors in the use and adoption of IPM (and biological control generally) in the past. Sometimes lack of effectiveness is a function of other agricultural practices. "Most agroecosystems are so disrupted by cultivation and chemical applications that biological agents such as microbials, antagonists, predators, and parasitoids become separated from their competitors, hosts, or prey (pests) in either space or time."[14] Changes in agricultural practices are more likely than genetic engineering to help overcome these impediments to IPM, at least in the near future.

On the other hand, the interaction of genetic factors with environmental factors such as temperature, moisture, and light also influences the ability of microbes to damage other organisms, and these factors are subject to manipulation using genetic engineering. If microbial biocontrol agents can be made less temperamental in regards to the timing and method of their application and use, this may help to permit wider use of IPM.

Even with genetic engineering, biological control is not likely to supplant the use of chemicals in agriculture.[12] Still, great benefits will be realized if genetically engineered biocontrol agents make the use of IPM more effective, economical, and feasible under a wide variety of agricultural conditions.

### 3. Protecting Ecosystems

In the future, biological control may be used to protect natural ecosystems, as well as agricultural systems, resulting in very direct environmental benefits. For example, research in molecular biology may help protect forests from pests; researchers at Iowa State University currently are using genetic engineering to help poplar trees resist insect damage.[32] Protecting woody plants yields direct benefits in the forms of preserving the soil from erosion and reducing carbon dioxide accumulation in the atmosphere.

# III. HOW TO DEVELOP SAFE BIOLOGICAL CONTROL AGENTS

A variety of factors influences the degree of safety that can be achieved when altering bacteria, viruses, and fungi to enhance their use as biocontrol agents. Scientists control many of these factors as they choose host organisms, vectors, donors, and environmental settings for testing. This section examines how those decisions can be made to minimize risks of encountering the environmental problems discussed in the previous section. It also briefly describes how economic and regulatory circumstances affect environmental risks.

Although it is important to take proper steps to minimize risks, equally important is to realize that many engineered biocontrol agents will present low risks to begin with. For the most part, biocontrol agents are intended to be introduced into controlled agricultural settings where they are unlikely to have widespread ecosystem effects. Furthermore, many will be closely related to domesticated species which are sufficiently familiar that likely risks can easily be identified.[2] The occasional exceptions to these rules, however, do deserve special attention.

## A. SCIENTIFIC CONSIDERATIONS

The best way to predict and minimize risks is to have extensive information on the parent organisms and the intended environments for use. Unfortunately, such information often is limited because knowledge about microorganisms and microbial ecology is not very advanced. For example, the vast majority of soil microbes cannot be cultured in the laboratory, and only a very few species have been studied extensively. Furthermore, because of their microscopic size, microorganisms are more difficult and expensive to monitor in the environment than most plants and animals.[8] For both of these reasons, using existing information on parent organisms and environments to avoid risks is difficult except for a very limited number of species and environments.

A second way to minimize risks is to fully understand the donated functions and the impact of vectors or other genetic transfer methods on the new organism. Both of these factors tend to be understood more fully at this time than information about whole organisms and environments. While insufficient to fully predict risks, information about these parameters can certainly be used to help minimize risks that may exist. For example, the choice of a transfer method can help contain organisms and their donated functions in the environments to which they are released.[32] This presents particular challenges to the scientist, described later in this section.

As mentioned earlier, a very important way to miminize environmental risks is to combine and integrate a number of pest management tactics rather than designing products that are intended to be relied on alone. Biocontrol agents designed to complement IPM systems will be safer and more reliable than those designed solely to replace chemical-intensive methods.

## 1. Selection of Parent Organisms

It is most important that parent organisms (both donors and recipients) be well characterized. The more that is known about the parent organisms from a genetic and physiological standpoint, the better able are scientists to predict and avoid negative effects from the modified organism.[8] For example, one of the few disadvantages of *B. thuringiensis* is that the physiology of its plasmids and bacteriophage are not well understood.[5] Similarly, lack of knowledge about the mechanisms of pathogenicity in many entomopathogenic fungi has restricted their use.[30]

When organisms with apparent potential for use as biocontrol agents are not well understood, tests can be conducted in contained conditions to learn more about their possible safety in the field. Tests have been developed to ascertain mammalian infectivity and toxicity, host specificity, growth conditions, ability to exchange genetic material with other organisms, and a variety of other relevant parameters.[34] In the absence of knowledge of these factors from historical experience, such tests are particularly relevant and important.

Certain features make bacteria, viruses, or fungi more desirable and safe for genetic engineering. One is specificity. The most successful and safest biocontrol agents tend to be those which are highly specific to their target pests.[7] Microbes that are generalists are more likely to cause unforeseen effects, so they should be avoided as a rule, or if they are used, they should be modified to enhance their specificity. For example, high host-specificity is one of the attributes of *B. thuringiensis* and *B. popilliae*, which has made them such attractive and popular choices for genetic engineering in the past decade.[30] In contrast, the lack of specificity of genera such as *Aerobacter*, *Cloaca*, and *Proteus* has inhibited their use.[5] Similarly, a proposed use of *Clavibacter xyli* subspecies *cynodontis* (Cxc) received objections from environmentalists because it does not have a narrow host range.[35]

A second attribute of safe parent organisms is lack of environmental persistence. Lack of persistence ensures that any unforeseen effects will have minimal effects. The ability to persist in the environment is often low among altered organisms due to their increased biological loads. Where persistence is likely, however, genetic modification may be needed to reduce it. For example, in 1986 the NERC Institute of Virology (Oxford) began a research project to genetically engineer baculovirus insecticides. They first modified the baculovirus to contain a genetic marker, and released it in order to determine environmental fate. The experiment showed that the virus was persistent, as expected. Then the Institute removed the protective viral coats of the viruses, thus limiting their persistence but not reducing their pathogenicity.[36] Researchers in the U.S. also are pursuing methods to reduce persistence of engineered organisms, such as the construction of phenotypes which are "conditionally lethal", that is, organisms that die when conditions subject to manipulation by the researcher are changed.[37]

Inability to exchange genetic material with other organisms is a third attribute of importance to the genetic engineer. The inability to exchange genetic material is desirable because it eliminates one avenue for undesirable effects on nontarget organisms. Mechanisms of genetic transfer must be identified and then should be studied in the laboratory[8] or, better yet, in microcosms which simulate the expected environmental conditions in which the organism will be used. If the expected rate of genetic transfer seems significant and if potential recipients exist in the environments where the organism will be used, the choice of a parent organism should be reconsidered or opportunities for transfer eliminated or disabled. One possible approach is to insert restriction systems (such systems are not yet well developed but they appear to hold "substantial promise").[8]

The ability to take mitigation action against an organism in the event of unexpected hazards is another important factor in its safety. For example, pathogenic organisms will be safer if they are dependent on hosts of restricted range and dispersal ability. Also, organisms of all types will be easier to mitigate if they can thrive only in habitats where chemical or physical control procedures are feasible.[32] Some of these traits may exist in the parent organism, but if they do not and mitigation is a concern, genetic modifications to achieve one or more of these characteristics may be appropriate.

A variety of other characteristics can make a parent organism more or less desirable as a subject of genetic engineering and for use as a biocontrol agent. A number of these were listed in a 1989 report by the Ecological Society of America. Besides those listed above, they include the following.[2]

- Origin — indigenous parents tend to be safer than exotic ones.
- Geographic range — narrow is safer than broad.

- Habit — free-living parent organisms are safer than those which are pathogenic, parasitic, or symbiotic (although note that most biocontrol agents will fall in one of these categories).
- Pest status — parent organisms which are neither themselves pests nor do they have relatives which are pests, are safer than pests.

Obviously, no parent organism can meet all of the criteria for "safest possible" organism, nor must it. For biocontrol purposes, organisms which are pathogenic, parasitic, or symbiotic often are needed for reasons of effectiveness. Such organisms need not be ruled out as hosts and donors for such uses. The existence of an effective combination of other characteristics that reasonably ensure their safety can counterbalance the existence of a few risk factors.

Even when parent organisms are well characterized through laboratory procedures, complete predictability cannot be achieved. For example, many species that would be expected to become established when introduced — because of their high population growth rates, dispersal capabilities, short life cycles, or other attributes — have not become established, while organisms lacking in these key traits sometimes have.[38] This points out the importance of evaluating modified organisms individually. One should begin with laboratory tests, and then incrementally move to less contained but more accurate test environments such as greenhouses, small-scale field plots, and then larger-scale tests. Results should be examined at each step and used to determine whether further, less contained testing is safe.

### 2. Selection of Donated Function

As with the parent organism, functions (controlled by genes) introduced from other sources should be well understood. Knowledge of the function of a gene in its natural host and environment is very helpful in predicting its effects in a new parent and a new environment.[8]

Testing the expression of a donated function in its new parent is important even for well-characterized genes, because knowledge of the natural history of a gene is helpful but not definitive. Proteins synthesized by identical DNA sequences may be chemically altered subsequent to the encoding process by enzymes in the parent organism. Protein activity also may be altered by intracellular factors such as pH.[39] Furthermore, it is possible that genetic modifications will unintentionally affect traits other than the intended ones.[1]

Pest resistance is an important issue to consider when choosing donated functions. For biocontrol purposes, genes encoding for pest toxins will often be donated to parent organisms that previously did not express those toxins, but which are useful parents for other reasons (e.g., they inhabit the part of the plant where most damage occurs). Because such new parent/function combinations are designed to concentrate the toxin at the site of pest damage, the problem of rapid pest resistance developing will be important to consider, and strategies for decreasing pest adaptation rates will be useful. In one example, Gould[27] has suggested linking the coding sequences for a toxin gene to tissue-specific promoters that limit the toxin expression to certain tissues in cotton plants. In principle, the same strategy could be used for limiting the expression of toxin genes in bacterial, viral, or fungal control agents to certain types of environmental stimuli.

It may be appropriate for environmental safety reasons to donate functions in addition to those needed specifically for biocontrol purposes. For example, donated genes encoding for sensitivity to light, oxygen, or high or low temperature may be useful ways of limiting an organism in the environment until more is known about its effects.[12] Also, marker genes for monitoring the spread and persistence of the organism may be useful to introduce; options include genes which permit the organism to catabolize substrates not normally used by that particular microbe (e.g., lactose), or nonfunctional genes which can be located using nucleic acid hybridization techniques.[1]

In general, safety of the engineered biocontrol agent should be enhanced as much as possible by trying to select donated functions that have the following characteristics.[2]

- The alteration is genetically stable.
- The alteration is either a deletion or a single gene change, rather than multiple genetic changes (except when additional genetic changes are made for environmental safety purposes, as noted above).
- The source of the donated function is closely related to the parent organism, except when either or both are pest species.

### 3. Selection of Vector and Genetic Transfer Method

Like the choice of the parent organism and trait, vectors or other methods used to insert or transfer a donated function to a parent organism can influence the likelihood that the altered organism will exhibit unintended environmental effects. They can affect the chances of unintentional genetic transfer to other organisms or may introduce genes which code for additional, undesired traits.

Certain features enhance the safety of vectors:[8,40]

- Inability independently to initiate horizontal gene transfer
- Resistance to being moved (excised or transposed)
- Limited host range
- Lacking additional genes beyond those needed for control, monitoring, and safety

There are a number of ways to minimize the chance that the genetic transfer method will permit genes to move beyond their intended parent. The most commonly used approach involves disabled vectors which have been altered specifically to render them incapable of moving or being moved.[8] A newer alternative is to use "gene cassettes", which act like transposons but produce stable genetic changes.[1] Removal rather than insertion of genes provides a third option which in certain cases·can produce the desired phenotypic changes, and always eliminates the risk of subsequent, unintended genetic transfer. Finally, scientists are developing "suicide plasmids" which soon should be available for carrying genetic material into cells yet "remaining viable only under the specific environmental conditions of [their] intended use . . . and [dying] if [they] escaped that location or if the environmental condition requirement for its survival changed."[41]

In addition to minimizing chances for genetic transfer, it is important to minimize the chance that additional traits inadvertently are introduced by the vector. One approach is to ensure that the plasmids that were used to transfer genetic material into the parent microbe fail to replicate, thereby eliminating from the final product genes that the plasmid carries. Whether or not this is necessary requires consideration of the traits that might be expressed and their relevance to the intended use of the organism.[1] Another option in some cases is to avoid the use of vectors altogether, and instead rely on transfer methods such as transformation or bioballistics. However, this carries risks of its own because it entails less control of whether and how the genetic function is expressed in the final product.

While concern about the method of insertion and the nature of introduced vectors is important, some people feel it needs to be tempered by the knowledge that genetic transfer among microorganisms appears to be fairly widespread in nature.[42] Unfortunately, most of our knowledge about genetic transfer in "nature" actually is derived from laboratory experiments. Thus, until a great deal more is known about the scope of such transfer under natural conditions, it is impossible to determine whether transfer of recombinant genes is either likely or potentially unsafe. In the meantime, it seems prudent to take steps to restrict such events wherever feasible.

## 4. Selection of Experimental Use Sites and Experimental Conditions

In many cases, the environment is as important as the organism itself in influencing the chance of environmental impacts. Therefore, attention to the choice of experimental use sites and conditions is critical.

The central principle for ensuring safe environmental use of genetically engineered biocontrol agents is to conduct research in a systematic progression. Testing first needs to take place in contained laboratories and/or microcosms (small-scale replicates of the environment). Using information obtained from these contained tests, experiments can safely move into semicontained greenhouses, small-scale field plots, and large field sites. Data relevant to environmental behavior should be collected at each point and used to identify possible risks and appropriate safety precautions to take at the next level of testing. Each level of testing will provide much relevant information, but since highly scale-dependent effects may not be discernible at the smaller-scale levels, monitoring is still appropriate even if at the larger-scale stages no unexpected effects previously have been noted.[1]

The likelihood of environmental impacts — especially the chance that organisms will disseminate or become established — is affected by the intrinsic characteristics of the site and the conditions of experimentation established by the researcher. Unfortunately, the factors that make ecosystems susceptible to disturbance and invasion are not well understood,[13] but potential test sites can be selected with some consideration for the characteristics that may help them resist invasion. That knowledge suggests that sites should be, as a general rule:[2]

- More complex (i.e., genetically heterogenous) and stable than genetically homogenous and disturbed[8,38]
- Suitable for the use of effective monitoring techniques
- Suitable for the use of effective mitigation techniques in case of unwanted effects (although in most natural environments it is difficult to completely eliminate any introduced organism once it is established)[1]
- Lacking in selection pressure for the engineered trait

In addition to choosing an intrinsically safe site, researchers can take certain steps to further enhance the site's safety. Such steps include:

- Controlling access from human, wildlife or environmental agents that could disseminate the introduced organisms.[2]
- Timing the application to miminize dispersal, e.g., avoiding climatic conditions that would enhance dispersal such as rain or wind.
- Minimizing the density of the introduced biocontrol agent and its host species because these are positively correlated with the rate of genetic transfer.[32]
- Ensuring that technologies for monitoring gene transfer, and for mitigation in case of an emergency, are planned and tested in advance to be sure they are effective.[8] [Although the techniques for monitoring and mitigation are limited at this point, research is progressing to improve the quality and quantity of available methods. For example, the U.S. EPA is currently supporting research to develop a range of detection methods for use during field trials.[19]]

It may be difficult to meet all of these criteria in any given test site, especially because chosen site(s) need to be reasonably representative of ecosystems where the biocontrol agents will eventually be used in order for the results to be meaningful. For example, biocontrol agents need to be tested in agricultural environments, which typically are homogenous and disturbed (undesirable) but at the same time relatively easy to monitor and suitable for mitigation techniques (desirable). Similarly, biocontrol agents often need to be introduced at high densities to achieve the desired effect.[1] Such competing considerations are unavoidable, so researchers simply must choose which of the above factors is most important, based on knowledge of the altered organism, the stage of research, safety considerations, and practical considerations.

### 5. Selection of Scope of Wide-Scale Use

Some organisms may be safe for small-scale experimentation yet may present unacceptable risks if used on a large scale because of scale-dependent effects, delayed effects, or effects that occur over a long time.

Concern over widespread use of antibiotic resistance genes is a well-known example of a possible scale-dependent effect. Antibiotic resistance genes are useful for monitoring organisms, but scientists are concerned that if used extensively, such genes could mutate and be transferred to pathogens, giving pathogens the ability to resist therapeutic antibiotics.[1] Another example involves a microbial biocontrol agent made by inserting the toxin gene from *Bacillus thuringiensis* into *Clavibacter xyli* (Cxc). Initial field trials in 1988 showed that the organism has a wide host range. Ecologists pointed out that in large-scale use, natural selection would promote spread of the microbe into native plants susceptible to lepidopteran damage, thus exposing and probably killing many nontarget moths and butterflies.[42] Thus, the product is judged risky for large-scale use, even though small-scale experiments are considered safe.

Another reason that large-scale uses sometimes can present problems not apparent in small-scale tests is that certain effects do not occur immediately. For example, experiments have shown that natural selection tends to increase fitness of engineered agents over time.[9] Other effects that may occur only after many years of large-scale use include bioaccumulation, species displacement, and perturbations of geochemical cycles.[2] These types of effects generally can be avoided by reducing populations to very low levels after experimental uses; this strategy is usually not appropriate for large-scale or commercial use, however.

As mentioned earlier, some of the problems that might occur with large-scale use probably can be avoided by using molecular techniques to limit the spread or survivability of introduced organisms, e.g., by inserting conditionally lethal genes.[44] In the case of biocontrol agents, however, it often is most useful to establish permanent populations that persist at low levels until a pest outbreak occurs.[2] Therefore, the desirability of large-scale use and of persistence will vary with the specific case and the degree of confidence that unintended large-scale effects will not occur.

### 6. Summary: How to Develop Safe Biological Control Agents

In 1989, the U.S. National Research Council (NRC) developed a risk evaluation "framework" for organisms intended for field testing (the framework does not cover large-scale uses). The NRC framework suggests that risk evaluation consider three major criteria: (1) familiarity with introductions similar to the proposed introduction, (2) control over persistence and spread of the introduced microorganism and its exchange of genetic material with other organisms, and (3) environmental effects.[1] The preceding sections show that, to a large extent, researchers can ensure that the organisms they develop will meet these criteria through their choice of parent organisms, genetic material, vectors, and circumstances of use. Attention to the NRC guidelines and the points discussed above should help to minimize risks.

### B. OTHER CONSIDERATIONS: ECONOMICS AND REGULATIONS

It is beyond the scope of this paper to address in any depth the economic factors that affect biocontrol agents or the government regulations that apply in many countries. Still these nonscientific issues are important determinants of the ultimate use and safety of engineered biocontrol agents. Therefore, a few brief words on these subjects and some references to other sources of information follow.

The economics of biocontrol agents obviously affects their environmental impact, because to the extent they are used or not used, their impacts will vary. In the past, biocontrol agents have had difficulty competing with other pest control methods due to their narrow host range, their sometimes difficult storage requirements, and the labor and attention necessary to determine the correct time to release them. Genetic engineering may overcome some of these impediments and hence increase the chances that such agents will indeed have environmental impacts — either negative or positive, as previously discussed.

Government regulations also affect the chance of environmental impacts, hopefully by reducing the likelihood they will occur. Regulations are proliferating in many countries of the world, particularly in North America, Europe, and Japan. Excellent summaries of these regulations as of 1990 are found in OTA (1989)[8] for the U.S., and Royal Commission (1989)[45] for overviews of most of the developed countries and some developing countries.

## IV. CONCLUSIONS

Safe use of genetically engineered bacterial, fungal, and viral biocontrol agents is achievable — particularly when the organisms and environments involved are well understood — but it does require attention to safety issues by those who are designing and using new products. Environmental safety requires knowledge of genetics, molecular and organismal biology, population and community ecology, evolutionary biology, and numerous other disciplines.[2] Because few if any scientists have training in all these areas, both companies and individual researchers who are designing such organisms should plan to obtain advice from other specialists on points which may be less familiar to them. In addition to interdisciplinary consultation, much more research in the area of microbial ecology will be necessary to expand the currently very limited options to work with organisms and environments that are well understood. As the scientific base of knowledge regarding organisms' survival, persistence, and effects grows, the chances of safe use will increase commensurately.

## REFERENCES

1. National Research Council, *Field Testing Genetically Modified Organisms: Framework for Decisions,* National Academy Press, Washington, D.C., 1989.
2. **Tiedje, J. M., Colwell, R. K., Grossman, R. K., Hodson, R., Lenski, R. E., Mack, R. N., and Regal, P. J.,** The planned introduction of genetically engineered organisms: ecological considerations and recommendations, *Ecology,* 70, 298, 1989.
3. **Dahlsten, D. L.,** Control of invaders, in *Ecology of Biological Invasions of North America and Hawaii,* Ecological Studies, Vol. 58, Mooney, H. A. and Drake, J. A., Eds., Springer-Verlag, New York, 1986, chap. 16.
4. **Huffaker, C. B.,** Where we are and what we need to do in biological control: the entomologist's view, in *Proceedings of the National Interdisciplinary Biological Control Conference,* Battenfield, S. L., Ed., Cooperative State Research Service, U.S. Department of Agriculture, 1983, 9.
5. **Faust, R. M., Reichelderfer, C. F., and Thorne, C. B.,** Possible uses of recombinant DNA for genetic manipulations of entomopathogenic bacteria, in *Genetic Engineering in the Plant Sciences,* Panopoulos, N. J., Ed., Praeger, New York, 1981, 225.
6. **Simonsen, L. and Levin, B. R.,** Evaluating the risk of releasing genetically engineered organisms, *Tree,* 3, S27, 1988.
7. **Howarth, F. G.,** Classical biocontrol: panacea or Pandora's Box, in *Proc. 1980 Hawaiian Entomol. Soc.,* 24, 239, 1980.

8. **U.S. Congress, Office of Technology Assessment,** New Developments in Biotechnology — Field-Testing Engineered Organisms: Genetic and Ecological Issues, OTA-BA-350, U.S. Government Printing Office, Washington, D.C., 1988.

9. **Lenski, R. E. and Nguyen, T. T.,** Stability of recombinant DNA and its effects on fitness, *Tree,* 3, S18, 1988.

10. **Sharples, F. E.,** Spread of organisms with novel genotypes: thoughts from an ecological perspective, *NIH Recombinant DNA Tech. Bull.,* 6, 43, 1983.

11. **Regal, P. J.,** Models of genetically engineered organisms and their ecological impact, in *Ecology of Biological Invasions in North America and Hawaii,* Ecological Studies, Vol. 58, Mooney, H. A. and Drake, J. A., Eds., Springer-Verlag, New York, 1986, chap. 7.

12. **Jutsum, A. R.,** Commercial application of biological control: status and prospects, *Phil. Trans. R. Soc. London,* 318, 357, 1988.

13. **Levin, S. A. and Harwell, M. A.,** Environmental risks and genetically engineered organisms, in *Biotechnology: Implications for Public Policy,* Panem, S., Ed., The Brookings Institution, Washington, D.C., 1985, 58.

14. **Battenfield, S.,** Ed., Proceedings of the National Interdisciplinary Biological Control Conference, Cooperative State Research Service, U.S. Department of Agriculture, Washington, D.C., 1983.

15. **Tolin, S. ad Vidaver, A.,** Guidelines for research with genetically modified organisms: a view from academe, *Annu. Rev. Phytopathol.,* 27, 551, 1989.

16. **Barr, B.,** unpublished memorandum dated Oct. 11, 1989.

17. **Fuxa, J. R.,** Environmental risks of genetically-engineered entomopathogens, in *Biotechnology, Biological Pesticides and Novel Plant-Pest Resistance for Insect Pest Management,* Roberts, D. W. and Granados, R. R., Eds., Boyce Thompson Institute for Plant Research at Cornell University, Ithaca, New York, 1988, 159.

18. **Pimentel, D.,** Biological invasions of plants and animals in agriculture and forestry, in *Ecology of Biological Invasions of North America and Hawaii,* Ecological Studies, Vol. 58, Mooney, H. A. and Drake, J. A., Eds., Springer-Verlag, New York, 1986, chap. 9.

19. **Pritchard, P. H.,** Review of Progess in Biotechnology-Microbial Pest Control Agent Risk Assessment, Report EPA/600/X-89/130, U.S. Environmental Protection Agency, Washington, D.C., 1989.

20. **Gould, F.,** Ecological-genetic approaches for the design of genetically-engineered crops, in *Biotechnology, Biological Pesticides and Novel Plant-Pest Resistance for Insect Pest Management,* Roberts, D. W. and Granados, R. R., Eds., Boyce Thompson Institute for Plant Research at Cornell University, Ithaca, New York, 1988, 146.

21. **Georghiou, G. P.,** Implications of potential resistance to biopesticides, in *Biotechnology, Biological Pesticides and Novel Plant-Pest Resistance for Insect Pest Management,* Roberts, D. W. and Granados, R. R., Eds., Boyce Thompson Institute for Plant Research at Cornell University, Ithaca, New York, 1988, 137.

22. **Marrone, P., Stone, T. B., Sims, S. R., and Tran, M. T.,** Discovery of microbial natural products as sources of insecticidal genes, novel synthetic chemistry, or fermentation products, in *Biotechnology, Biological Pesticides and Novel Plant-Pest Resistance for Insect Pest Management,* Roberts, D. W. and Granados, R. R., Eds., Boyce Thompson Institute for Plant Research at Cornell University, Ithaca, New York, 1988, 112.

23. **McGaughey, W. H.,** Insect resistance to the biological insecticide *Bacillus thuringiensis, Science,* 229, 193, 1985.

24. Rural Advancement Fund International, Microbial insecticides: special focus on *Bacillus thuringiensis,* RAFI Communique, January 6, 1989.

25. **Lewin, R.,** In ecology, change brings stability, *Science,* 234, 1071, 1986.

26. **Sagoff, M.,** On Making Nature Safe for Biotechnology, unpublished paper dated January 10, 1989, University of Maryland Institute for Philosophy and Public Policy, College Park, MD.

27. **Gould, F.,** Genetic engineering, integrated pest management and the evolution of pests, *TIBTECH,* 6, S15, 1988.

28. National Research Council, *Alternative Agriculture,* National Academy Press, Washington, D.C., 1989, chap. 2.

29. **Carlton, B. C.,** Genetic improvements of *Bacillus thuringiensis* as a bioinsecticide, in *Biotechnology, Biological Pesticides and Novel Plant-Pest Resistance for Insect Pest Management,* Roberts, D. W. and Granados, R. R., Eds., Boyce Thompson Institute for Plant Research at Cornell University, Ithaca, New York, 1988, 38.

30. **Miller, L. K., Lingg, A. J., and Bulla, L. A., Jr.,** Bacterial, viral and fungal insecticides, *Science,* 219, 715, 1983.

31. **Weiss, S. A., DeGiovanni, A. M., Goodwin, G. P., and Kohler, J. P.,** Large scale cultivation of insect cells, in *Biotechnology, Biological Pesticides, and Novel Plant-Pest Resistance for Insect Pest Management,* Roberts, D. W. and Granados, R. R., Eds., Boyce Thompson Institute for Plant Research at Cornell University, Ithaca, New York, 1988, 22.

32. **Anon.,** Toward a "more subtle approach", *Ag Bioethics Forum,* 1, 1, 1989.
33. **Glaser, D., Stotzky, G., and Watrud, L.,** Prospects for containment of genetically engineered bacteria, in *Prospects for Physical and Biological Containment of Genetically Engineered Organisms,* Proc. Shackelton Point Workshop on Biotechnology Impact Assessment, Gillett, J. W., Ed., Institute for Comparative and Environmental Toxicology and Ecosystems Research Center, Cornell University, 1987, 32.
34. U.S. Environmental Protection Agency, Subpart M: biorational pesticides, National Technical Information Service, Springfield, Virginia, PB 83-153965, 1983.
35. **Goldburg, R.,** unpublished memorandum dated 2/25/89.
36. **Bishop, D. H. L.,** The release into the environment of genetically engineered viruses, vaccines and viral pesticides, *TIBTECH,* 6, S12, 1988.
37. **Cuskey, S. M.,** Lethal genes in biological containment of released microorganisms, unpublished article.
38. **Orians, G. H.,** Site characteristics favoring invasion, in *Ecology of Biological Invasions in North America and Hawaii,* Ecological Studies, Vol. 58, Mooney, H. A. and Drake, J. A., Eds., Springer-Verlag, New York, 1986, chap. 8.
39. **Goldburg, R.,** unpublished memorandum dated 6/7/89.
40. **Rissler, J.,** personal communication.
41. **Bej, A. K., Perlin, M. H., and Atlas, R. M.,** Model suicide vector for containment of genetically engineered microorganisms, *Appl. Environ. Microbiol.,* 54, 2472, 1988.
42. **Miller, R. V.,** Potential for transfer and establishment of engineered genetic sequences, *TIBTECH,* 6, S23, 1988.
43. **Goldburg, R.,** unpublished memorandum dated 2/25/90; see also *Christian Science Monitor,* April 29, 1988, and *Wall Street Journal,* April 29, 1988.
44. **Cuskey, S.,** unpublished data.
45. Royal Commission on Environmental Pollution, Thirteenth Report: The Release of Genetically Engineered Organisms to the Environment, Her Majesty's Stationery Office, London, July 1989.

*Index*

# INDEX

# C

transgenic, 73
Insecticidal crystal proteins (ICPS), see Crystal proteins
Insecticides, see also Biopesticides
  chemical, 54, 60, 256—257
  engineered activity, 100—101
  management strategies, 61—62
  mosquito control, 150—151
Integrated Pest Management (IPM), 54, 61, 62, 257
Integrative transformation, 83
International Cooperative Program, 197
Inversion markers, 185
IPL-41 medium, see Media, IPL-41
IPM, see Integrated pest management
Iridovirus, 142—143
Irradiation, 31
Isozymes, 58, 141
Ivermectin, 185

**J**

JHE, see Juvenile hormone esterase
Jinguan et al. process, 204—205
Juvenile hormone esterase (JHE) gene, see Gene(s), juvenile hormone esterase (JHE)

**K**

Kanamycin, 111
Kapow selection, 80
Kellen K, see *Bacillus sphaericus*, toxin
Killing volume hypothesis, in insect cell culture, 229
K$^+$-ATPase, 11

**L**

Lactalbumin hydrolysate, 225, 226
*Laodelphax striatellus*, 121
Larvae, baculovirus susceptibility, 70—71, see also Baculovirus
Laser interferometry, 107
Latex test, 122—123, 125, 126, 127
Lepidoptera, see also genera by name
  cell lines, 224
  crystal proteins, 4, 5, 6, 10, 97—98
  diptera cell fusion, 138
*Leptinotarsa decimlineata*, 57
Levamizole, 87
Ligand blots, 10
Light, see Irradiation
Linkage analysis, 58—59
Low input sustainable agriculture (LISA), 246
Low molecular weight (LMW) antigens, 182—183
Luciferase (LUX), 109
LUX, see Luciferase
*Lymantria dispar*, 6, 70, 80, 142
  nuclear polyhedrosis virus of, 219, 220
Lysolecithin, 136

**M**

mAb, see Monoclonal antibodies

Macroprojectiles, 106—107, see also Particle bombardment
Maize, see Corn
Mammals
  biopesticides, 42
  cell culture, 222—223
*Manduca sexta*, 10, 60, 101
*Mansonia uniformia*, 40
Marine impellers, 228
MCA, see Monoclonal antibodies
McAb, see Monoclonal antibodies
MCAP, 15, 45
MDR clones, 137, 141
Mechalas process, 204
Mectizan, see Ivermectin
Media
  BML/TC-10, 226
  CDC, 226—227
  Difco nutrient, 197
  Grace's, 224, 225
  HAT, 140
  Hink's TNM-FH, 225, 226
  insect cell culture, 225
  IPL-41, 226
  megna, 206
  MTCM-1601, 226
  pilot-scale *Bacillus* culture, 198—200
  semisolid fermentation, 204
  serum-free MM, 226
  submerged fermentation, 206
  ZH1%, 137, 141
Mediterranean fruit fly, see *Ceratitis capitata*
Megna medium, 206
Mendelian inheritance, 58
Mermithida, 88—89
Metabolic resistance, see Resistance, mechanisms
Metal, high-density, 107
Methylcellulose, 225, 228, 229
Microinjection, 82—83
Microorganisms, risks of genetically engineered, 157—159, 253
Microprojectiles, 107—109, see also Particle bombardment
Midgut, see Insect, midgut
Mitochondria, 113
MM-SF, see Media, serum-free MM
Monoclonal antibody, see also Polyclonal antibody
  advantages, 120
  cytoadherence-inhibitory, 169
  *Entamoeba histolytica*, 164—169
  insecticidal crystal genes, 14—15
  ochocerciasis immunodiagnosis, 182
  rice disease, 124—127
Monocotyledons, 17—18
Mosquito, see also *Aedes*; *Anopheles*; Biopesticides; *Culex*
  control
    duration, 43—44, 45
    problems, 150—151
  crystal proteins, 7, 8, 12, 26
  food sources, 30